Theory and Practice in Machining Systems

Yoshimi Ito · Takashi Matsumura

Theory and Practice in Machining Systems

Springer

Yoshimi Ito
Tokyo Institute of Technology
Yokohama
Japan

Takashi Matsumura
Tokyo Denki University
Adachi, Tokyo
Japan

ISBN 978-3-319-53900-3 ISBN 978-3-319-53901-0 (eBook)
DOI 10.1007/978-3-319-53901-0

Library of Congress Control Number: 2017932400

Printed on acid-free paper

This Springer imprint is published by Springer Nature
The registered company is Springer International Publishing AG
The registered company address is: Gewerbestrasse 11, 6330 Cham, Switzerland

Preface

The machine tool engineering technology can be, in wider scope, twofold: one is the production technology consisting of the design and manufacture, and the other is the utilisation technology. Importantly, the majority of people of machine tool concerns are interested in and related to the utilisation technology; however, the matters in production technology concerns are generally at issue especially in the academia, because the machine tool underpins all the industrial sectors as represented by a famous maxim, i.e. "*Only the Real Industrial Nations Can Produce the Machine Tool and Ordnance*". In fact, we cannot produce the necessary components for all the commercial and defence supplies without having the machine tool.

In use of the machine tool, we need, in principle, the synergic knowledge ranging from the form-generating movements possible by, and also structural design and numerical control of the machine tool, through the attachment and tool, to the machining technology. It is, however, regrettable that we cannot obtain the preferable materials enabling such knowledge to be learnt at glance. In fact, all the books having been and being publicised deal with each subject mentioned above separately with narrower scope.

As a result, it appears that the machine tool and cutting/grinding technologies belong to another engineering sphere each other. For example, we can observe one of the serious problems in the self-excited chatter vibration of regenerative type. In short, to suppress the chatter vibration, we must consider the corresponding problem related to the machine-attachment-tool-work system. It is, however, worth suggesting that nearly all academic, engineering and technical reports on the chatter vibration do not state anything about the chuck and tool holder, although they play very important roles in the suppression of the chatter vibration.

In this context, we have experienced a similar story in the thermal deformation, and importantly, we must be aware of the necessity of establishing such a new viewpoint even in the machining technology with the advance of the tool and attachment. In short, we can find a considerable number of novel cutting tools, which may innovate more process planning than ever before; however, for their effective uses, we need to establish also the innovative attachment to hold them. For example, it has been the common sense to hand-lap the sharp cutting edge (sharpness-killed cutting edge) of the end mill, so that the work surface can be finished with the better quality without chipping the cutting edge. With the advent of the innovative milling chuck, however, such a remedy becomes obsolete, and on the contrary, the sharp cutting edge itself is recently recommended.

In considering the preferable leverage among the related subjects and issues depending upon the machining requirements, we need now an informative book dealing with the machine-attachment-tool-work system from the viewpoint of the user. The book proposed herein aims thus at the holistic description for the technologies necessary to the component generation by placing the stress on the machining space. More specifically, the machining space is represented by the "*Linkage Diagramme*", one of the variant of the "*Flow of Force*", and visualised obviously the structural configuration entities and concerns surrounding the machining space together with their leading linkages among one another (see Chap. 2).

More importantly, we must be aware of the increasing need for such a book with the innovative development in each entity consisting of the machine-attachment-tool-work system as follows.

(1) Prevalence of the "*Platform Concept*" in the structural design of the machine tool.
(2) Upheaval in the development of modular attachments, especially in chuck.
(3) Prevalence of cutting tools of modular and combination types.

Of these, the most influential factor is the "*Platform Concept*", i.e. user-oriented modular design, especially in the machining method-integrated kind, e.g. "*Mill-turn*". In discussing the holistic utilisation technology mentioned above, thus, the book gives us some quick notes for the platform concept (see Chap. 7).

Obviously, it is furthermore desirable to extend the concept of the "*Linkage Diagramme*" to the raw material to be machined and NC information as follows.

(1) Enhancement of machinability of the raw material with wider scope, e.g. elimination of the scale in the hot-forged raw material and work preparation considering the directional orientation in strength caused by cold drawing.
(2) Development of the innovative machining method by facilitating the NC information, e.g. turbine blade machining by whirling and slot drill moving by helical programming like boring with planetary movement.

In the latter case, we must understand the fundamentals in removal processing of unnecessary allowance at the machining point, e.g. cutting and grinding mechanism, together with differing features in various machining methods. In Part III of this book, thus, we will discuss such issues extremely placing the stress on the advanced computational method being established.

As will be clear from the above, the book can be characterised by its challenging editorial work to give the machine tool user with core and synergic knowledge. Thus, the machine tool user benefits considerably and is able to conduct the lucrative business by reading this book. In addition, the book will provide the reader in the academia and research organisation in the enterprise with some clues to carry out the forerunning and noteworthy research into and engineering development in the machining technology.

Finally, the book consists of Parts I, II, III and Appendices, and Profs. Ito and Matsumura are in charge of Parts I, II and Appendices and also of Part III, respectively.

Yokohama, Japan
Adachi, Tokyo, Japan
December 2016

Yoshimi Ito
Takashi Matsumura

Contents

Part I
Fundamentals

Abstract

The machining technology should be learned, discussed and innovated with wider scope, i.e. that ranging from the manufacturing system to the cutting point in the machining space. To do so, we need to conduct something definite, and thus Part I will first depict an overall view for the desirable machining space in the manufacturing system. Then, we suggest the importance of the linkage of structural configuration entities within the machining space, i.e. machine-attachment-tool-work system, in discussing the leading engineering problem in the machining technology.

In Chaps. 1 and 2, primary concerns in the machining space will be detailed, for example, by using the hierarchical classification system of machining methods, closed-loop concept in the linkage of structural configuration entities and form-generating movement together with its description. Importantly, Chap. 1 provides us furthermore the knowledge about the supply of the raw material for machining and discharge of the swarf from the machining space. In short, metal-working, phase-change processing and additive processing are for the raw material and semi-finished component production.

Headnotes

In the production activities for the commercial and defence supplies, we use the factory system, which ranges from large-sized to cell configurations, depending upon the enterprise size. More specifically, primary concern in the production activity is the production morphology, which represents the life history of the product, starting from the market survey or user's order placement and terminating at the product disposal. Conceptually, the production morphology reveals the explicit and implicit representations of materials and information flows in the product history, and can be defined as follows:

"*A time series of 'Processes (Procedures)' dealing with the 'Product Creation and Disposal', ranging from the collection of market information of and order*

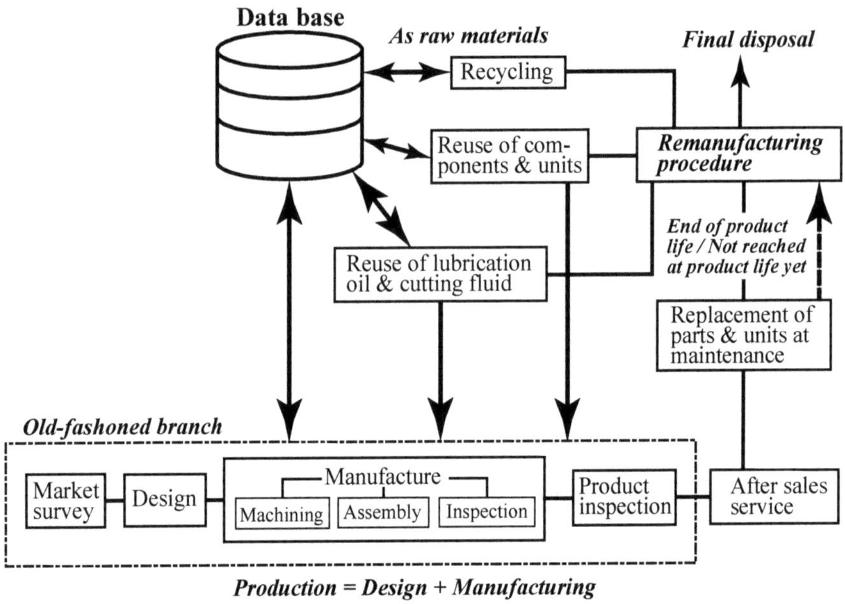

Fig. I.1 Concept of production morphology

placement by the customer, through the design and manufacturing, to the after sales service and remanufacturing".

Although depending upon the product kind and its production volume, the production morphology shows clearly the position of the machining procedure in the factory system, and thus Fig. I.1 illustrates the generalised production morphology in discrete type, and we used to specify the production morphology on the basis of the old-fashioned branch within Fig. I.1.[1] Importantly, we place now the stress on the remanufacture in Fig. I.1, in which the end-of-life product is, in general, at burning issue, in consideration of the social requirements for sustainability at present. We must, however, pay special attention to the following, because these will result in a new perspective in the production technology.

[1]In the discrete type, materials for manufacture concern, e.g. raw material, semi-finished work, finished component, cutting tool, jig and fixture, and swarf, are transferred in the pattern of "*Point-to-Point*," but not continuous flow. In contrast, in the process type, materials flow continuously, i.e. seamless material flow. The former can be observed, for example, in the motorcar, aerospace, industrial machines and machine tool industries, whereas the latter can be observed, for example, in the cigarette, petroleum refining, and cosmetic and detergent industries. In addition, we have a combination system of the discrete and process types, which can be exemplified by the large-scale integration (LSI) and sintered carbide tip manufacturing.

(1) Repair of the product in consideration of its operating history and by using the second-handed components (parts) and units, which are obtainable on the occasion of maintenance and repair.
(2) We may have some products, which are not yet reached at its life estimated at the design stage, but do not use anymore. We could obtain such products through (a) the renewal to the new product by the customer, (b) with the merchandise of new products by the competitor, and (c) by the function and performance deterioration in even new product due to the technology innovation. Of special note, we must be furthermore aware of the apparent product deterioration caused by the enactment of the new law.

Roughly saying, a group of the components can be finished and then assembled in the factory system. In general, we may finish the component by machining with various patterns, e.g. mass-production, production with medium volume and mediocre component variation, batch production and one-off production, in accordance with manufacturing requirements. In any patterns, the metal-cutting machine tool[2] can facilitate to finish the component to a larger extent. Importantly, without having the well-qualified manufacturing technology, the commercial and defence supplies required of the human society are not to be in reality, even though the product design is very outstanding. More importantly, the manufacturing technology involves much more professional subjects than our expectation, even when interpreting it with narrower scope and in relation to skills in the factory floor. In fact to understand correctly the manufacturing technology, we need to have the knowledge, at least, those ranging from the engineering material and its various processing methods, through machine tools and other industrial equipments to assembly and inspection technologies of the product.

Reportedly, the product involves two-pronged design attributes, i.e. fully quantified and ambiguous attributes, and a crucial issue in the product design is to rationally convert the ambiguous attributes into the quantified ones in consideration of the leverage between the quantified and ambiguous attributes. For example, in the motorcar, we can quantify the fuel efficiency of engine and traction force of tyre, whereas cannot quantify the roadability.

Now let us consider the machining technology from the manufacturing system point of view. In this context, we must be especially aware of the very importance of the grass root-like knowledge so far accumulated through the long-standing experience, and in part related to the culture and mindset background. Supposedly, such knowledge can be represented by the jargon in the factory floor.

With growing importance of flexible manufacturing systems (FMS) extremely in the 1980s and 1990s, we contrived various system configurations, and at present FMS of cell-integrated type are dominant across the whole industry. Obviously, the cell itself has enough flexibility, i.e. flexible manufacturing cell (FMC), and has been prevailed even in the very small-sized enterprise. As widely known, such

[2]Machine tools can be classified into "metal-cutting machine tools" and "metalworking machine tools". In this book, the machine tool represents shortly that of metal-cutting type hereafter.

system and cell have five functions, i.e. processing, transportation, storage, surveillance and maintenance, and also the software for computer-integrated manufacturing (CIM) in the lowest level (cell controller).

In flexible manufacturing, we must consider the three leading attributes, i.e. "*Flexibility*", "*Expandability*" and "*Redundancy*" in the design of system and cell, which is to be in reality by using the *modular design* (Ito 2008). In addition, when the processing function is for machining, we call it *flexible machining system* or *flexible machining cell*.

The machining function can, at present and in general, facilitate by the turning centre (TC) and machining centre (MC), and with the further contrivance of the machining function-integrated kind, e.g. "*Mill-Turn*" and MC with grinding function, compact and cubic-like flexible machining cell becomes dominant. Such a cell appears as to be a machine tool as a whole, and thus we must pay the special attention to the machining space, even when discussing the flexible machining system.

Now let us discuss the configuration changes in flexible machining system and cell for the motorcar industry for ease of understanding of time series-like variation in the machining space. In this context, primary concerns are the flexible transfer line (FTL), a variant of FMS, and its compact type, i.e. "*Transfer Center*".

Figure I.2 shows the typical FTL for cylinder block machining with 4 min. in cycle time, resulting in machining volume of 50,000 units per year. Importantly, the line is designed by the modular principle, where an MC is of basic module for machining, and the machining space within MC itself is very important from the

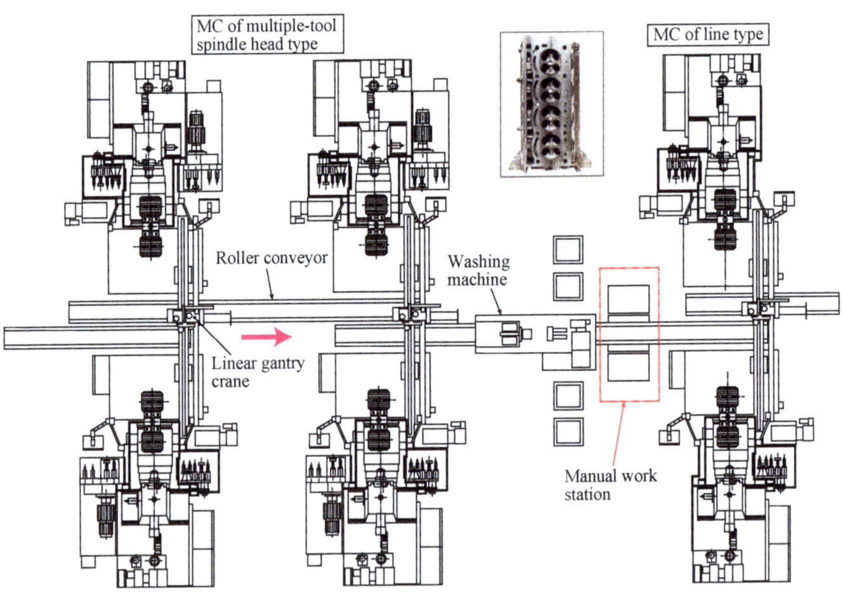

Fig. I.2 Flexible machining line for cylinder head—Fritz Werner, in 1993

viewpoint of machining technology. It is thus necessary to understand the dimensional and performance specifications of MC. For example, MC of multiple-tool spindle head type has the following specifications.

Maximum size of machining space (X, Y, Z): 1000 × 950 × 800 mm
Maximum rotational speed of main spindle: 6000 rev/min
Output power of main motor: 37 kW
Size of multiple-tool spindle head: 500 × 630 mm
Tooling system: Hollow shank system (HSK)

With the advance of the related technologies, FTL has been replaced by a group of the flexible machining cell apart from the very special machining requirements. In a certain case, the cell of stand-alone operation is employed. This aims at the further economisation of manufacturing activity, for example, by saving the installation cost and floor space for manufacturing facilities.

Figure I.3 shows a flexible machining cell complex, where a flexible machining cell is of basic module, and we can first install only a cell with the automatic pallet changer (APC). With growing volume for machining requirements, then, we can expand the cell to the cell complex by installing furthermore the new cells necessary together with the load/unload station, rail-guided vehicle and buffer, i.e. the cell having the *expandability* function. After having the two cells, we can operate the cell complex, even when one cell is down, i.e. the cell having the *redundancy* function. As will be clear from the above, we can provide this cell complex with flexibility, expandability and redundancy by using the modular design.

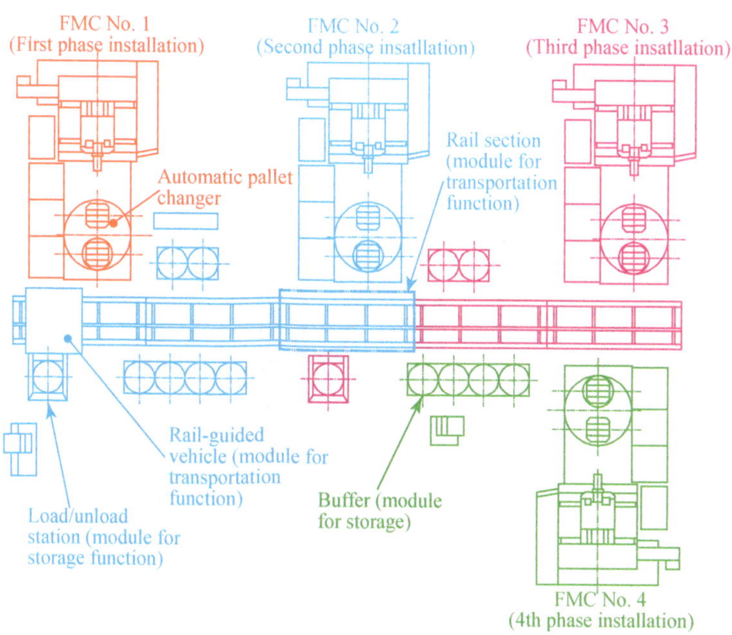

Fig. I.3 Flexible machining cell complex (by Cincinnati Milacron, 1990s)

In the middle of the 2000s, a facing issue was the compactness of the flexible machining cell, and in due course, its variants have been contrived. Of those, the "*Transfer Center*", i.e. FTL of compact type, is now in leading position. Importantly, the "*Transfer Center*" may facilitate the batch production, which is being prevailed in the motorcar industry. Typically, Fig. I.4 shows a "*Transfer Center*" of ANGER-brand, whose machining capability is compatible with those carried out by 2–5 MCs. In contrast to the head changer in the past, the work spindle can travel and rotate, and the work can be machined by single-tool spindle head, multiple-tool spindle head and various tools mounted on the turret head and tool cassette. Paraphrasing, the "*Transfer Center*" is one of the modernised head changers.

In the past, a leading variant of FTL was furthermore the dial machine, i.e. transfer line of rotary indexing type, and for example Witzig und Frank has merchandised its advanced type as shown in Fig. I.5. In this case, the machining space consists of not only the machining function, but also the transportation and storage functions, as will be clear from top and bottom-right of Fig. I.5. In contrast, the machining space of flexible machining cell consists of only the machining function.

To this end, it emphasises that we must pay certain attention to the machining space within the machining station, even when discussing the flexible machining system.

Now let us discuss furthermore the leading engineering problems in the machining space.

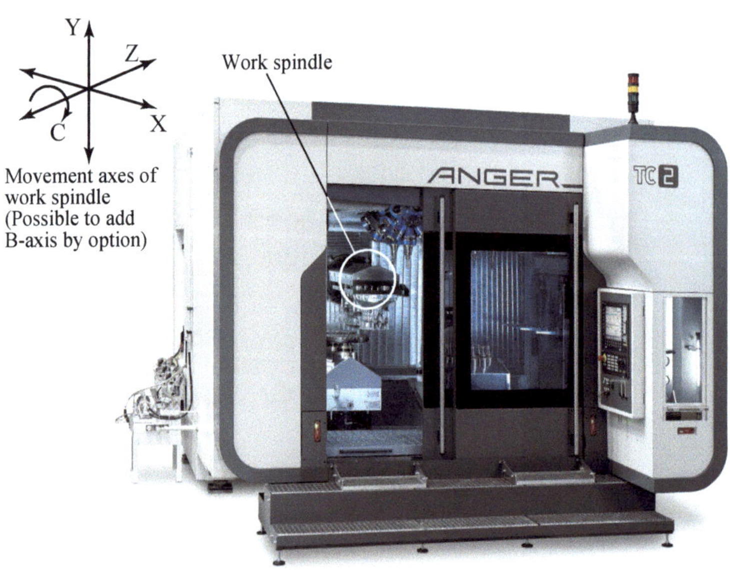

Fig. I.4 Appearance and machining space of "Transfer Center" (by courtesy of Anger 2014)

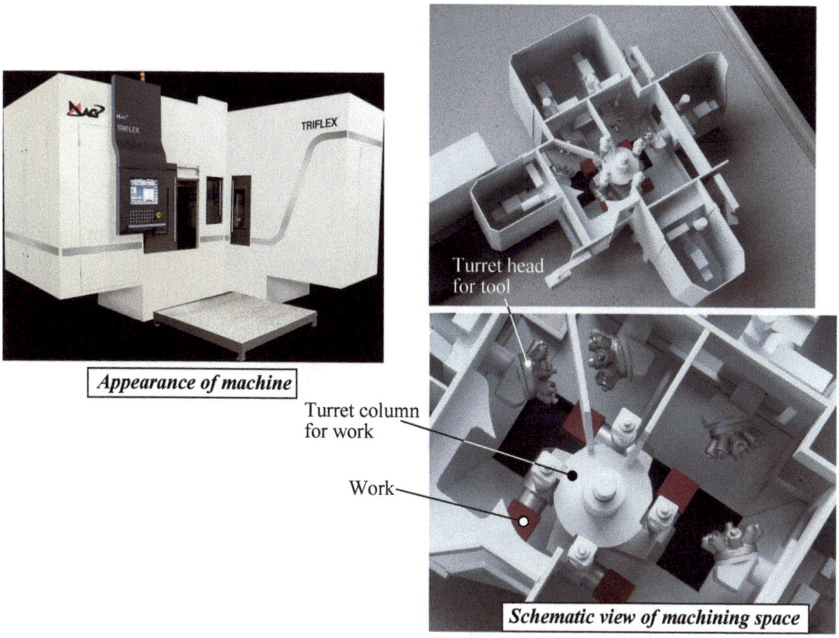

Fig. I.5 "Transfer Centre" of rotary indexing type (by courtesy of Witzig und Frank 2009)

In the functionality and performance contexts of the machine tool, there are the three leading design attributes, i.e. "*Better Machining Accuracy*", "*Machining with Higher Linear and Rotational Speeds*" and "*Heavy-duty Cutting Capability*". From the viewpoint of machining space, these result in the leading engineering subjects for the *chatter vibration* and *thermal deformation*.

Figure I.6 is the scenery of the machining space within TC, which consists of the structural body components of the machine tool, attachments, cutting tools and work. As can be readily seen, we must deal with the engineering problems mentioned above by using the systematic and synergic points of view, i.e. those in machine-attachment-tool-work system. Importantly, with the advent of the machining function-integrated kind like "*Mill-Turn*", we must solve more unimaginable and uncountable problems ever than before. Thus, we will discuss quickly the present perspectives for the chatter vibration and thermal deformation, placing the stress on the machining space-dependent aspects.

Chatter Vibration

In nearly all research papers and engineering reports, it is first stated in the headnote that the chatter vibration is one of the dynamic behaviours in the machine-attachment-tool-work system. In contrast, they do not state, for example, the detail for work and tool holding in the body text. In this context, Doi et al. (1982, 1985) investigated the self-excited chatter vibration of regenerative type in turning with the three-jaw chuck and unveiled the mingling effect of parametric vibration. More

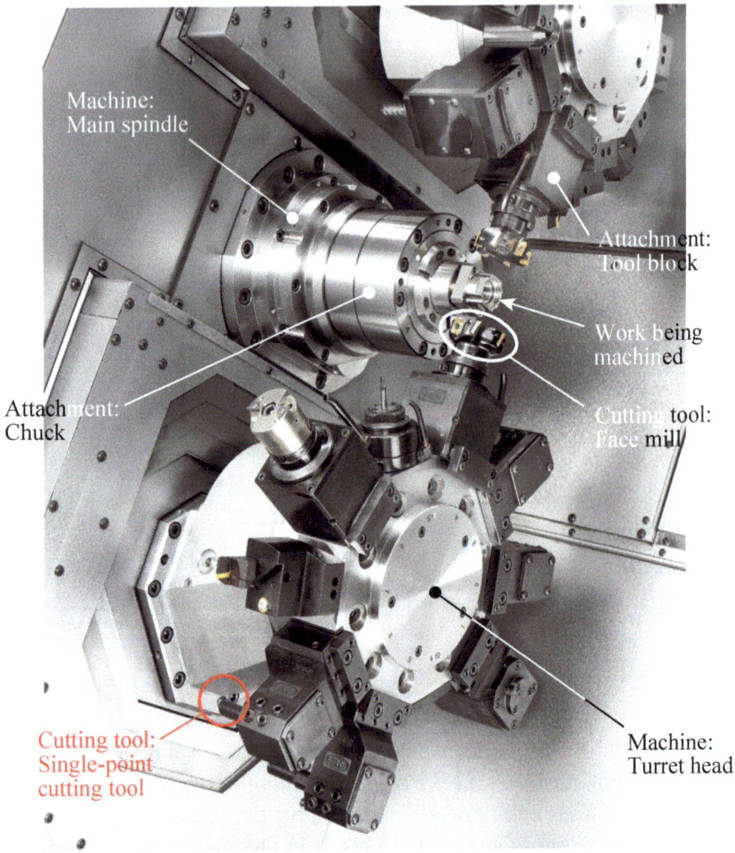

Fig. I.6 Scenery of machining space (by courtesy of Traub 2009)

specifically, the work held by the three-jaw chuck shows the directional orientation in its stiffness, which causes a parametric vibration, and deteriorates considerably the anti-chatter capability.

Another typical evidence is the lower-speed stability. The stability limit decreases theoretically in the lower-speed range of the main spindle; however, in practice, the stability limit increases considerably in the same speed range. Since 1960s, it has been believed that the lower-speed stability is caused by the interference between the flank (relief face) of the cutting tool and the wavy surface of the work, resulting in the increase of damping. Admitting the validity of such a belief in part, we must be aware that there are various damping sources in the machine-attachment-tool-work system not only at the cutting point but also other joints within the system. Of special note, we must produce the "*Damping Capacity Distribution Diagramme*" of the system, at least, at the machining space, as similar as the "*Rigidity Distribution Diagramme*" of the machine tool structure as a whole (Koenigsberger and Tlusty 1970).

Table I.1 Dimensional specifications of milling cutters (by Monnin, Dissertation of ETH Zürich)

Tool-holder assemblies	Type	Diameter mm	Free length mm	Number of teeth	Helix angle deg.	Engaging angle deg.
T1	Insert end mill	40	175	5	10	+90
T3			100			
T8	End mill (sintered carbide)	10	103	2	30	
T10	Face mill	63	90	5	20	+45
T12	End mill (sintered carbide)	16	124	4	40	+90
T15		12	171	2		
T16	Round insert end mill	40	116	5	4	—

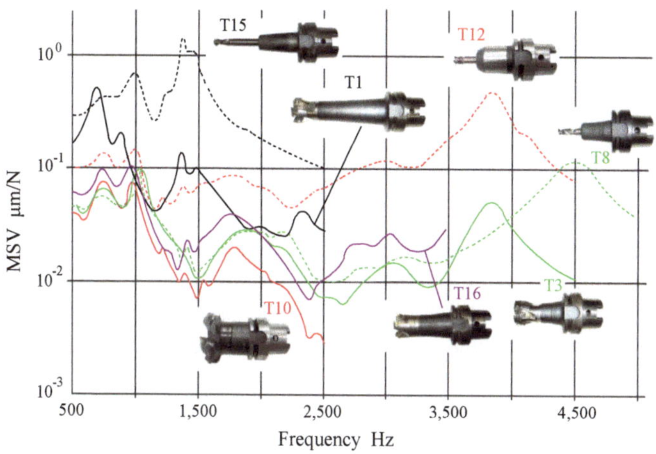

Fig. I.7 Receptance functions at tool tip for various tool-holder assemblies (by Monnin)

Importantly, Monnin (2013) has measured the frequency response in the first bending mode of the main spindle in still stand, when varying the milling cutter and its holding device, as shown in Table I.1 and Fig. I.7. The experiment was carried out by using the quinaxial-controlled MC of Mikron Agie Charmilles brand (Type HPM) and by milling the work made of either carbon steel (Ck 45) or Al-alloy (EN AW-6082 T6). In addition, the frequency response was measured by impulse excitation and also by detecting the corresponding signal with the accelerometer. In Fig. I.7, furthermore the maximum singular values (MSVs) are an index to represent the maximum amplitude of receptance function. As can be readily seen, we can observe the remarkable differing feature when varying the milling cutter and its holding device.

With the growing importance of the higher-speed machining in the aircraft industry, we face a new problem regarding the availability of the stability chart,

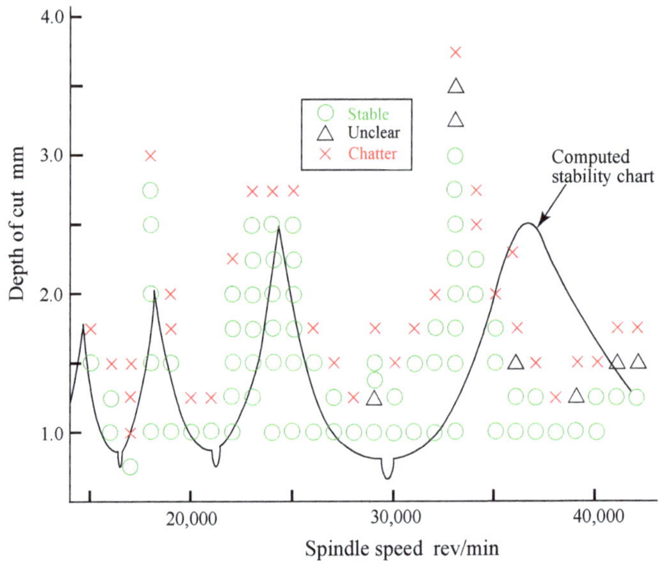

Fig. I.8 Comparison between computed stability chart and experimental values (by van Dijk et al.)

when investigating the preferable cutting condition without the chatter vibration. In this context, Van Dijk et al. (2008) reported the discrepancy between the stability charts obtained theoretically and experimentally as shown in Fig. I.8. It is very interesting that the computed stability chart differs considerably from the experimental one at the higher-speed range of main spindle, i.e. up to 27,000 rev/min. They suggested that such a discrepancy is caused by the speed-dependence behaviour of the spindle dynamics.

More specifically, the theoretical stability chart was computed for the milling machine of Mikron brand (Type HSM700) with the two-flute end mill (10 mm in diameter, 57 mm in length), which was held by shrink-fitting. Of note, the necessary data for computing the stability chart were obtained by the impulse hammer method, and also the cutting test was carried out for Al-alloy 7075 with slot (full circular) milling.

Importantly, Schmitz et al. (2004) publicised, earlier than that of Van Dijk et al., a similar result, i.e. considerable difference between the theoretical stability chart and the experimental data in the higher-spindle speed while end milling. Apparently, we can observe such an interesting behaviour when the rotating speed is more than 17,000 rev/min. In addition, a machine tool manufacturer has an experience, in which the blur-like or micro-beat-like chatter marks can be clearly observed while wet-machining the Al-alloy with 3-flute end mill, i.e. side cutting pattern of full length, and more than 11,000 rev/min in cutting speed.

To this end, thus, it emphasises that the chatter vibration problem should be investigated in consideration of the state of machining space as well as the cutting point.

Fig. I.9 Total enclosure for TC—Type SPRINT 50 linear (by courtesy of Gildemeister 2009)

Thermal Deformation

With prevailing the conventional TC and MC, the total enclosure becomes popular to enhance the marketability from the industrial design point of view. As can be readily seen from Fig. I.9, the total enclosure affects the thermal behaviour of the machine tool as a whole from both the heat dissipation and the accumulation to some extent.

Jędrzejewski et al. (2007) conducted an interesting research into the influence of the enclosure in the lathe for higher-speed machining. Although they did not show the detail of the model having been used in the research, they proposed first a mathematical model for the thermal deformation together with considering the environment temperature surrounding the machine. Then, they identified the thermal deformation of the main spindle by measuring it with the straight edge made of quartz.

Importantly, they suggested extremely that the temperature of the structural body component changes with certain time delay due to the variation in the ambient temperature, and then they eyed the heat accumulation due to the guard placement. Fig. I.10 reproduces some typical influences of the cover to the displacement in main spindle, where the Z-axis is identical with the spindle axis. As can be readily seen from the measurement, the semi- and total enclosures have considerable and certain influences on the displacements in the directions of Y and Z axes.

In general, we can observe various heat sources in the machining space, and the heat generated is dissipated in part to the environment through the enclosure. In contrast, the temperature distribution in the machining space is influenced by the change in the environment temperature.

Of special note, we must thus understand the mutual effects in the 3-dimensional heat flows between the machining space and the surroundings. Accordingly, it is necessary and inevitable to identify each heat source and its influencing region, and thus a concept of "*Thermal Volumetric-Space*" is proposed herein as shown in Fig. I.11. This concept is to visualise a cell-like room, in which the temperature

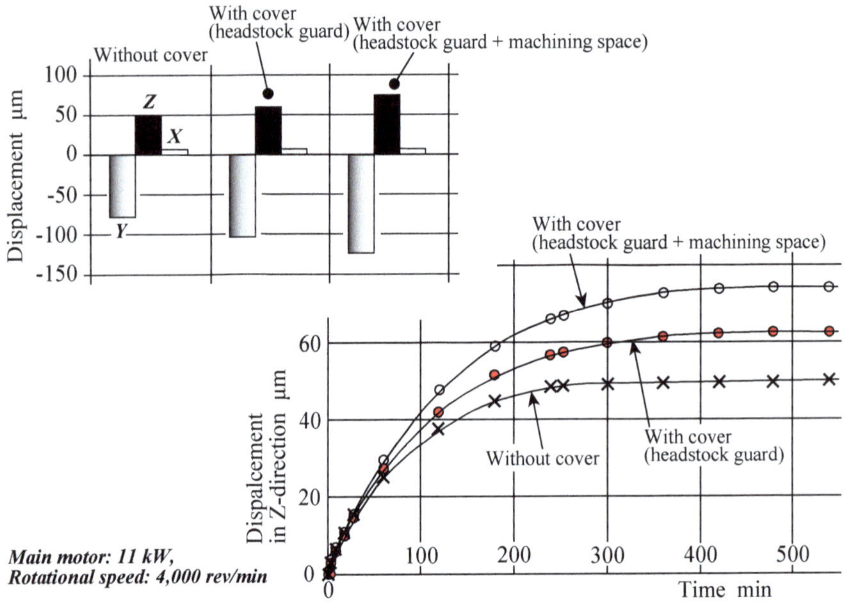

Fig. I.10 Influence of enclosure on thermal deformation in main spindle of lathe (by courtesy of Jędrzejewski et al.)

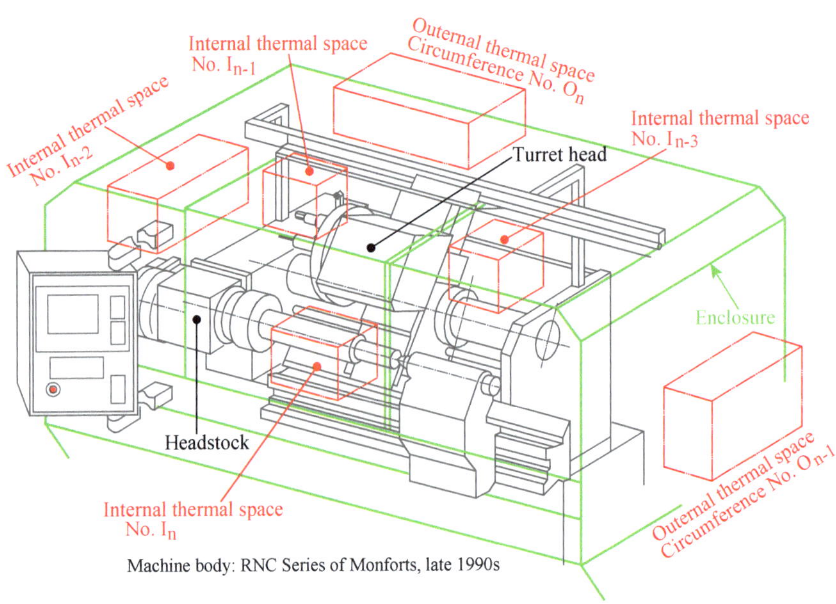

Fig. I.11 Concept of "Thermal Volumetric-Space"

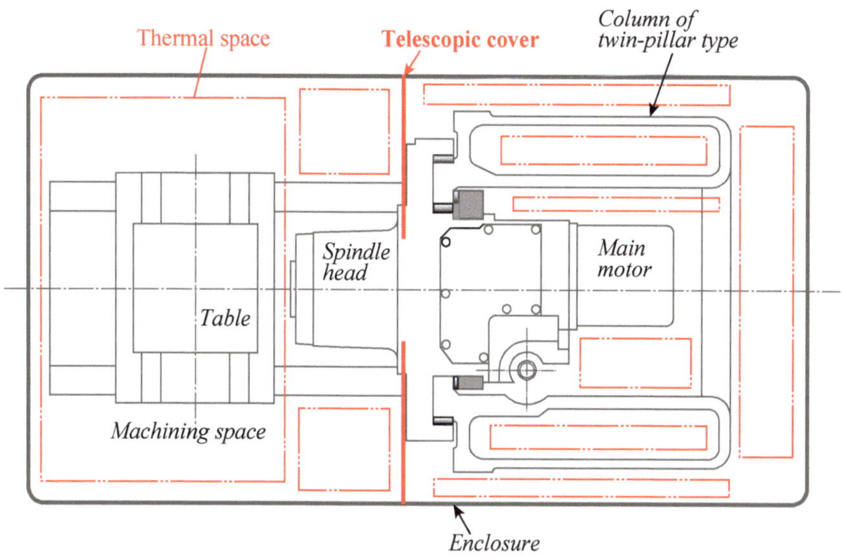

Fig. I.12 Allocation example of "Thermal Space" within enclosure

distribution is nearly constant, and in general the machine body may be partitioned into several rooms by the structural body components and configuration entities. Obviously, we can detail the heat flow among volumetric spaces by using their distribution diagram. Conceptually, the "*Thermal Volumetric-space*" may be simplified into the two-dimensional thermal space as shown in Fig. I.12, provided that the machining and factory spaces are in the constant temperature distribution in the vertical plane, and we may benefit from it to some extent.

In the thermal deformation, it is furthermore important to determine correctly the dynamic and thermal boundary conditions, and to do so we must consider the "*Dynamic and Thermal Closed Loop*". In fact, the thermal state of the machining space reaches at stable after repeating several times the closed-loop effect between the dynamic and thermal changes. For example, Fig. I.13 reproduces the temperature rise of the model spindle supported by the tapered roller bearing with high-speed adapter (Lee 1991). As well known, the high-speed adapter was contrived by Timken. More specifically, the thermal deformation of the outer ring is quicker in radial direction than that in axial direction, and thus small clearance at the land B can be facilitate the release of the dynamic constraint of the outer ring, resulting in the reduction of heat generation.

As can be readily seen, the high-speed adapter can reduce the temperature as expected, whereas we can observe the spike-like temperature rise by placing the adapter as shown also in Fig. I.13. This implies that the high-speed adapter shows a stick-slip like movement in both the radial and axial directions, and induces the complexity in the closed-loop effect. Importantly, the more increase the rotational speed of spindle and loading, the much more apparent is the closed-loop effect

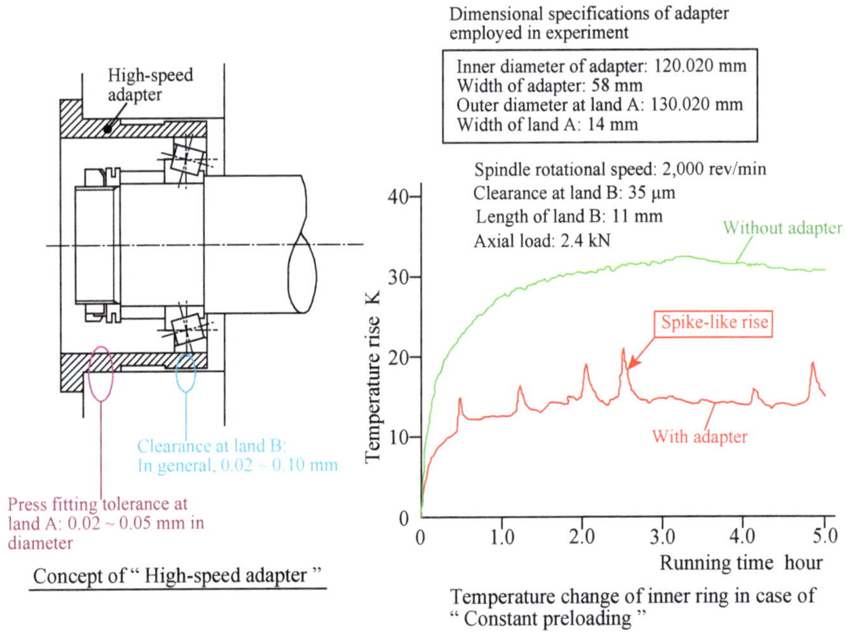

Fig. I.13 Temperature rise in spindle with and without high-speed adapter—implication of "closed-loop effects"

In many respects, it is thus necessary to integrate the "*Thermal Volumetric-space*" and "*Dynamic and Thermal Closed Loop*" as shown in Fig. I.14. Eventually, we will be able to scrutinise the thermal behaviour in the machining space, extremely discussing the mutual effect between the machining space and the environmental field through the enclosure. Importantly, it is furthermore possible to consider the "*Dynamic and Thermal Closed Loop*" within each "*Thermal Volumetric-space*", and we can expect the authentic prediction of the machining accuracy.

Within the thermal behaviour context, we must consider another leading issue, i.e. uncertainty in the determination of the thermal boundary condition. Such an uncertainty is, in general, caused by the flows of air and cooling media within the machining space and structural body component, around the machine body as a whole and within the enclosure, and furthermore around the outer surface of enclosure and factory floor.

In fact, we cannot determine the heat dissipation capability from the surface of machine body and heat transfer pattern in the machining space without considering such an uncertainty.

Figure I.15 reproduces such flows of air in the rotating jaw chuck and grinding wheel (see Sect. 5.4, Ito 2010). In both the cases, we can visualise the flow pattern by using both the smoke wire method and the *Taft Method*, and observe the very complicated flow pattern caused by the protruded jaw from the chuck body and the

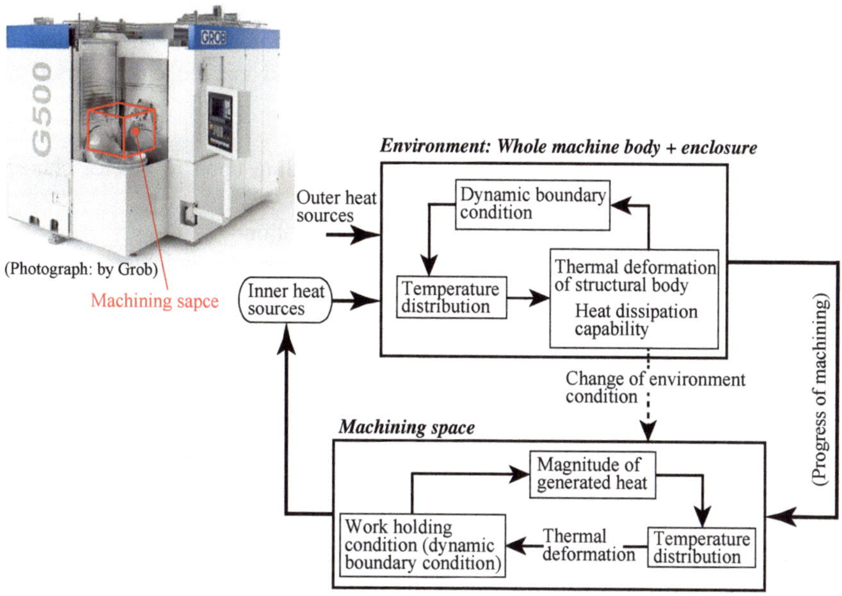

Fig. I.14 Closed-loop effect of dynamic and thermal behaviour within machining space and environment

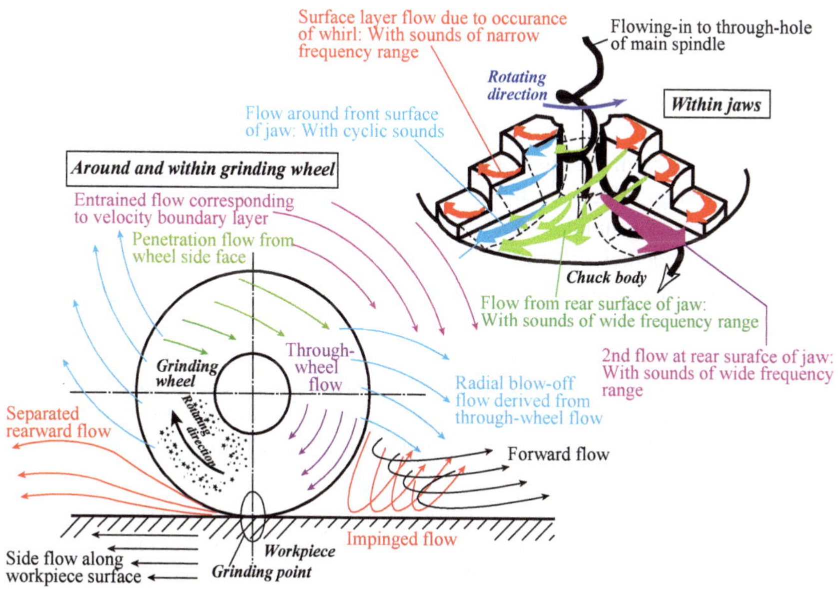

Fig. I.15 Very complicated flows of air for rotating chuck and grinding wheel

porosity of the grinding wheel. Obviously, these flow patterns may affect the thermal boundary condition of the machining space. Importantly, the flow pattern in chuck is closely related to the jaw configuration and can be characterised as follows:

(1) Flow-in to through-hole of main spindle: Depending upon resonance characteristic of air column
(2) Surface layer flow: Depending upon jaw configuration
(3) Flow around front surface of jaw: Depending upon jaw height along radial direction and jaw rotating speed
(4) Flow from rear surface of jaw: Depending upon chuck body configuration
(5) Second flow at rear surface of jaw: Depending upon cross-sectional shape of jaw along radial direction

Of course, the flow pattern of grinding wheel is related to the abrasive grain size, kind of binder and rate of porosity, and more importantly, the flow pattern can facilitate the magnitude of the local heat transfer coefficient as will be discussed below.

In the flow pattern in the grinding wheel, primary concern is the entrained flow, the layer thickness of which is equal to the radius of the grinding wheel or much more than the radius.

It is very interesting that the flow of air is in nonlinear relation to the lower rotational speed and becomes linear at the higher rotational speed, respectively. Supposedly, various flows seem to be relevant at the lower-speed range, where the porosity of the grinding wheel is very effective. As can be readily seen from the flow behaviour of Al-alloy disc, with the increase in the rotational speed, the laminar flow becomes dominant even in the grinding wheel.

In addition to the flow pattern, Saito et al. (1983) reported some measured results for the local heat transfer coefficient on the work while surface grinding as shown in Fig. I.16. Obviously, we can observe some interesting behaviour as follows:

(1) In comparison with the case of Al-alloy solid disk, the porosity of the grinding wheel has large effects on the local heat transfer coefficient. Importantly, the more porosity a grinding wheel is, the more increase and wider range are the heat transfer coefficient and its distribution, respectively.
(2) The larger heat transfer coefficient appears at the area, where the impinged flow is dominant.

It emphasises again that the thermal boundary condition should be determined in full consideration of the flow of air within the machining space. Of special note, we must be also aware of the importance of the measuring method for the three-dimensional distributions of both the temperature distribution and the thermal deformation within the machining space.

Summarising, the thermal deformation should be estimated in consideration of (1) the mutual interference among various heat sources within and around the machine tool as a whole including the factory floor environment, (2) closed-loop

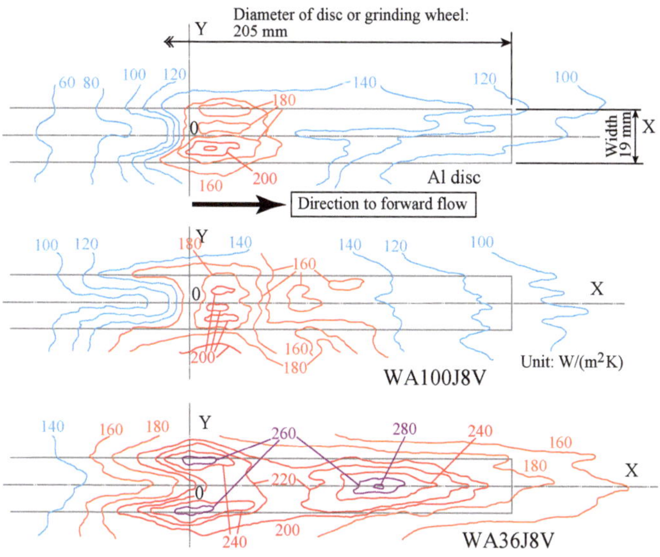

Fig. I.16 Effects of abrasive grain size on local heat transfer coefficient and its distribution in grinding wheel

effects, (3) flows of air within the machining space, enclosure and factory floor and also (4) coolant within the machining space.

References

Doi M, Masuko M, Ito Y (1982) Re-observation of the chatter vibration in chuck works—effects of jaw number on the appearance of a scale-like chatter mark. In: Proceedings of 10th NAMRI of SME, pp 409–416

Doi M et al (1985) Study on parametric vibration in chuck work. Trans JSME (C) 51(463):649–655

Ito Y (2008) Modular design for machine tools. McGraw-Hill, New York

Ito Y (2010) Structural materials and design for preferable thermal stability (Chapter 3). In: Ito Y (ed) Thermal deformation in machine tools. McGraw-Hill, New York

Koenigsberger F, Tlusty J (1970) Machine tool structure, vol 1. Pergamon Press, p 10

Lee SK et al (1991) Thermal behavior of machine tool spindle system based on the thermal closed-loop concept (Remedies for Thermal Deformation by Using the High-Speed Spindle Adapter). Trans JSME (C) 57(538):2085–2091

Monnin J (2013) Active structural methods for chatter mitigation in milling process. Dissertation of ETH Zürich

Jędrzejewski J et al (2007) Precise modeling of HSC machine tool thermal behaviour. J AMME 24(1):245–252

Saito Y et al (1983) Thermal behaviour and flow pattern around the workpiece surface during dry surface grinding. J Precision Eng 49(10):1421–1427

Van Dijk NJM et al (2008) Real-time detection and control of machine tool chatter in high speed milling. 2nd International Conference "Innovative Cutting Processes & Smart machining", Cluny

Schmitz TL, Ziegert JC, Stainslaus C (2004) A method for predicting chatter stability for system with speed-dependent spindle dynamics. Trans North Am Manuf Res Inst SME 32:17–24

Chapter 1
Metal Processing Technology in General—Importance of Hierarchical Classification

Prior to discuss the machining technology, we must understand the *bird's-eye view* of the metal processing technology, which may handle other non-metallic materials, e.g., plastics, glass and ceramics, in certain cases. Of note, the metal processing technology can be, in general, classified into the (1) machining (metal removal processing, swarf-generating type), (2) metalworking (metal non-removal processing), (3) phase-change processing, e.g., welding, casting and plastic injection moulding, and (4) additive processing (rapid prototyping) technologies. In addition, we must use the surface treatment technology, such as coating, sand blasting, shot peening and so on.

More specifically, the material as a whole or in part is first melted into the liquid state, and then solidified again in the phase-change processing. It is also notable that the additive processing has been rapidly improved its technological standard. For example, the Airbus Defence and Space has tried to produce the rocket component by using 3D printer of laser processing type, i.e., one of additive processing methods, which is made of platinum-rhodium alloy (Pultarova 2015).

Importantly, the metal removal processing is, as literally shown, to remove the unnecessary allowance, i.e., undue proportion, from either the raw material or semi-finished component as the swarf, whereas the metalworking can deform all the raw material into the finished or semi-finished components without generating the swarf. More importantly, the availabilities of these processing technologies depend upon the properties of product, and thus the production engineer should learn first their overall views together with their advantages and disadvantages from both the technological and economic aspects. Obviously, the machine tool manufacturer and user must learn extremely metal removal processing, especially the machining technology in details.

To ease the following discussion, several turning methods for generating the outer surface of the cylindrical component are shown in Fig. 1.1a–c. At a glance, we have certain difficulties in understanding the differing features of each method. Eventually, we face a problem for choosing the preferable method, when providing

Y. Ito and T. Matsumura, *Theory and Practice in Machining Systems*,
DOI 10.1007/978-3-319-53901-0_1

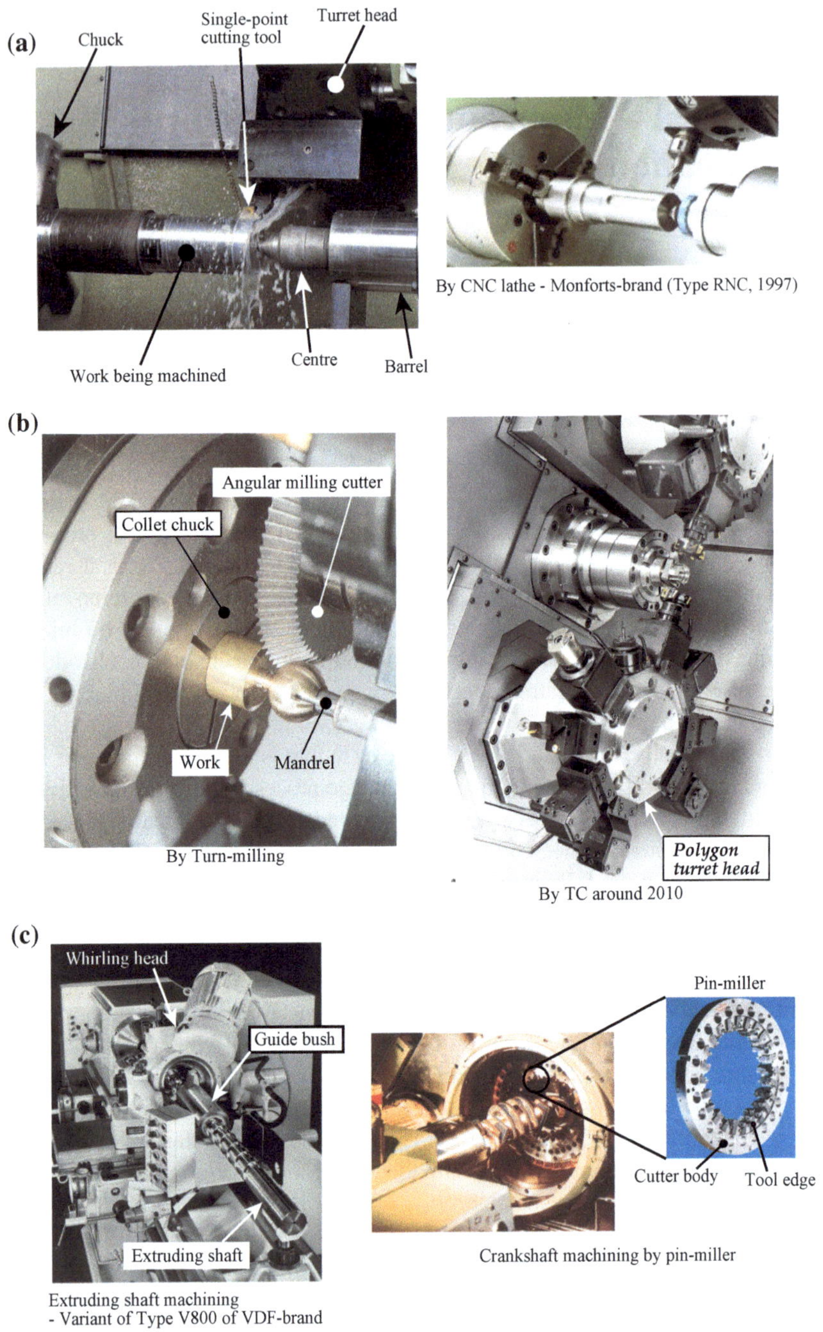

Fig. 1.1 Various turning methods to finish outer surface of work: **a** Centre work. **b** Chuck work (by courtesy of Traub). **c** Whirling and its variant in 1970s

the part drawing. To overcome such difficulties and to solve facing problems, a leading remedy is to systematically classify each processing method by using its *hierarchical structure*. Obviously, it is furthermore desirable to give simultaneously the information for the kind and type of machine tools available for each machining method.

1.1 Various Metal Removal Processing Methods and Available Kinds of Machine Tool

Figure 1.2 delineates such a hierarchical structure in metal removal processing methods, and for the sake of meaningful discussion, we classify furthermore in detail for "Turning" , "Gear cutting" and "Special-purpose grinding" as the samples. Conceptually, Fig. 1.2 is a first-hand view for the metal removal processing as a whole, and very effective in understanding the machining technology.

In general, we used to systematise machining by employing 3rd and 4th tiers in the classification. In this context, Fig. 1.3a–f show some components (parts) in commercial and defense supplies, their machining methods and kinds of machine tool available, if necessary. In addition, the characteristic features of each product will note quickly in the following.

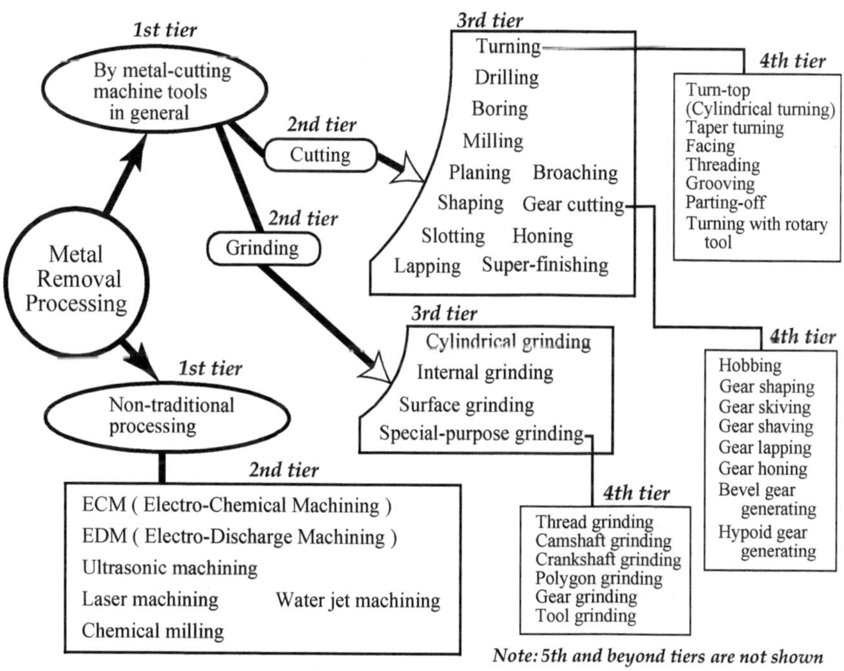

Fig. 1.2 Hierarchical classification for representative metal removal processing

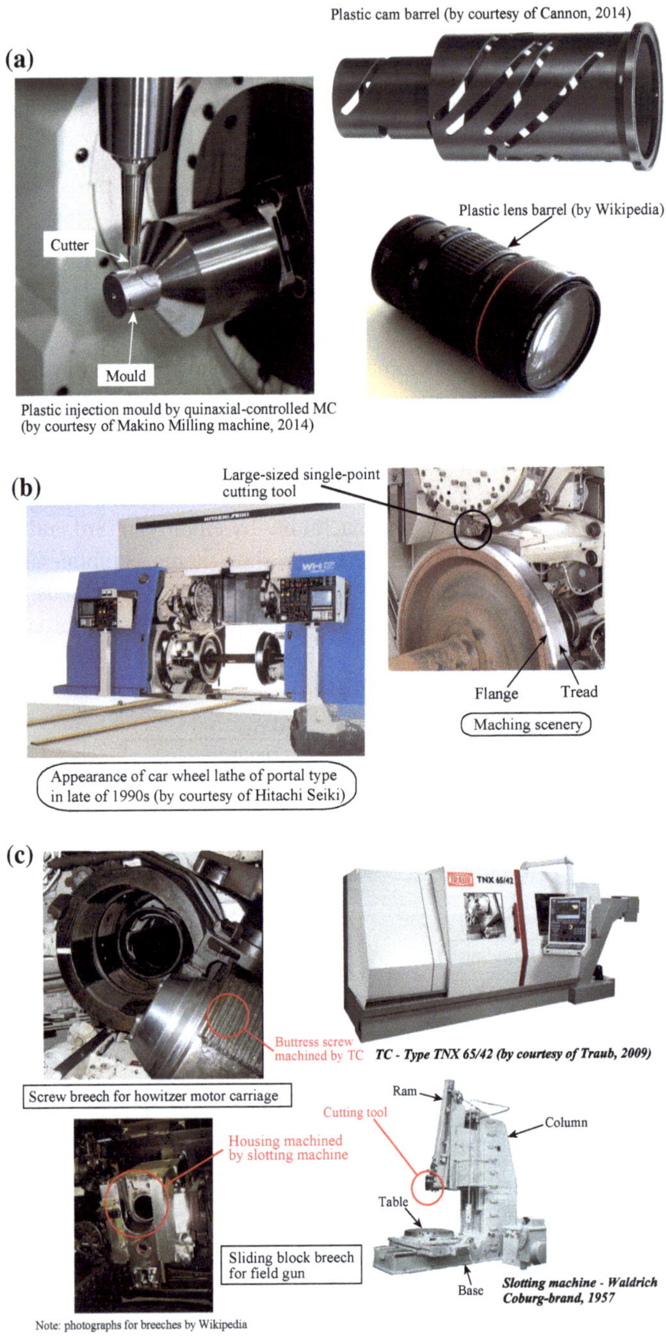

Fig. 1.3 Machining various components for cpommercial and defence supplies: **a** Plastic barrel for telephoto lens of camera. **b** Tread and flange of car wheel. **c** Breech finished by TC and slotting machine. **d** Various machining methods for motorcar components (by courtesy of Mitsubishi Materials, 2007). **e** Friction-welded camshaft (by Izumi Machine Mfg, 2014). **f** Gears in power train

(d)

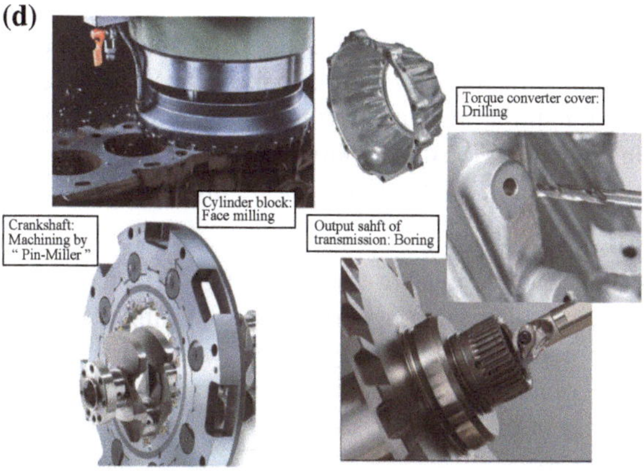

(e)

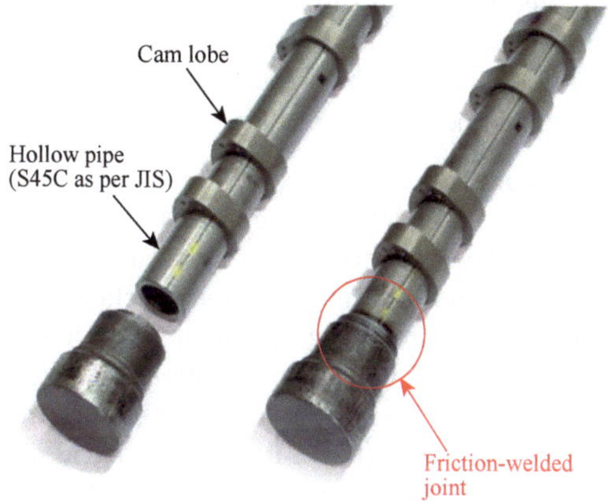

(f)

Fig. 1.3 (continued)

Cam barrel of camera
With increasing use of the plastic barrel in the camera, we need to machine the plastic injection mould. Importantly, the cam barrel becomes very complicated, because of providing multifarious functions to the camera. In general, the details of cam barrel are in company's confidentiality, and thus we can know only that the required machining time is more than 10 h in the ordinal mould.

Car wheel of railway
To prevent the derailment, we must re-true the tread and flange of the car wheel to the regulated shape and dimension at certain interval. Importantly, the car wheel lathe is designed to carry out heavy-duty cutting, and nowadays the under-floor installation type is dominant to render the disintegration and re-assembly of the train useless, resulting in the saving in the maintenance time and cost.

Breech of gun
For the sliding block breech, we use still the slotting machine (slotter); however, this kind is obsolete. Up to 1960s, the slotting machine was one of major players (kinds) in manufacturing, but nowadays we install it at the factory floor only for simple re-machining the part necessary to adjust neatly in the assembly procedure.

Various motorcar components
As widely known, the motorcar industry is the utmost leading user of the machine tool, and thus it is said that we may observe nearly all machining methods once visiting the turnery and concerns in the motorcar manufacturer. In contrast, we must recognise the new waves for producing the lightweight component with less energy loss, so that the fuel efficiency can be improved. We can produce the friction-welded camshaft by welding the hollow shafts with cam lobes separately finish-ground. Obviously, the camshaft is of lightweight and as can be readily seen, the cam lobe can be made of the more suitable material than ever before, which benefits to reduce the frictional energy loss between the cam and tappet. In addition, we ease camshaft machining, whereas the camshaft grinding machine, one of representative kinds in the machining facilities of the motorcar industry, is not necessary any more. We must thus be ware of the rapidly ongoing integration of machining and metalworking, and also its marked influences on the drastic changes in the kinds of the machine tool being demanded.

Gears for power train in motorcar
With growing importance for the improvement of fuel efficiency, one of primary concerns is to finish the lightweight gear with hardened tooth flank. We must thus use gear grinding instead of gear shaving or gear honing. In addition, it is necessary to develop the distortion-less heat treatment for the tooth flank and also chatter-less machining for the thin-walled gearbox.

More specifically, the tooth profile (flank face) is of Involute curve in nearly all cases, and thus the hob or pinion cutter and work move relatively to generate the tooth profile in spur and helical gear cutting. In spiral bevel gear cutting, the quasi-Octoid tooth profile has been used due to certain difficulties in form-generating movement in the past. There is, in principle, no interchangeability between both the straight or spiral bevel gears generated by the machines of different-brands. Importantly, bevel gearing should be generated in a couple by using the machine of same-brand.[1]

In addition, other machining sceneries will be displayed below, so that we can obtain the overall views for machining. Figure 1.4a, b show dry and wet drilling, and also cylindrical and internal grinding, respectively. For the sake of further convenience in the description in Part II, we indicate the typical work and tool setting devices in these machining sceneries. In addition, the description for form-generating movement is also shown in each machining method (refer to Chap. 2), and from such the description, we benefit considerably to understand the movement of the work and tool. For example, drilling can be represented by "*/CZ*", where the drill can travel in Z axis direction while rotating around C axis to generate the hole.

In relation to Fig. 1.4a, b, furthermore, we can suggest the following.

(1) We must eye especially the structural body components facing to the machining space in discussing the machining technology, although all the structural body components and their joints govern the rigidity and damping capacity of a machine tool as a whole.
(2) As will be discussed in Chap. 3 and Part III, we can benefit considerably by supplying the cutting and grinding fluids, whereas we must refine such a fluid harmless to the environment in the final disposal.
(3) The form-generating movement in grinding may be regarded as that of turning apart from the differing properties of the tool.

Apparently, the objective work can be classified roughly into those with axi-symmetrical and box-like shapes, which can be machined by the turning centre (TC) and machining centre (MC), respectively. Importantly, TC can carry out centring, turning, drilling, reaming, tapping, recessing, undercutting, chamfering, knurling and so on. In short, learning TC and turning technologies are first mandatory to understand metal removal processing. In contrast to TC, in which the work rotates while machining, the box-like works including the flat plate—and beam-like shapes are accommodated and fixed on the table of MC, and they can be finished using drilling, tapping, face milling, end milling, boring and so on.

More importantly, nearly all components for the commercial and defence supplies can be finished by the conventional TC and MC nowadays; however, we must use ultra-precision cutting, grinding, honing, lapping and super-finishing, when the machining accuracy should be in the order of better than 1 μm. Of these, grinding has been widely used to improve the machining accuracy after machining the work by TC and MC, and the grinding method can be characterised by its applicability to

[1]Octoid curve is one of tooth profiles like Involute curve. Even in the case of the straight bevel gear, which is easier to generate than spiral bevel gear, there are three ways to determine the tooth depth and face width, and these impede the interchangeability. The representative manufacturers of the bevel gear generator are Gleason, Klingelnberg, Oerlikon and Reinecker.

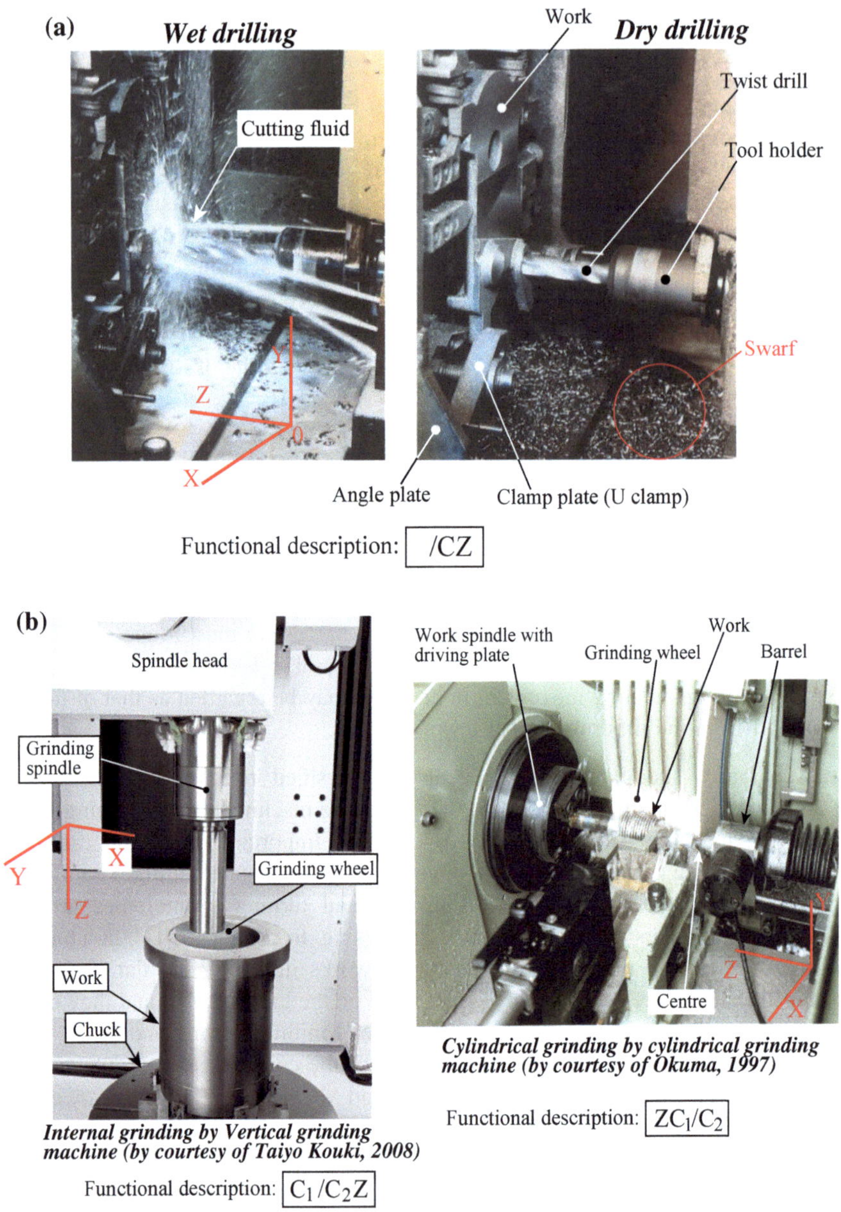

Fig. 1.4 Sceneries of drilling and grinding: **a** Dry and wet drilling with work and tool setting devices [by BMBF (Federal Ministry of Education and Research), 1990s]. **b** Grinding outer and inner surfaces of cylindrical components

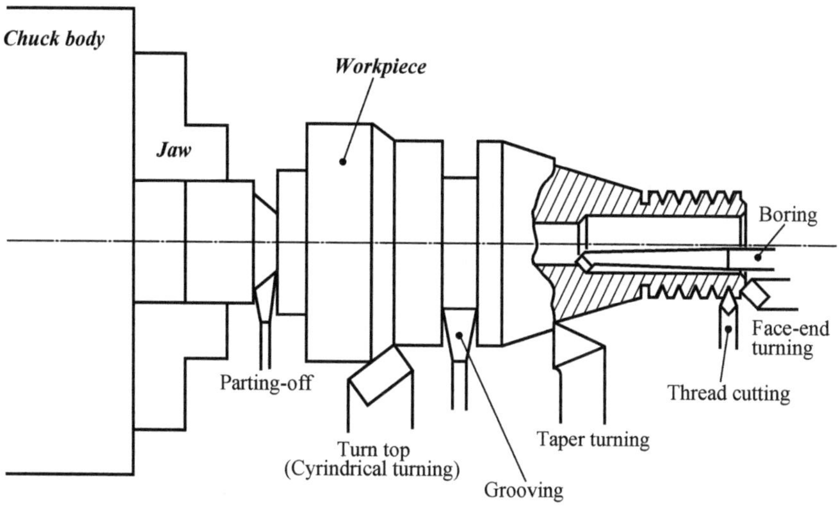

Fig. 1.5 Representative turning methods—4th tier classification of turning

the hardened work by the heat treatment. In fact, TC and MC are only applicable to the non-hardened work apart from that using the special cutting tool.

Admitting that we used to discuss the machining method by 3rd and 4th tiers in the classification, it is natural that machining should be deployed duly into those for 5th and beyond tiers in consideration of a myriad of machining requirements with uncountable varieties in the human society. Importantly, Fig. 1.5 illustrates some methods in turning of 4th tier, and the *turn top* can be detailed and represented by the form-generating movement as shown in Fig. 1.6. More specifically, the turn top can be classified into the fundamental type observed almost daily in turnery and its variants, in which the cutting tool is of single-point in general. In the variants, we use the special cutting tool like the surface broach in certain cases. Of note, skiving was for machining the axel of rolling stock in the past.

As can be readily seen from Fig. 1.6, we may use various machining methods to generate the cylindrical part, and which method we must employ is in dependence upon the machining requirements, and furthermore upon the technological, economic and human resources available at the factory.

With growing importance to cope with the new machining requirement on each occasion, we contrived so far a handful of interesting methods even in turning with rotating tool as shown in Fig. 1.7. In addition, we must be aware of the necessity for differentiating *turn-milling* furthermore as shown in Fig. 1.8.[2] In fact, there are two variants in *turn-milling* by the relative allocation of the rotating axes for the work and tool, i.e., form-generating movement, and obviously the obtainable benefits

[2]*Turn-milling* sounds like new, but National Broach & Machine Tool in USA contrived already this method in the 1950s, and called "Roto mill".

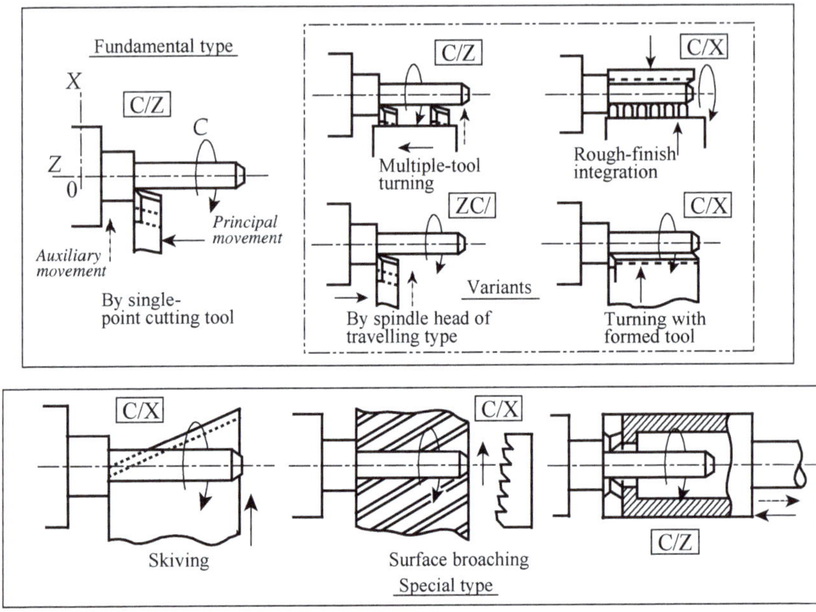

Fig. 1.6 Fundamental type, its variants and special types in turn top

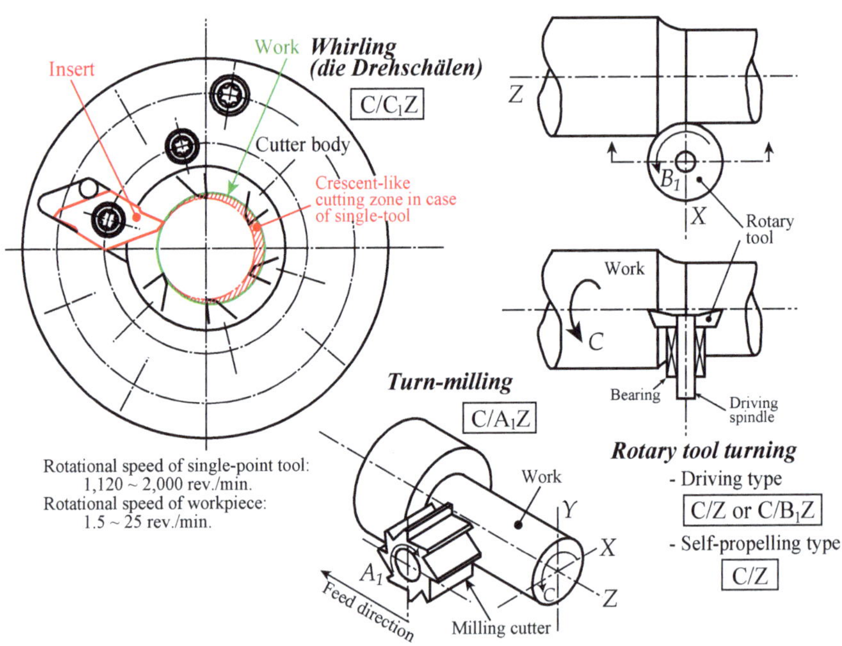

Fig. 1.7 Turn top with rotating tool

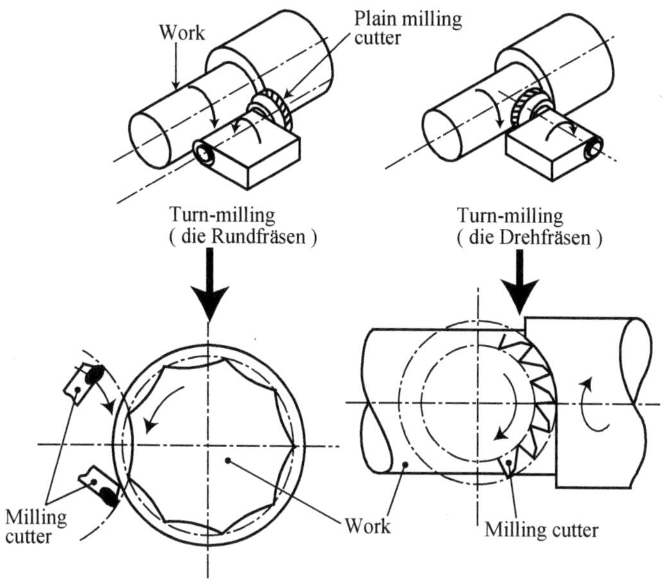

Fig. 1.8 Variants in turn-milling

differ from each other as will be shown in Fig. 1.9. Reportedly elsewhere, it is very interesting that the German technical term discriminates clearly these as shown together in Fig. 1.8.

As widely known, the rotating tool system in turning grows recently its importance, and necessitates the clarification of the comparative advantage and disadvantage. As can be readily seen from Fig. 1.7, there are obvious differences among *whirling*, *rotary tool turning* and *turn-milling*. More specifically, Fig. 1.9 depicts the differing features by choosing the three attributes, i.e., configuration complexity of component manageable, shape geometry possible to generate and type of swarf. In fact, there are considerable differences in each form-generating movement. In addition, we can suggest further characteristic features as follows.

(1) Whirling: The cutting tool is of single-point, and it can rotate around the work by centring the differing position from that of work. As a result, the swarf of crescent-like shape can be generated. This method has been very famous as the *Mach high-speed screw cutting* to produce the lead screw of lathe in the past.
(2) Rotary tool turning: The cutting tool is half cup—or cup-like shape with monolithic edge, which is propelled either freely driven by the cutting force (self-propelling type), or rotates by external driving power. As a result, it appears that the succeeding cutting edge approaches to the cutting point continuously, and thus we can expect the cooling effect of cutting edge after its disengagement from the cutting point (Thomas and Lawson 1976). In principle, this method is applicable to turning of the *difficult-to-machine material*, e.g., heat-resistance steel.

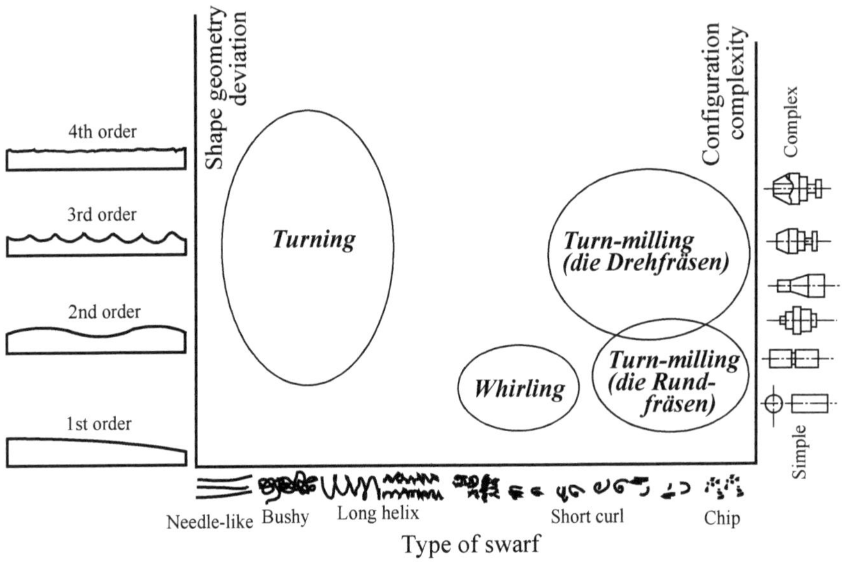

Fig. 1.9 Comparison for available ranges of turning in general, whirling and turn-milling in consideration of three attributes (by Sorge)

(3) Turn-milling: The cutting tool is milling cutter, and there are a handful of variants depending upon the relative position of the rotating axes of the milling cutter and work, and also the kind of the milling cutter. In fact, both the down— and up-cut patterns in milling are available, and as can be easily imagined, *turn-milling* is for heavy-duty cutting (Sorge 1984, Schulz 1990). On the strength of beneficial features of *turn-milling*, Heisel and Ruziczka (1994) applied it to boring and drilling, provided that the milling cutter should be rotated inversely to that of work to prevent the strong torsional vibration.

In short, we may use various turning methods in accordance with the machining requirement, and thus we must produce the choosing flow or diagramme for the preferable method. As can be easily imagined, the classification system of hierarchical structure is very convenient as will be clear from the above.

Admitting the hierarchical characteristics in the machining method, we must discuss furthermore whether structural configuration entities, e.g., structural body components surrounding the machining space, have such properties or not. In fact, the machine tool and attachment have similar characteristics, although involving uncertainty to some extent. Figure 1.10 shows a trial for classifying the kinds of machine tools at present (1st and 2nd tiers), and for example, TC may be furthermore detailed as shown in Fig. 1.11. In short, the classification has been, in part, established, but is not in fruition as yet. This is because the machining function-integrated kind has been and being prevailed, resulting in the complexity

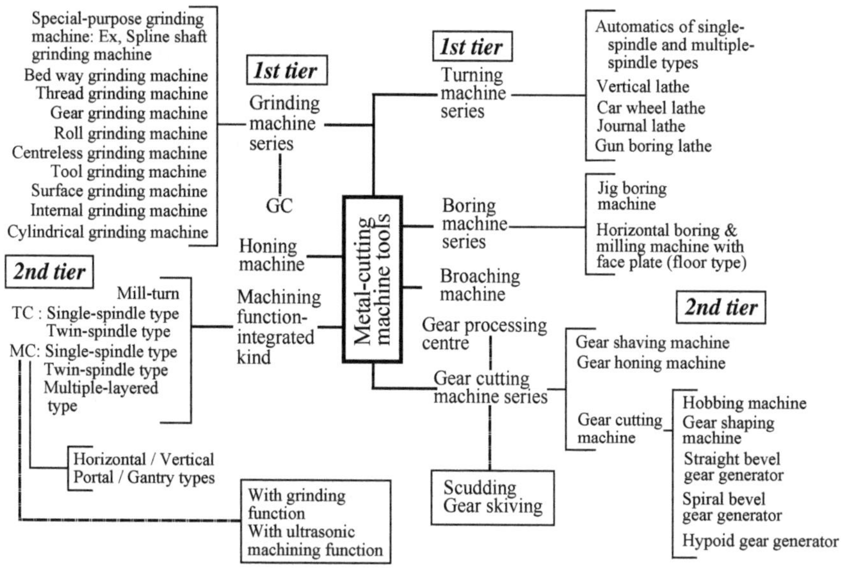

Fig. 1.10 Machine tool classification by structural configuration in 2010s

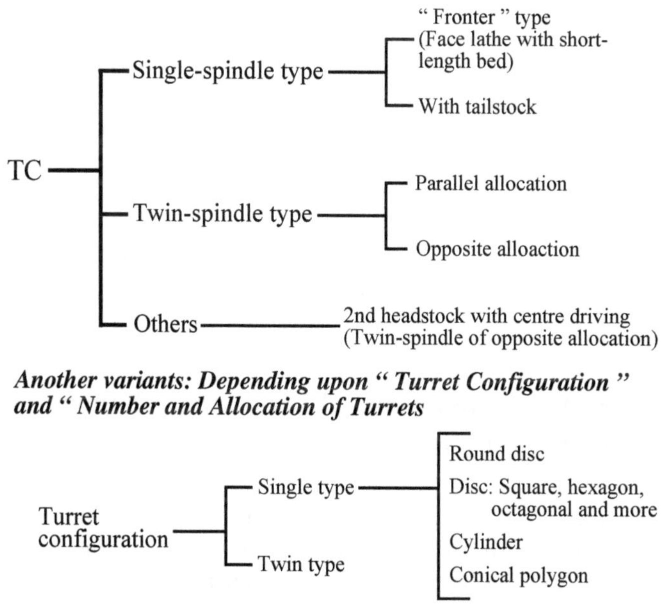

Fig. 1.11 Classification of TC in 3rd and beyond tiers

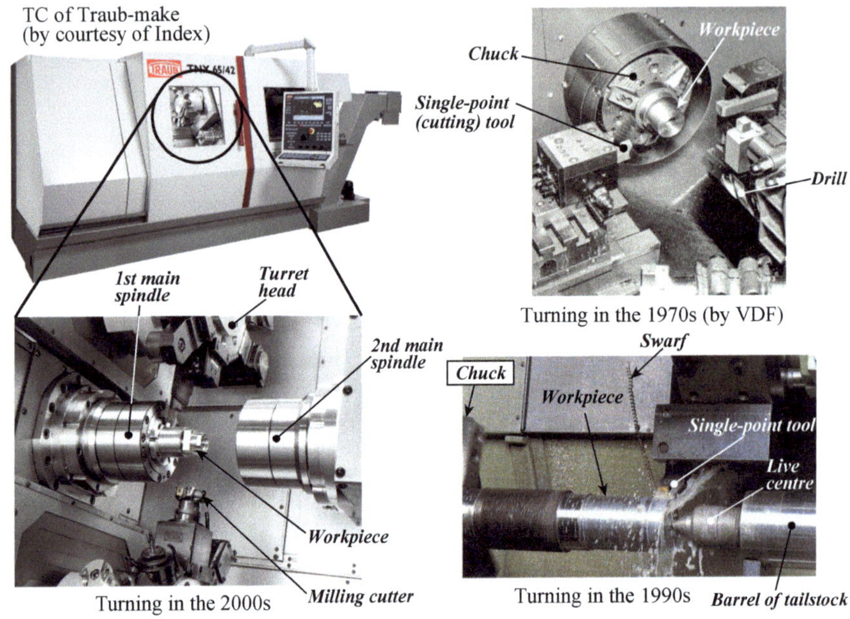

Fig. 1.12 Various turning methods in old—and new-fashions—for axi-symmetrical workpiece

in structural configuration. In contrast, the cutting tool is capable of systematising by the hierarchical structure when using the modular principle (refer to Part II).

To this end, it is worth suggesting that the machining technology maintains its essential features even when the technology itself and concerns advances in due course. Figure 1.12 shows various turning operations of old—and new-fashions, which are for axi-symmetrical work and carried out by using the engine lathe in the past and TC at present. Although fundamentals of turning are the same, various method-integrated turning have been prevailed with the advance of NC technology.

1.2 Various Metal Non-removal Processing Methods and Their Characteristic Features

Figure 1.13 shows a classification system for the metal non-removal processing available at present in accordance with the same idea applied to Fig. 1.2.[3] Actually, cold forming and welding are furthermore classified in detail. As already stated,

[3]Figure 1.13 includes welding from the ease of understanding for the supply of semi-finished component to the machine tool. Welding is, in principle, one of the phase-changing processing methods with material additive function. For example in arc-welding, a part of raw material is

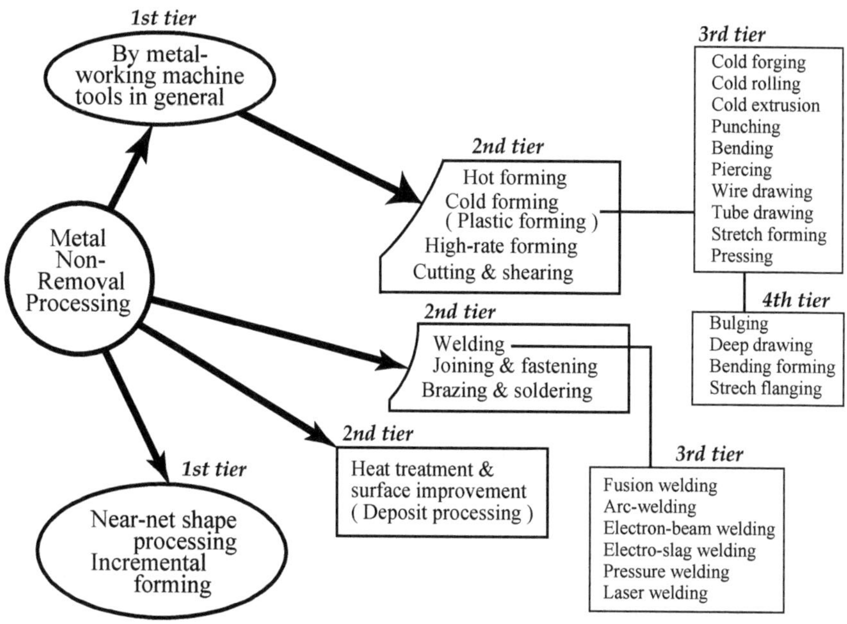

Fig. 1.13 Representative metal non-removal processing

machining is dominant, because nearly all components in the product can be finished by machining, whereas metal non-removal processing can produce the raw material, semi-finished and finished components, depending upon the processing requirements. Then, both the raw material and the semi-finished component can be, in general, supplied as the work to the machine tool, and finally finished by the machine tool. Having in mind such a relationship between the supply of either the raw material or semi-finished component and the machining technology, some quick notes will be given in the following.

(1) There is an inevitable exchange for using either machining or metalworking in accordance with the advent of the new technology, and often by reason of reducing the production cost, even when processing the same work.
(2) In machining, the integration of several different methods has been prevailed as exemplified by turn-milling, in order to reduce the preparatory work, to positively use one-set machining and to shorten the machining time. Following the machining technology, such the integration of processing methods is under way in the metalworking.

(Footnote 3 continued)

melted and solidified after forming the beat, where the electrode provides the welding portion with small volume of material necessary.

(3) Metalworking has already been supported by the NC technology following NC machine tool, and such NC metalworking includes forging, bending, rolling, drawing out, hammering and so on.
(4) In metalworking, it is vital to enhance the processing accuracy and also the compatibility with batch manufacturing. In retrospect, it is said that the former was first tried within a Japanese Big National R&D Program called Flexible Manufacturing System Complex Provided with Laser (FMSC) launched out in the late 1970s (MITI 1980s). In contrast, a leading subject related to the latter was proposed by Verein Deutscher Werkzeugmaschinenfabriken e. V. (VDW) around 2005.

As a result, metalworking is applicable to batch manufacturing as well as large volume and a less variation manufacturing nowadays, and obviously, the processing accuracy has been improved amazingly. Figure 1.14 shows a schematic view of step-land bearing for the CD player, which should be machined to generate the sharper edge at the step. Although being in the trial stage, we can produce it by using cold pressing or coining. Geminately with this example, the practical use of the friction-welded crankshaft implies again the difficulty in establishing the choosing guide for the preferable processing method as will be discussed in Chap. 2 (refer to Fig. 1.3e).

Figure 1.15 shows an example of typical integration of processing methods across the border between metalworking and machining. In fact, the turret punching press can finish the sheet metal by punching, tapping after embossing, and also

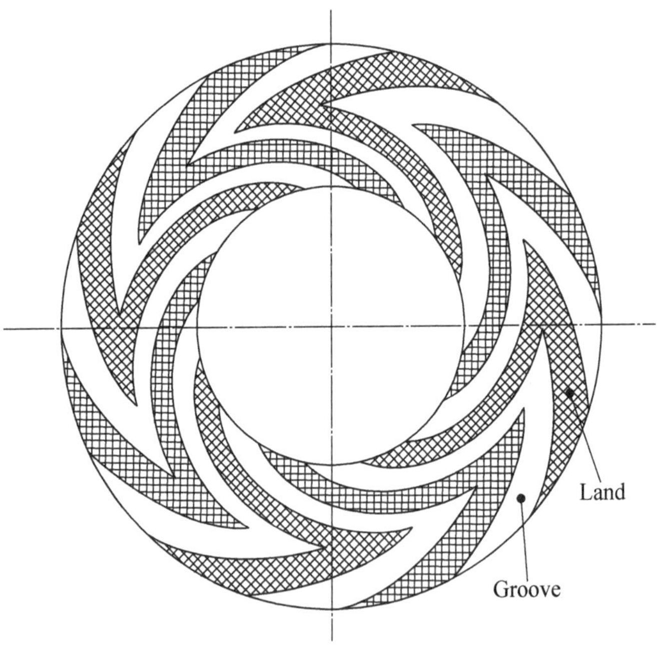

Fig. 1.14 Thrust bearing of spiral groove type possible to cold pressing

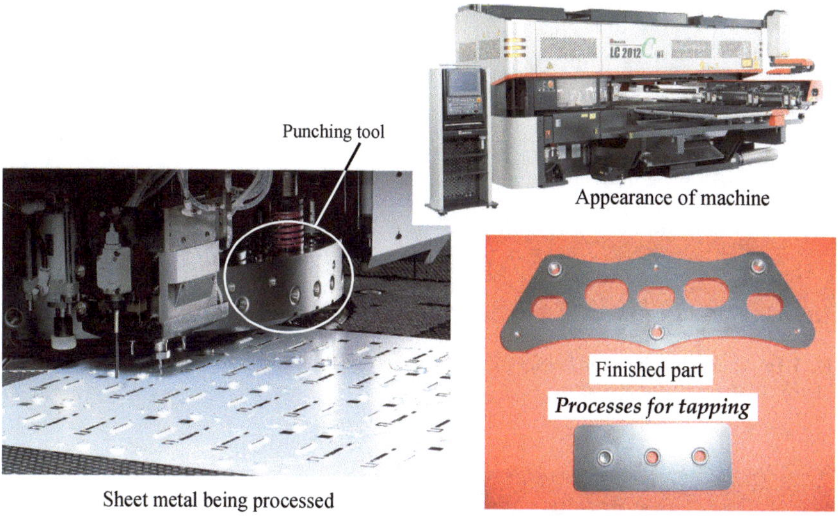

Fig. 1.15 Sheet metal processing with turret punching press (by courtesy of Amada, 2010)

routing by the laser. More specifically, the punch can pierce small holes with various shapes, whereas the large hole and aperture can be produced by means of routing with the laser.

Admitting that metal non-removal processing has been advanced amazingly during the last few decades, it is necessary and inevitable to understand its essential features. In this context, we must first be aware that metalworking can be, in general, classified into two-pronged spheres, i.e., cold and hot working. Cold working means that the metal undergoes plastic deformation below its lower critical temperature, e.g., that of maintaining the *Pearlitic structure* unchanged in the case of steel (just over 700 °C). In contrast, hot working is carried out when the metal is above the lower critical temperature (Chapman 1972).

More specifically, cold working occurs the distortion of crystal structure in the metal, rendering it harder and more brittle. Such a distortion may, however, be recovered by tempering or normalising. In contrast, cold working can facilitate finish processing with acceptable part accuracy, and also render the post-cutting process useless with the advance of the related technology to some extent.

Against to this context, hot working can deform the metal by using the mould even with relatively deep cavity, when its temperature is above the upper critical, together with refining the crystal structure. Importantly, hot forging benefits to save a great deal of time and expense, because it may be forged to the raw material of near-net-shape, i.e., that with shape very close to the finished one, as shown in Fig. 1.16. More importantly, hot forging may control the flow of the metal and result in the increase of material strength. In the past, the friction drop stamp and

Fig. 1.16 Crankshaft being in hot forging (by courtesy of Prof. Emeritus Kiuchi, the University of Tokyo)

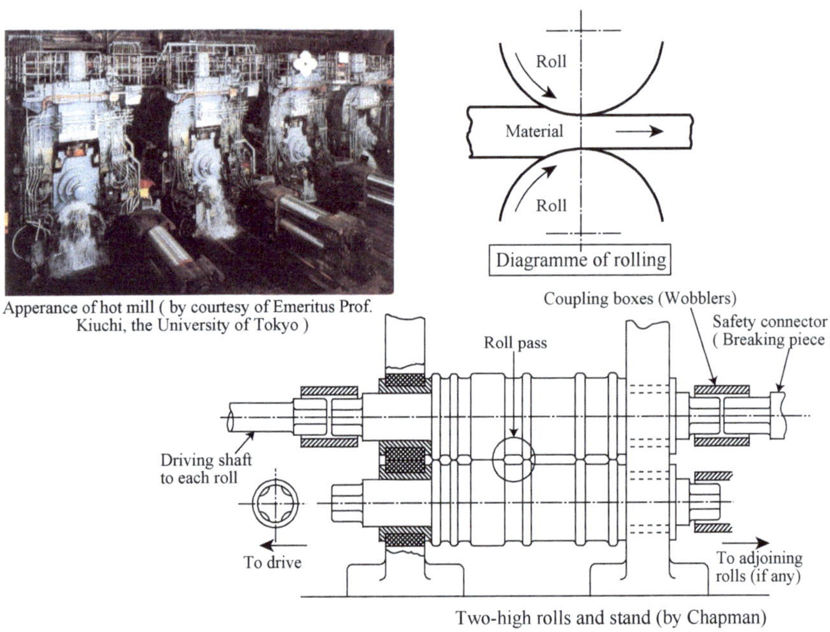

Fig. 1.17 Schematic views of hot-rolling

pneumatic hammer were prevailed, but now the high-speed mechanical forging press is popular for hot forging.

Figure 1.17 shows a group of machines for hot rolling to produce the strip and sheet. In hot rolling, the thicker sheet can be, in principle, stretched out to thinner one such as shown in the right above of Fig. 1.17. In addition, we can produce the bar with various cross-sectional shape by changing the roll pass, which can be

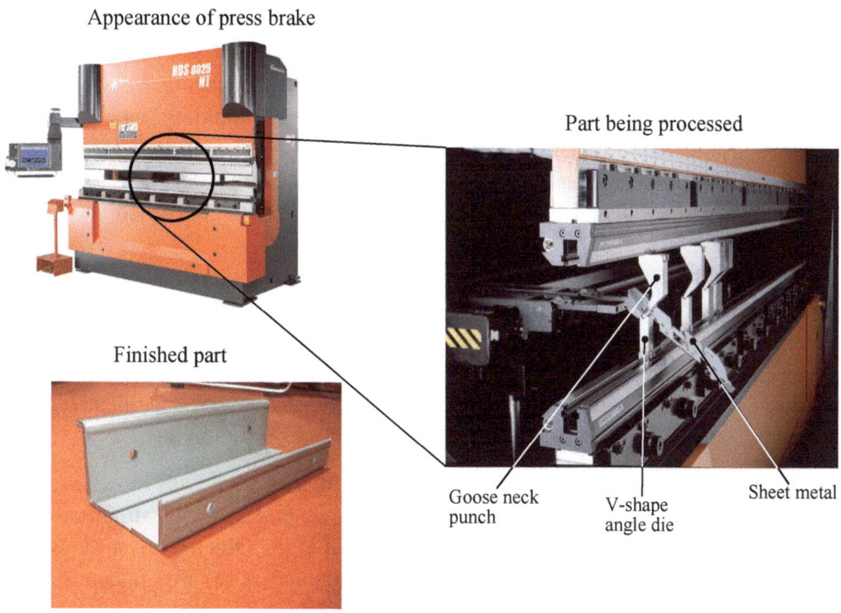

Fig. 1.18 Cold sheet metal bending and press braking (by courtesy of Amada, 2010)

formed between the top and bottom rolls as shown in right below of Fig. 1.17. In due course, hot rolling has a handful of variants depending upon the allocation of rolls, e.g., three-high rolls and four-high rolls. Figure 1.18 demonstrates furthermore cold bending (bent cold), in which the sheet metal can be bent by using a pair of punch and die. In fact, the sheet metal can be placed between the die and the punch of various shapes and duly bent by step-wise way.

To this end, we suggest extremely that metalworking can be characterised by its deformation mechanism with free-boundary problem including uncertainties. In contrast, machining can be, in general, carried out under constraint boundary condition. As a result, metalworking has been so far believed very experienced-based engineering, although having been endeavouring to sublimate the experience-based knowledge into the engineering theory. Such a belief is, in part, correct, and in retrospect, a trial to interpret material processing from theoretical point of view is, for example, credited to Thomas of Technical College Worcester (1972).

1.3 Phase-Changing Processing

Within phase-changing processing, the utmost representative is casting, and casting can be classified depending upon the material to be processed and casting method. In general, the available material ranges from the cast iron and cast steel, Al-alloy and Mg-alloy, through plastics, to the ferrite resin concrete.

Within various methods, the sand mould casting is very typical; however, another methods are now available including the variants of sand mould casting, e.g., shell moulding, investment casting, die casting and centrifugal casting.

Apart from the component made by plastic injection moulding, nearly all castings are provided the machine tool with the raw material and then machined in accordance with the indication in the part drawing. It is thus desirable to learn, at least, their fundamental properties, especially those concerned with the machinability.

For example, Fig. 1.19 illustrates objective components, their moulds, wooden patterns necessary, and so on. As will be clear from Fig. 1.19, a cast component or structure can be produced by pouring the molten metal into a mould made to its shape, where the mould is produced beforehand by means of the wooden pattern having the shape similar to the component required. More specifically, some characteristic features of the casting will be stated in the following.

(1) The top surface of a casting is liable to be rather spongy with blowholes, and thus the important portion of the casting, e.g., guideways of structural body component in machine tools, should be placed at the underneath of the bottom box in the mould. This spongy zone is unwelcome as the raw material and is caused by the entrapped gases unable to escape before the solidification of metal.
(2) Because of occurring the contraction while cooling, the pattern should be a little bit larger than the finished casting. In case of cast iron, the approximate

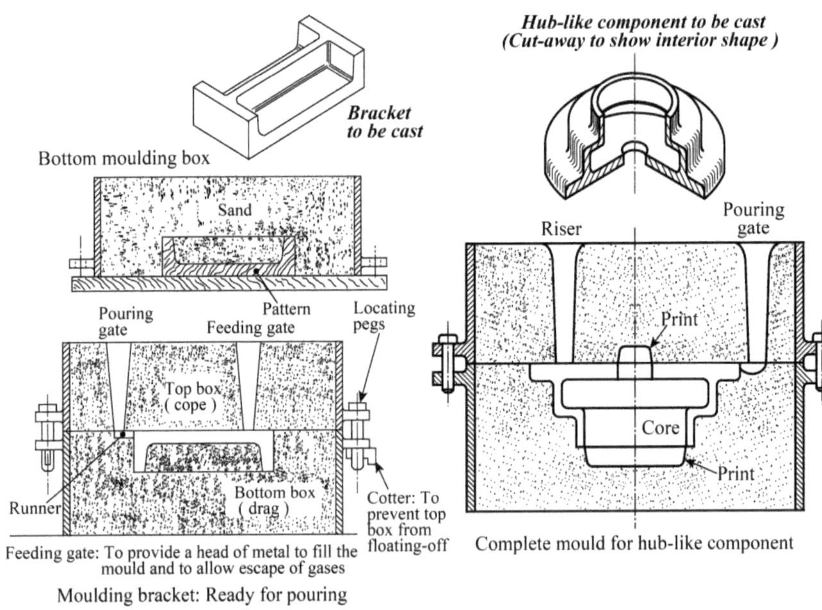

Fig. 1.19 Representative methods for sand mould casting (by Chapman)

contraction per unit meter is between 8 and 9 mm (around 1/120). Thus, the patternmaker employs the "*Contraction rule*"; however, the contraction states depend upon the complexity in the component shape and also the structural configuration to a large extent. In addition, we must consider the machining allowance. These imply that we need to have the skilled patternmaker to produce correctly the casting in accordance with the requirements.

(3) The wooden pattern is, in general, made of well-seasoned soft wood, e.g., yellow pine.
(4) We must use the core when the casting is more complicated with holes, aperture and cavities.

Obviously, we must consider the experienced knowledge at the factory floor, which is often called by the jargon and differ from one another by casting procedure. In addition, we have various kinds of the cast iron, e.g., ranging from the grey cast iron to alloy cast iron. In many respects, Fig. 1.20 shows some castings together with specifying the material and casting procedure.

Importantly, we have employed the cast or steel welded structure for the structural body component in the machine tool. It is interesting that the casting includes that made of concrete. Of these, the cast iron structure is even now prevailed, and in this context, we use the grey, alloy including that with low thermal expansion, and spheroidal graphite cast iron, i.e., ductile (Mg additive) and Meehanite (Ca additive). Of these, the grey cast iron is dominant from viewpoint of

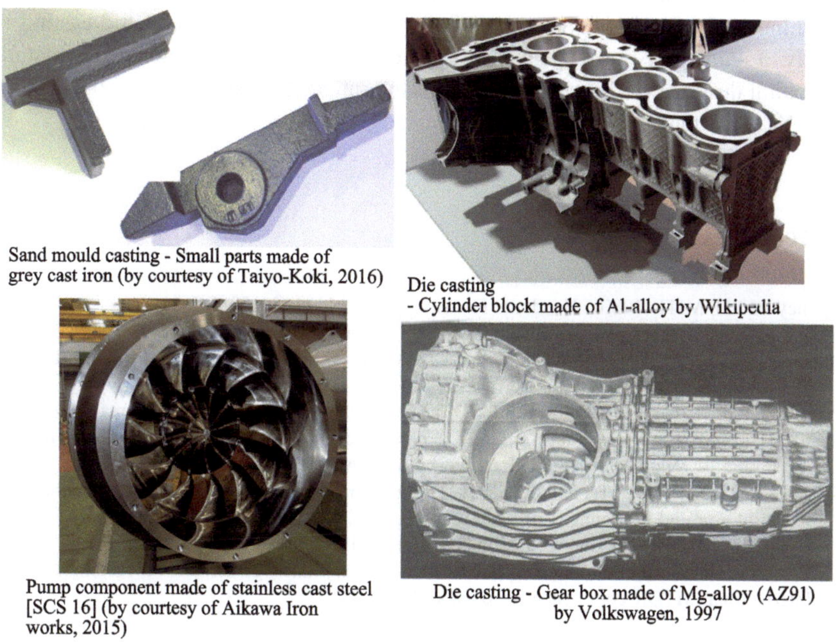

Sand mould casting - Small parts made of grey cast iron (by courtesy of Taiyo-Koki, 2016)

Die casting - Cylinder block made of Al-alloy by Wikipedia

Pump component made of stainless cast steel [SCS 16] (by courtesy of Aikawa Iron works, 2015)

Die casting - Gear box made of Mg-alloy (AZ91) by Volkswagen, 1997

Fig. 1.20 Some examples of castings

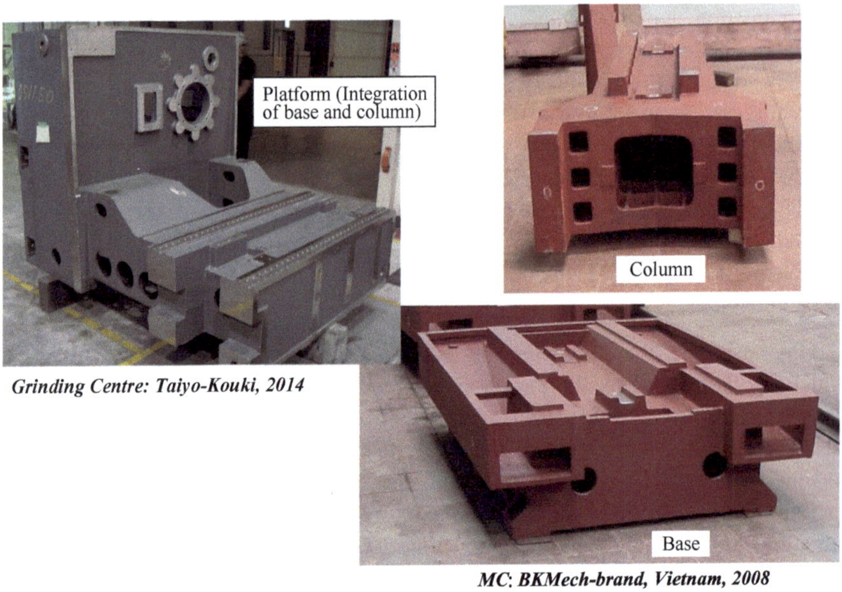

Fig. 1.21 Structural body components made of grey cast iron for machine tools (Photographs taken by Ito)

economisation, and the spheroidal graphite cast iron is often employed for the machine tool with the higher performance and quality, although such a casting is costly. Figure 1.21 shows some cast structures, and in due course, the mature structural designer of the machine tool has the satisfactory knowledge about casting as well as its machining technology.

References

Agency of IST MITI (1980s) Handouts for tsukuba test plant with FMSC

Chapman WAJ (1972) Workshop technology part 1, 5th ed. Edward Arnold, London: 49–54

Heisel U, Ruziczka G (1994) Machining bores with milldrilling and millboring. Prod Eng I–2:67–70

Pultarova T (2015) World's first (partly) 3D printed rocket engine tested. E & T magazine

Schulz H (1990) High speed turn-milling—a new precision manufacturing technology for the machining of rotationally symmetrical workpieces. Ann CIRP 39–1:107–109

Sorge K-P (1984) Die Technologie des Drehfräsens. Darmstädter Forschungsberichte für Konstruktion und Fertigung. Carl Hanser

Thomas GG (1972) Production technology. Oxford University Press

Thomas RM, Lawson RLJ (1976) Application of a rotary turning tool. In: Proceeding of 17th International MTDR conference 1976, MacMillan

Chapter 2
Choosing Guide in Form-Generating Movement and New Approaches to Machining Technology—Correlation Diagramme Among Machining Method—Tools—Attachments—Machine Kinds

In discussing the machining technology, a facing crucial issue is to establish a choosing guide for the machining method, when providing us with the part drawing. For such a purpose, we must conduct the meticulous investigation into the mutual correlation among a group of the structural body components allocated around the machining space (hereafter, call it "*Structural Configuration Entities*"), attachments like tool bracket and chuck, cutting tool, work and so on. These machine body and facilities can feature the machining space to various extents.[1]

In this context, it is worth suggesting that nearly all the books so far publicised are far from fruition to respond such requirements of wider scope. Reportedly, we have discussed the machining technology so far with narrower scope by only eying the machining method and its mechanism. In addition, it has been, as reported elsewhere, believed that machining technology is very experienced-based engineering problems, although having been endeavoured to sublimate the experience-based know-how into theoretical-based konwledge. Such a belief is, in part, correct, but not correct in certain extent. In retrospect, a trial to interpret material processing from theoretical point of view is, for example, credited to Thomas of Technical College Worcester in (1972) as already introduced.

Conceptually, we can interpret the machining space as to be a cross point of the production and utilisation technologies of the machine tool. From another viewpoint, we can envisage both the material and information flows in the machining space as shown in Fig. 2.1.

From Fig. 2.1, we can obtain duly necessary information according to the machining requirement. It is thus very convenient to discuss, with wider scope, the leading issues in the machining technology by placing the stress on the machining

[1]The description, discussion and concerns in this Chapter will be done, provided that the machine tool available for corresponding machining has acceptable functionality and performance specifications. The basic knowledge about the structural design principle of the machine tool will be stated quickly in Appendix B. Such the design principles are mandatory even for the machine tool user.

Y. Ito and T. Matsumura, *Theory and Practice in Machining Systems*,
DOI 10.1007/978-3-319-53901-0_2

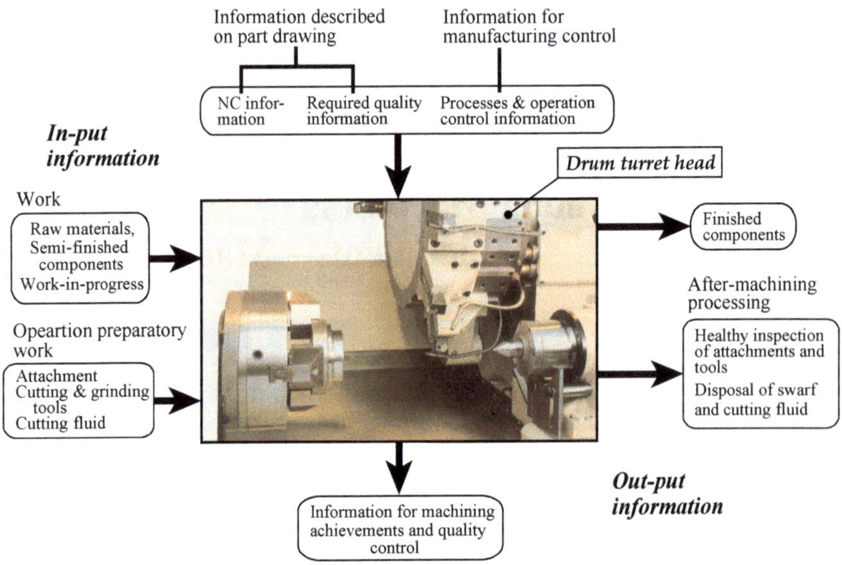

Fig. 2.1 Material and information flows in machining space

space. Of special note, flexible manufacturing means simultaneous management for both the material and information flows within due time, i.e., no redundancy in time and space domain, in the machining space (Weck 1974).

2.1 Facing Problems in Choosing Guide for Preferable Machining Method

Obviously, the choosing guide for the recommendable machining method is one of the leading engineering problems related to the machining space, and for such a purpose we can propose an idea as shown in Fig. 2.2. As can be readily seen from Fig. 2.2, it is possible to enumerate the potential processing, constraint conditions and evaluation indices; however, a choosing guide for practical use is far from completion. One of the hindrances is the amazing advance of the metalworking technology, which may often replace or force to improve the machining technology so far used as exemplified by the friction-welded crankshaft (see Fig. 1.3e).

To simplify and clarify our discussion, thus, we focus now our objective on a choosing flow and diagramme applicable to machining in practice. In this context, we must first suggest the involving problems within Fig. 2.2 as follows.

(1) No evaluation functions are established as yet.
(2) Leading constraints are in reciprocal relation to one another, and as a result, we cannot determine the optimum constraint condition, but may use the preferable

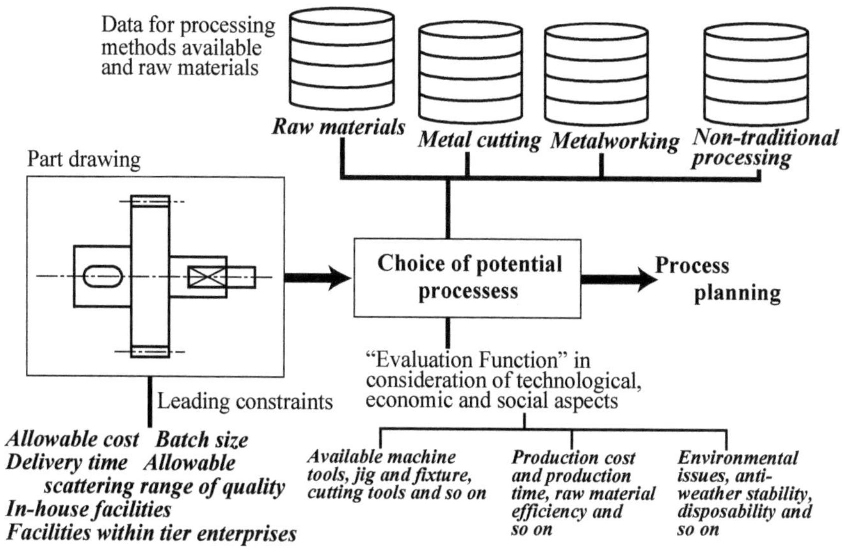

Fig. 2.2 Searching concept of rational processing method

constraint only, resulting in a handful of constraint conditions possible in certain cases.

(3) Reportedly, there are the three leading attributes in the design of machine tools, i.e., better machining accuracy, higher rotational and travelling speed and heavy-duty cutting capability. These three attributes are also in reciprocal relation, and thus each machine tool can be characterised by its own design features. In consideration of primary compatibility of machine tool available with the machining requirement, we must determine the machining method.

(4) With the advance of the technologies related to the machine tool, attachment and tool, the machining method available changes to various and large extents as exemplified by the advent of the machining function-integrated kind, e.g., "*Mill-turn*". In certain cases, the machining method changes drastically by the improvement of the work material itself.

Of these, the novel contrivance for the machine tool affects the choosing guide for the preferable machining method to large extent. For example, Fig. 2.3 shows the interesting gear cutting methods being prevailed, i.e., *gear skiving* and "*Scudding*", where the scudding is a variant of gear skiving. The gear skiving was already contrived in the 1980s, but has not been positively used so far (Spur and Stöferle 1980). In the 2000s, the gear skiving has been re-considered its effectiveness, e.g., high productivity by cutting the gear blank with high hardness and availability for cutting the stepped gear like gear shaping (Schmidt and Bechle 2003).

More specifically, the gear skiving is a synergy of hobbing and gear shaping, and characterised by the tooth profile-generating movement as follows.

Gear skiving for work of high hardness
(by courtesy of Pfauter)

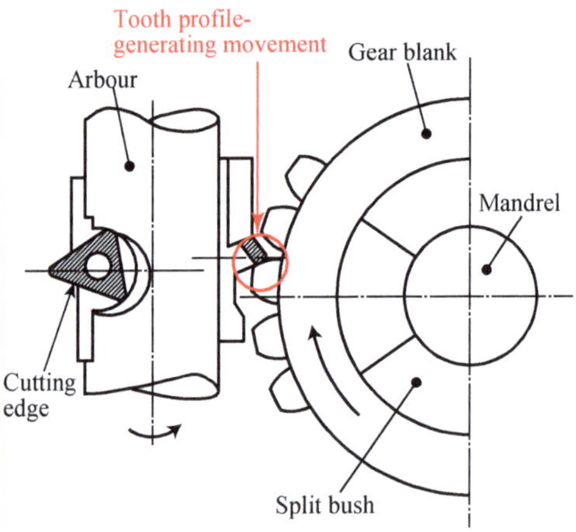

Concept of "Scudding"
(Depicted from publicised material of Wera)

Fig. 2.3 New proposal and resurrected method in gear cutting

(1) The rotating axis of pinion cutter engages with certain angle to the rotating axis of the gear blank.
(2) Maintaining such the engagement of the axes, the pinion cutter removes unnecessary allowance along the rotational axis of the gear blank.

Importantly, the gear skiving can be recently carried out by the machining centre (MC), and thus may accelerate the development of the gear production centre.

In addition to the technological aspect, a root cause of difficulties lies in the saving of the machining cost, i.e., economisation, provided that we can guarantee the finished quality of the component as indicated as well as maintaining the delivery time. Importantly, the machining cost depends upon especially the machining pattern, e.g., either batch or large volume and a less variation productions.

Following such conceptual discussion mentioned above, we must then detail the choosing instruction for the preferable machining method from the viewpoint of factory floor. In fact, we must consider the following factors and their complicated combination while machining the component within the in-house and also at subsidiary.

(1) Machining facilities, their capabilities and healthy conditions, and also expectable machining cost.
(2) Available cutting and grinding tools, jig and fixture, and so on.
(3) Machining accuracy and surface quality to be finished, necessity of heat treatment, properties of work material and so on.
(4) Batch size, allowable delivery time and cost.
(5) Skill qualification of operator and requirements for extremely special machining.

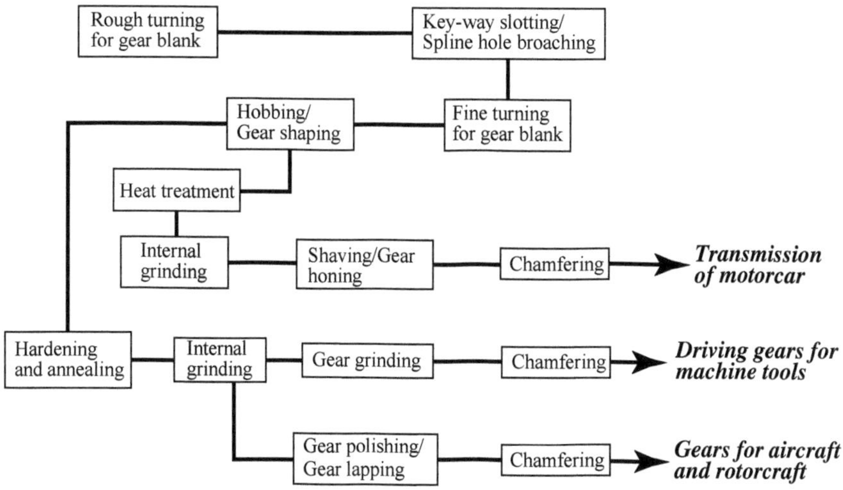

Fig. 2.4 Various machining procedures of spur gear depending upon product kind

Now let us show the two examples, which cause the difficulties in choosing the preferable machining method.

Machining Procedures in Spur Gear

Figure 2.4 illustrates the various machining procedures of the spur gear, which depend upon the product kind. Up to gear cutting, there are no differences; however, after then the gear can be finished by various methods depending upon the product kind. Of course, this is because of fulfilling the function and performance required by each product.

Importantly, the gear for the transmission of motorcar is being finished by gear grinding to reduce the frictional energy loss at the tooth profile (refer to Fig. 1.3f).

Surface Roughness and Machining Method

Figure 2.5 shows a sample work to learn the sensory flair for the quality of the finished work. The surface roughness is, as widely known, one of the indices to evaluate the work quality, and there are various choices in machining methods even when the surface roughness required is the same.

In short, more versatile in the machining requirements, the more difficult is the determination of the utmost preferable machining method.

Against to this context, it is expectable to establish a choosing guide for the suitable machining method by applying the design methodologies in the machine tool structure, i.e., concept of "*Flow of Force (der Kraft Fluß)*" and description of form-generating movement (functional description). In many respects, the *flow of force* is very effective to represent the machining space, i.e., linkage among structural configuration entities, by the direct graph as will be discussed later. Conceptually, we may clarify the correlation between the machining method and the machine tool by such a directed graph.

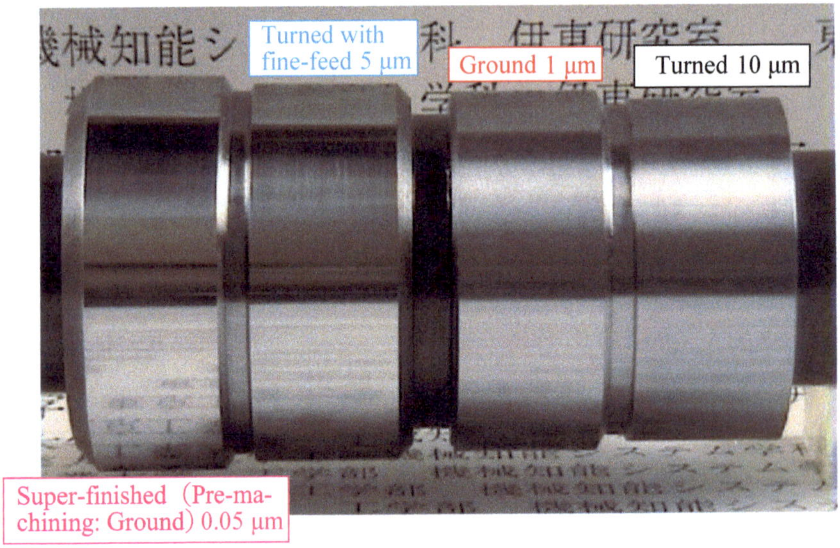

Fig. 2.5 Relationships between surface roughness and machining method

In the following, we will thus discuss the concept of *flow of force*, description of form-generating movement and its application to the choice for suitable machining method, and also the representation of linkage in machining space.

2.2 *Flow of Force* and Description of Form-Generating Movement

It is believed that the concept of *flow of force* was German-originated supporting tool (design methodology) for the structural design of machine tools. In short, the *flow of force* can visualise all the structural body components within a machine tool as a whole, which are being burdened the cutting force, and in the case of large-sized machine tool, its weight.

As can be readily seen from Fig. 2.6, the cutting force acts first on the main spindle in the headstock and barrel of tailstock through the work, and also on the turret head through the cutting tool. These structural body components are mounted on others, e.g., cross slide, bed and base. For the ease of understanding, we display such 3-dimensional loading states by the arrowed line as shown in Fig. 2.6, and thus call it the *flow of force*.

The concept of *flow of force* is very simple, but very effective in conducting the design methodology for the structural design, because the *flow of force* can be converted into the directed graph in the mathematics, where we can represent the

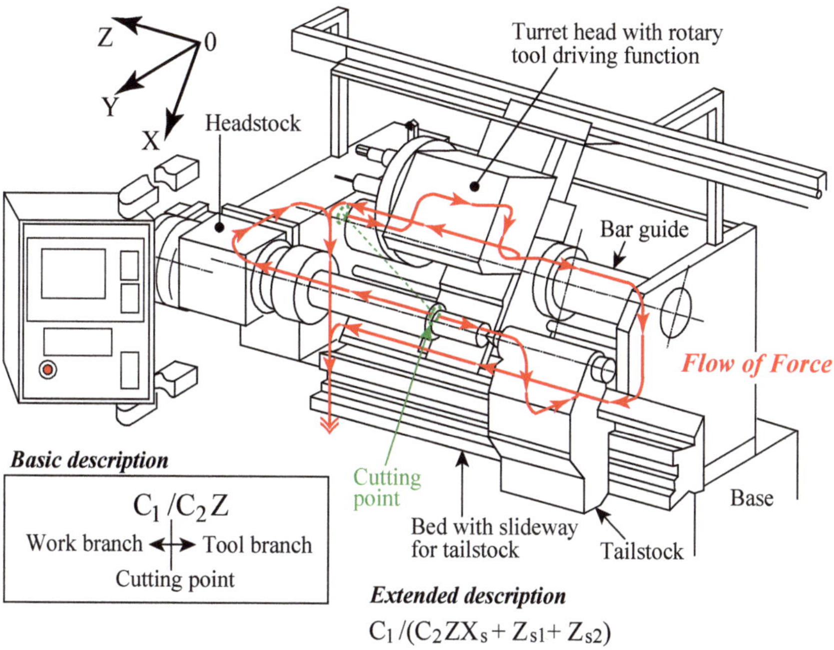

Fig. 2.6 "Flow of Force" in and "Functional Description" of CNC lathe (Modified RNC Series of Monforts-brand in 1990s)

structural body component and flow of force by the "*Vertex*" and "*Directed Edge*", respectively (Ito 2008).

In the design methodology for the machine tool structure, another issue is the description of the form-generating movement, i.e., functional description. More specifically, we can represent all the rotational and liner movements in a machine tool as a whole by using the co-ordinate system, preferably that in accordance with NC axes, as shown also in Fig. 2.6. We can thus describe form-generating movements possible by the combination of several axes. Reportedly, we have already established the functional description without ambiguity, and importantly, the *flow of force* can facilitate it implicitly by giving absolute order of the alphabetical letter in the description.

As will be clear from the above, we can apply the concept of *flow of force* and functional description with narrower scope to the machining space. Conceptually, it is possible to delineate the linkage of structural configuration entities by using the *flow of force* as shown in Fig. 2.7a, b. Importantly, the machining method being carried out in the machining space can be represented by using a variant of functional description. In addition, we may thus discuss the machining technology in detail from the viewpoint of machine-attachment-tool-work system.

Of special interest, the *flow of force diagramme* is very convenient to grasp at glance the mutual relationship between the production (design and manufacture)

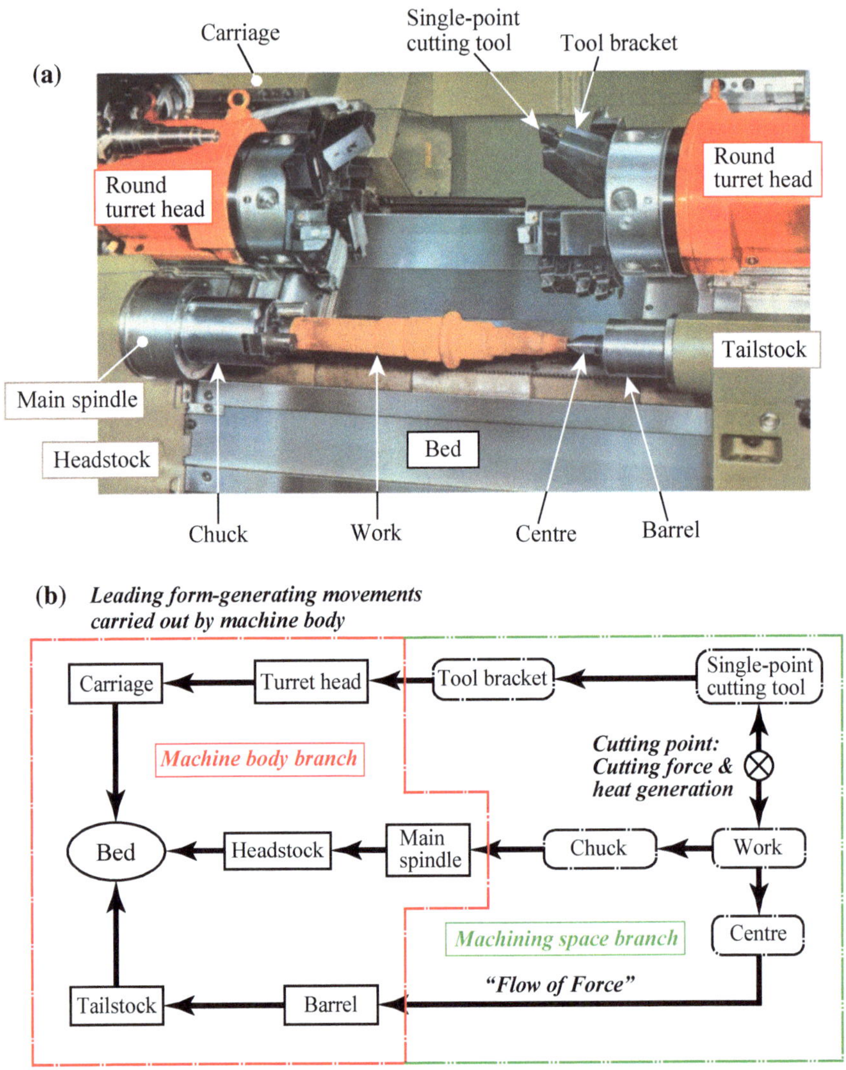

Fig. 2.7 Machining space in NC lathe and its "Flow of Force" diagramme: **a** Typical machining space—NC lathe of twin-turret type (Heyligenstädt-brand 1979). **b** "Flow of Force" diagramme

and utilisation technologies in the machine tool. As already shown in Fig. 2.7b, the branches for the machine body and machining space correspond closely to the production and utilisation technologies, respectively. In general, the machine tool manufacturer pays the special attention to the machine body branch, whereas the machine tool user eyes only the machining space branch; however, it is recommended to understand the total view of the *flow of force diagramme* with the preferable balance between the machine body and machining space branches with the advance of both the machine tool and its application technologies.

Importantly, similar stories are available for the "Linkage Diagramme" (refer to Fig. 2.24), and also for the illustration to represent the stiffness definition (refer to Fig. B.3). In the former case, the machining method (left-hand side) and the machine kind (right-hand side) belong to the machining space and machine body branches, respectively. In the latter case, we can obviously display the differing design principle of the machine tool from that of other industrial machines.

2.2.1 *Functional Description of a Machine Tool as a Whole*

In the functional description, we must first classify the rotational and linear movements into three kinds, i.e., principal, auxiliary and supplementary (supporting) movements. Of these, the supplementary movement is for setting the depth of cut, indexing of turret head, clamping motion of barrel and so on.

Importantly, we can shape, in principle, the component by the principal and auxiliary rotational and travelling movements, and in general, the former and the latter are carried out by the structural body components allocated around the machining space, and also the attachment and tool, respectively.

In the functional description proposed herein, thus, we use the following description rules (refer to Fig. 2.6).

(1) Indicating the cutting point by the symbol "/"(slash), and then the movements of work and tool are placed at the "left-hand" and "right-hand" branches, respectively.
(2) Starting from the cutting point, the alphabetical letter is placed in the closer order to the cutting point, in other words, placement of alphabetical letters along with the *flow of Force*.
(3) When having primary movement designated by the same alphabetical letter more than plural, we use the numerical subscript for the differentiation, and also the symbol "+", if it belongs the same branch.
(4) In cases of auxiliary and supplementary movements, we use the subscripts "a" and "s", and also apply the same rule as per the item (1).
(5) When the structural body component can move simultaneously more than two axes, the corresponding alphabetical letters are marked by the symbol "⌒" (hat).
(6) If necessary, we can place the secondary co-ordinate (U, V, W), which is especially required of describing the machining method-integrated kind.

In the extended functional description in Fig. 2.6, the supplementary movement is given; however, in general, we use the basic functional description, which consists of primary and auxiliary form-generating movements. Figure 2.8 shows a functional description for "*Mill-Turn*", which is the upmost advanced type within

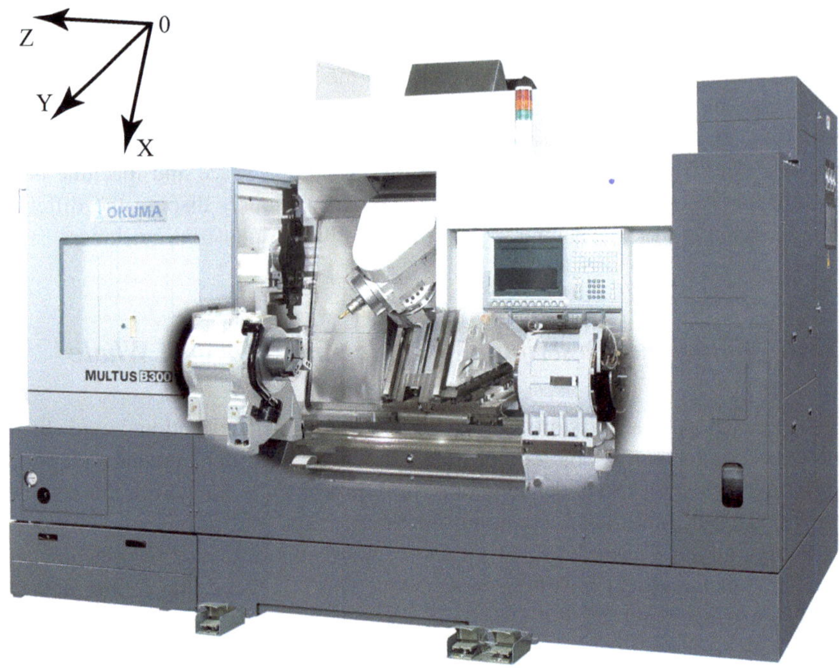

$(ZC_2 + C_1)/(Z + AX + C_3Z)$

Fig. 2.8 A representative "Mill-Turn"—Type MULTUS B 300 (by courtesy of Okuma 2004)

the machining function-integrated kind at present. This kind is extremely characterised by the built-in tool driving mechanism in the tilting head. In the conventional form-generation, the primary movement can be facilitated by the configuration and allocation of the main spindle. Against to this context, the tilting head can facilitate various machining methods and often the auxiliary movement is converted to primary one as can be seen from the functional description shown also in Fig. 2.8.

Of note, we can use the numerical controlled (NC) turning machine like the turning centre (TC) or *Mill-Turn* by attaching the rotating tool spindle to the turret head, and in such a case, the rotating tool plays the role of the auxiliary form-generating movement as shown in Fig. 2.9.

2.2.2 *Form-Generating Movement in Machining Function-Integrated Kind*

In the modern machine shop, the conventional TC and MC are doubtlessly major machining facilities, and thus we often forget that TC and MC belong to the

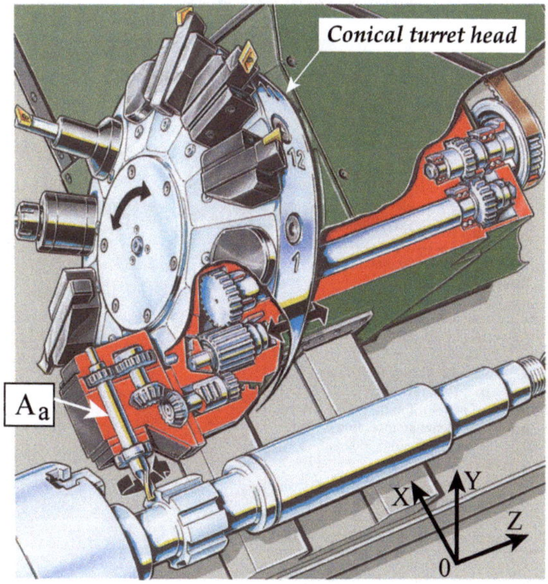

Auxiliary movement: A_a

Fig. 2.9 An auxiliary form-generating movement—rotary tool spindle unit mounted on turret head in CNC lathe (Type TNA, Traub-brand 1993)

machining function-integrated kind (see Fig. 7.1).[2] In fact, on the strength of beneficial features of TC and MC, the machining function-integrated kind has been deployed across the whole kinds of the machine tool. For example, Fig. 2.10 shows the vertical turning machine with milling function, where turning and milling are primary and secondary functions, respectively, as can be also assumed from the machine specifications. In short, milling can be carried out by the built-in spindle, which is placed within the ram, and the milling cutter is accommodated within the automatic tool changer (ATC) together with the single-point cutting tool for turning as shown in left top of Fig. 2.10.

In contrast, *Mill-turn* can carry out various machining methods without any discrimination in form-generating movement (refer to Fig. 2.8).

Having in mind that there are a considerable number of variants in the machining function-integrated kind, in general, the primary form-generating movement is, in principle, to be in reality with a handful of configurations in the main spindle and concerns as follows.

---

[2]The idea of the machining function-integrated kind is not new, but old. In the 1960s, Ramo merchandised the engine lathe, which is capable of broaching in addition to the ordinary machining function as the lathe. As compared with this lathe, TC and MC are the very advanced type and characterised by the automatic tool and tool cassette changer. Although staying a trial stage, in the Big National Project of Japanese Government called *Flexible Manufacturing Complex Provided with Laser*, we already produced a processing function-integrated kind.

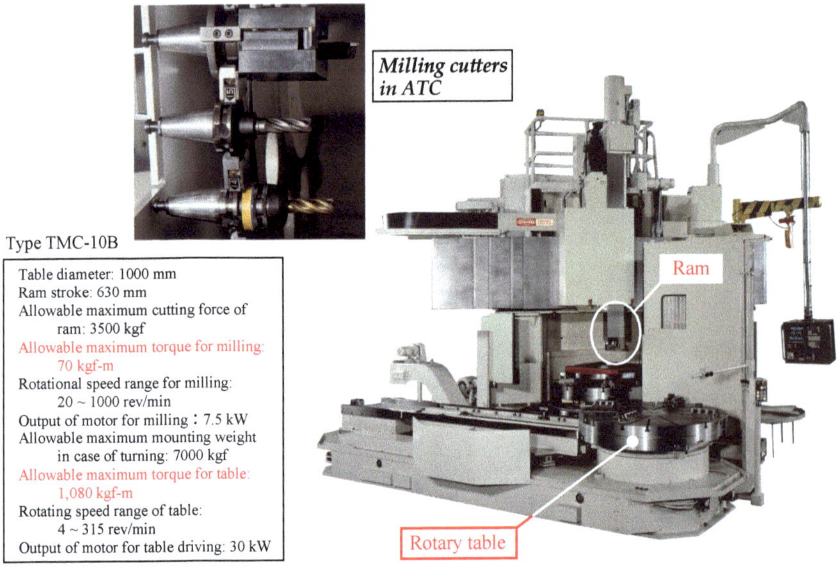

Fig. 2.10 Integration of milling function to vertical TC (Type TMC, around 2005, by courtesy of Toshiba Machine Mfg.)

(1) Twin-spindle, e.g., vertical and horizontal allocation (refer to Fig. 1.11).
(2) Main spindle with multiple-layered type, and in certain cases with eccentric type.

Figures 2.11 and 2.12 show two typical configurations of the main spindle. One is for twin-spindle type with travelling function, and the other is for two-layered spindle with also travelling function. For references, Fig. 2.12 shows furthermore the appearance of NC horizontal boring and milling machine with three-layered spindle in its right top. Obviously, the form-generating movement is reinforced much more than that in two-layered spindle, although NC commands become complex.

Intuitively, we can see the capability of carrying out various machining methods from the functional descriptions. With the growing importance of the quinaxial-controlled MC, furthermore, both the tilting spindle head and trunnion table can facilitate the primary form-generating movement as already exemplified in Fig. 2.8.

Paraphrasing, the structural body component allocated around the machining space is closely related to the primary form-generating movement, and a facing issue is to reinforce the versatility in the machining method by adequately incorporating the auxiliary form-generating movement. One of the remedies to such a requirement is to employ a variant of the modular designs, i.e., "*concept of Platform*" (Ito 2015).

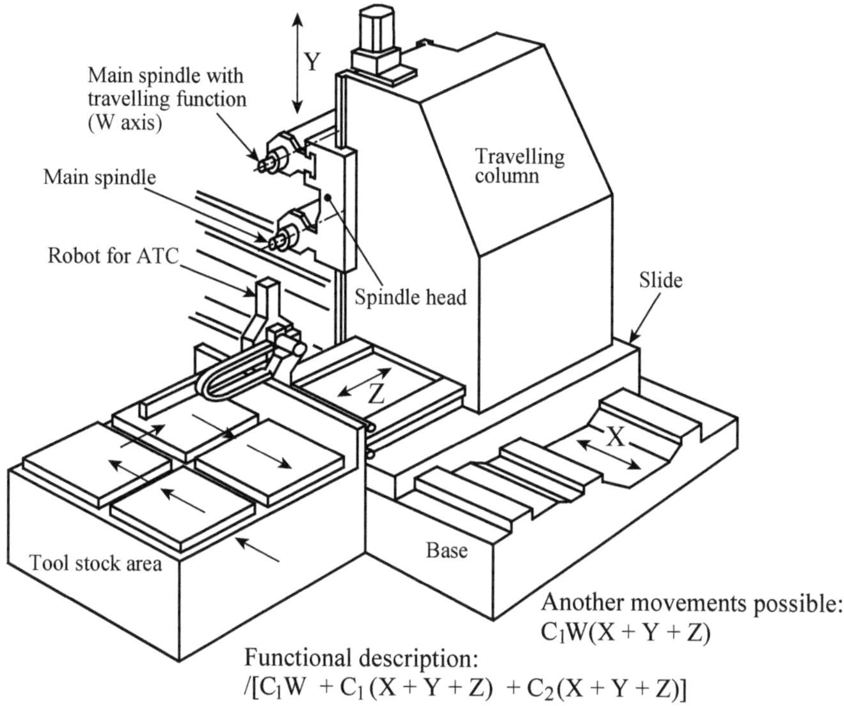

Fig. 2.11 Spar milling machine of twin-spindle type—a variant of AUTOMAX I, MPM-brand, middle of 1980s

In the *concept of platform*, a group of the structural body components allocated far from the machining space is integrated into one large-sized module, and we call it *platform*. Apart from the necessity of pains-taking work in the structural design, we can expect considerable advantages in both the production and the utilisation of the machining function-integrated kind by employing the *platform* as will be discussed in Chap. 7. Supposedly, the user can obtain the lucrative benefit without any awareness by installing, for example, *Mill-Turn*, which is designed on the basis of *platform concept*. Although there is not a relevant symptom at present, a *user-oriented platform design* should be developed hereafter on the strength of vigorous contrivances in the attachment and tooling.

In fact, we may expect more marked economisation by establishing the *user-oriented platform* than ever before, where the auxiliary form-generating movement is to be in reality by the attachment and tool. It is furthermore better to use the cutting function-integrated tool, e.g., formed tool and gang tool. Accidentally, the tooling layout in TC becomes recently more sophisticated, and thus we may elevate the synergy effect of the *user-oriented platform* by combining it with the advanced tooling layout as shown some examples in the following.

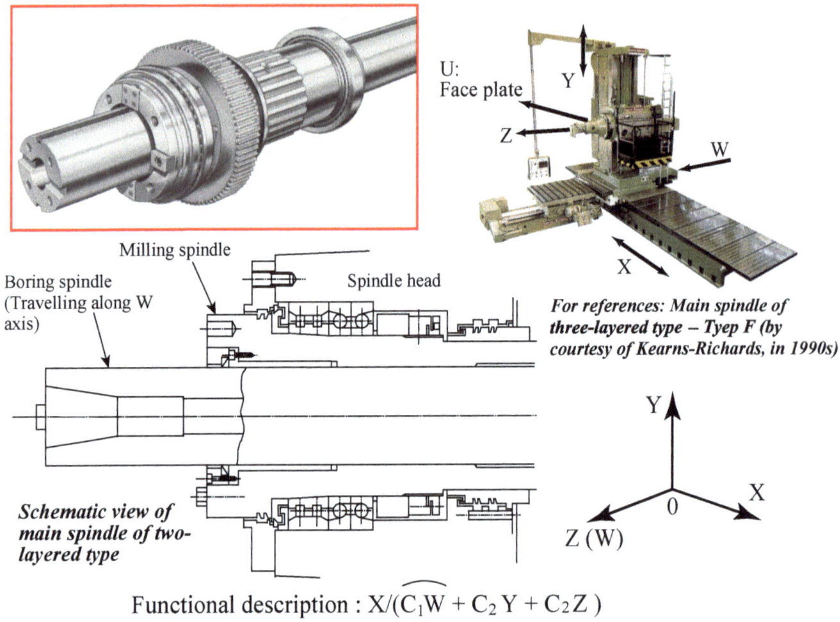

Fig. 2.12 Main spindle of two-layered type in horizontal boring and milling machine

Figure 2.13 shows a machining space in the computerised numerical controlled (CNC) automatic of twin-spindle type of Traub-brand (Type TNL). As can be readily seen, metal sawing is applicable as one of machining methods. In fact, this CNC automatic has several special cutting functions, e.g., gear cutting and gun drilling as like as TC. Reportedly, it is now common senses that MC can facilitate grinding, ultrasonic machining and laser processing as shown in Fig. 2.14, where the mould for the denture (artificial tooth) is being ground by the high-speed quinaxial-controlled MC. In addition, Fig. 2.15 shows the machining scenery in the five-face processing machine, and as can be seen, the machine can generate various work configurations by changing the attachment mounted on the main spindle, i.e., snout and angle head. Of note, MC of gantry type is the successor of the five-face processing machine.

In the integration of the grinding function to MC, there are two facing issues at the spindle-end and front main bearing: one is to maintain the oil and water tights, and the other is to clean the tapered hole by the air cleaning system. It is however said on the basis of our long-standing experience that such an air cleaning system cannot eliminate the debris of worn abrasive grain and micro-particle removed from the work while grinding. Supposedly, MC with grinding function shown already in Fig. 2.14 is well equipped something necessary, although it is not publicised, because of company's confidentiality.

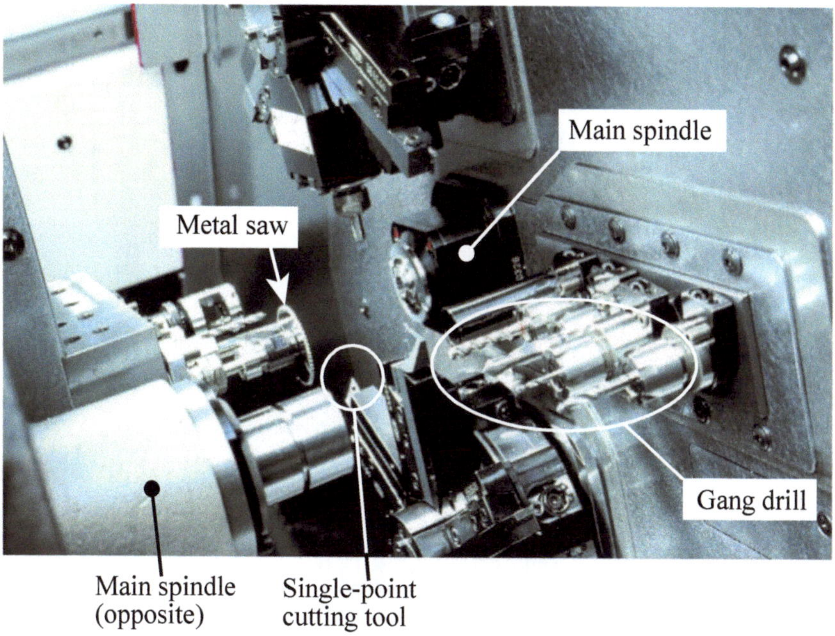

Fig. 2.13 Much more integration of machining methods (by courtesy of Traub 2009)

Another idea is thus to provide the grinding centre (GC) with cutting function, and probably we will be able to solve the problems mentioned above. Figure 2.16 shows a GC, where the internal grinding spindle is for both grinding and milling after stiffening its rigidity and increasing the allowable torque at the lower rotational speed. It sounds very easy to mill the work by GC; however, we will face another new problems as follows.

(1) GC has two spindles, i.e. grinding and work spindles, and thus GC is, in principle, capable of conducting various form-generating movements as compared with MC. In contrast, there are considerable difficulties in determination of the design specifications, e.g., rotational speed range and allowable maximum spindle torque, which are derived from the essential difference between cutting and grinding as exemplified the three components in the machining force.
(2) It is desirable to provide GC with hard-turning function, e.g., that for hardened bearing steel (100Cr6, hardness around 62 HRC) by the Poly-crystal Cubic Boron Nitride (PcBN) with TiC coating.[3]

[3]Based on the on-the-spot investigation by Ito at Technical University of Aachen (TH Aachen) in 1997. On that occasion, cutting speed, feed rate and depth of cut are 150 m/min, 0.06 mm/rev and 0.1 mm, respectively.

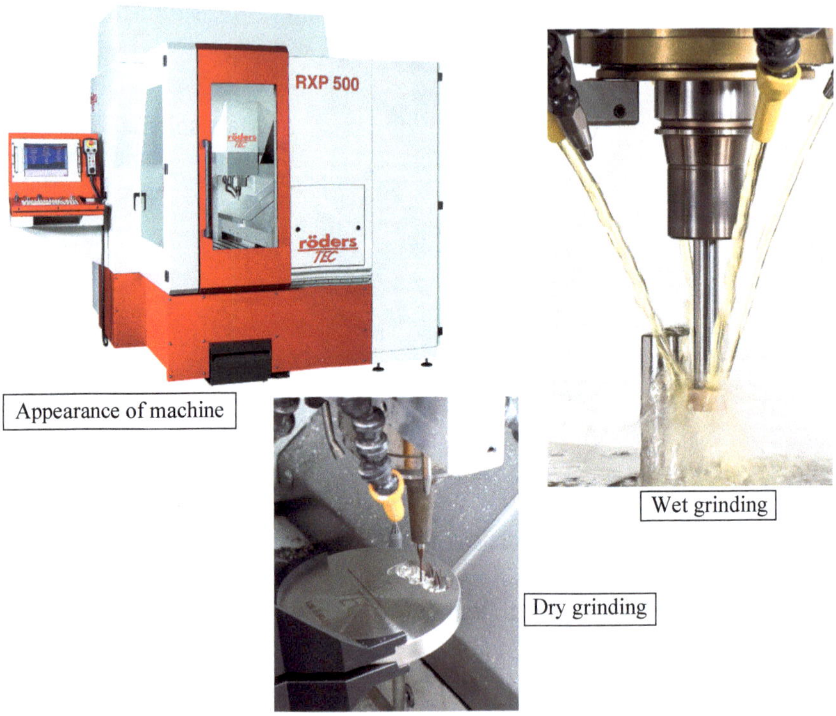

Fig. 2.14 Grinding carried out by high-speed quinaxial-controlled MC—type RXP 500 (by courtesy of Röders 2013)

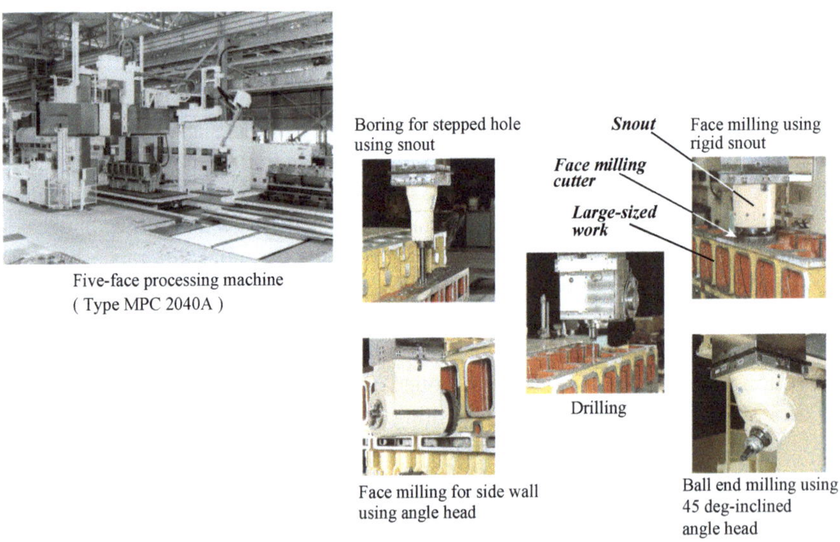

Fig. 2.15 Various machining methods capable of carrying out by five-face processing machine—for box-like work (by courtesy of Toshiba Machine Mfg., late 1990s)

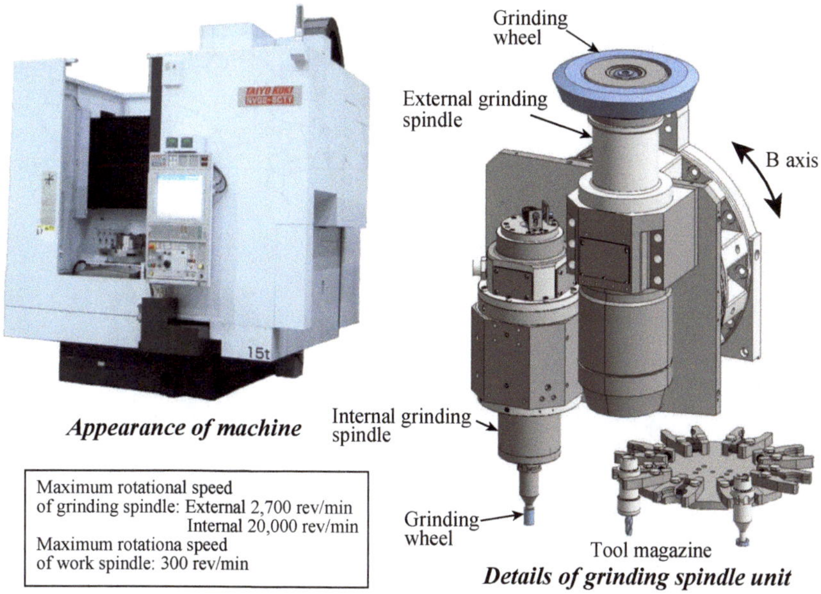

Fig. 2.16 GC with milling function (Type NVG II-5CTY, by courtesy of Taiyo Koki 2013)

As will be clear from the above, it is recommended that we should reinforce the auxiliary form-generating movement by the cutting and grinding tools in the *user-oriented platform*, and accidentally, we can purchase, without any difficulties, such tools, e.g., gang tool, combination tool and special boring tool for stepped holes from the market across the whole world. In the following, thus, some quick notes will be given.

(1) The multiple-spindle head shown in Fig. 2.17 is for MC and "*Transfer Centre*" and suitable for gang drilling and boring. In addition, gang milling can be regarded as a variant of formed milling cutter, and as well known, consists of a group of single milling cutters of various configurations so as to economise the tool cost than that for formed milling cutter. In retrospect, these have been in practice since 1960s.

In the case of gang cutter shown in Fig. 2.17, furthermore, we can see an interesting contrivance, i.e., staggered pattern mounting on the arbour, to suppress the chatter vibration. More specifically, the milling cutter has two keyways, one of which is placed with a half of tooth pitch in shifting from the other.

(2) Trister Engineering has merchandised a combination tool, i.e., "Drill Reamer" as shown in Fig. 2.18. As literally shown, this tool is for drilling and reaming of Al-alloy and of custom-made. In addition, we can purchase another combination tool, e.g., drill and boring bar with chamfering at the entrance of hole and counter-boring. In this context, an interesting contrivance is shown together in

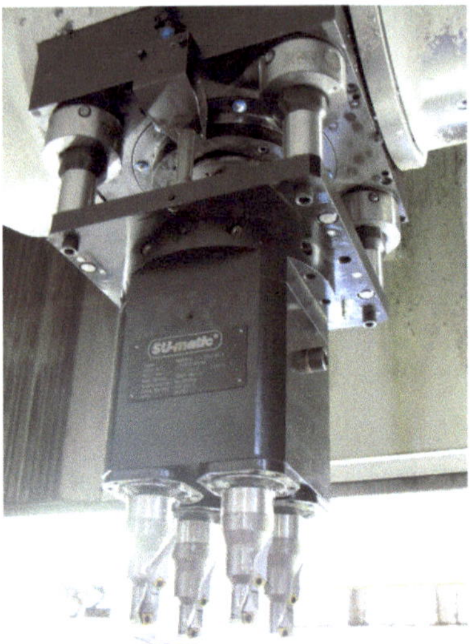

Multiple-spindle head for drilling and boring (by courtesy of SU-matic, 2015)

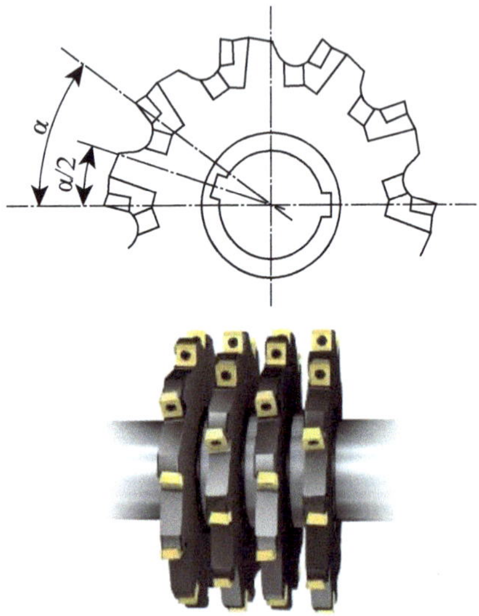

Gang milling in staggered pattern mounting (by courtesy of Sandvik, 2015)

Fig. 2.17 Gang drilling, boring and milling

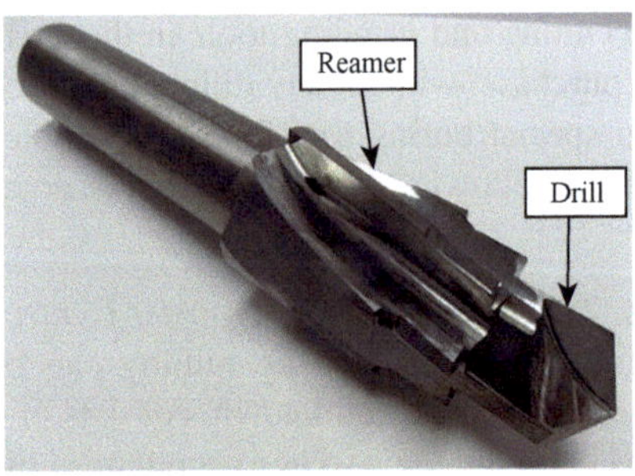

Drill reamer (by Tri Star Engineering, 2015)

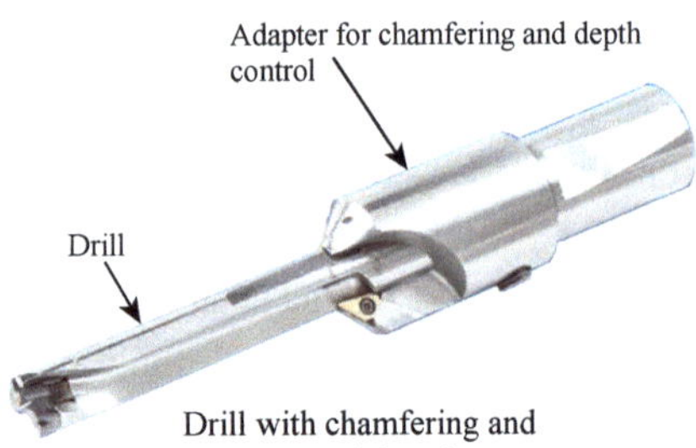

Drill with chamfering and depth control function (by courtesy of Iscar, 2015)

Fig. 2.18 Examples of combination tool

Fig. 2.18, where the drill and adapter with depth control and chamfering function are assembled. Reportedly, in the Al-alloy die casting for the transmission and cylinder block of the motorcar, we have often the hole with unacceptable taper, bending, undersized diameter, and even the deviation of the hole position. Tanoi Co. has thus been on the market a tap called "Multi-Tap" since 2005, in which the end milling function is integrated at the front and side of the tap to provide a remedy for finishing such holes.

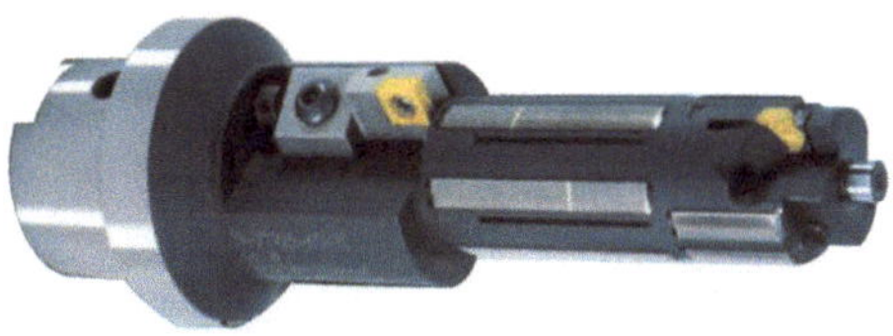

Typical boring bar for stepped hole (by Mapal, 2015)

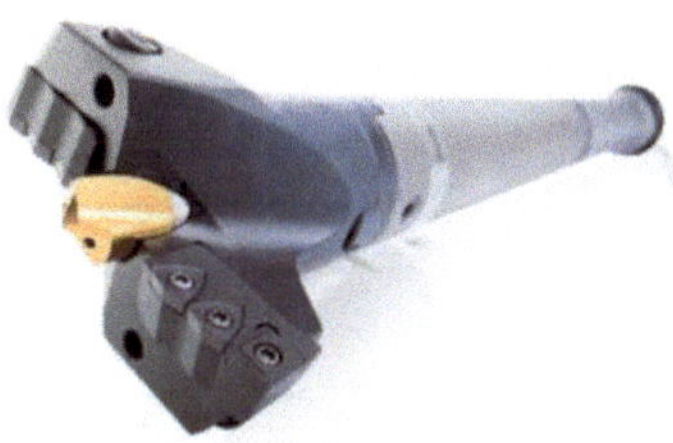

Drill for deep-hole (by Kennametal, 2015)

Other tooling systems by Mapal (2015)

Fig. 2.19 Tools for deep hole and stepped hole

(3) Kennametal has proposed a special deep hole drilling system as shown in Fig. 2.19, which can be characterised by the pilot drill placed at the centre of the drill. The tool is of insert type, available for the work made of steel, stainless steel, cast iron or non-ferrous metal, and capable of drilling up to 10 times of drill diameter. Of special interest, Mapal supplies the special boring tool for machining the motorcar component made of Al-alloy such as cylinder head, cylinder block, transmission box and so on as shown also in Fig. 2.19. In this case, we can see also the improvement in machining accuracy by providing the pad as compared with that in the past. In addition, it is worth suggesting that the drill and boring tool for stepped hole are now very popular in the cutting tool sphere.

As a matter of course, we must pay special attention to the usability of these gang and combination tools especially from the viewpoint of the machining volume. Equally, we must be aware of the amazing advance in tool technology, and often some novel contrivances render the design technology for machining system and process planning so far generalised useless.

As will be clear from the above, we can expect the potentiality of the *user-oriented platform* available for the machining function-integrated kind in not distant future, and on that occasion, there may be three choosing ways for the machine tool to carry out the machining requirements.

(1) Placing stress on primary form-generating function
(2) Placing stress on auxiliary form-generating function
(3) Synergy of primary and auxiliary form-generating functions.

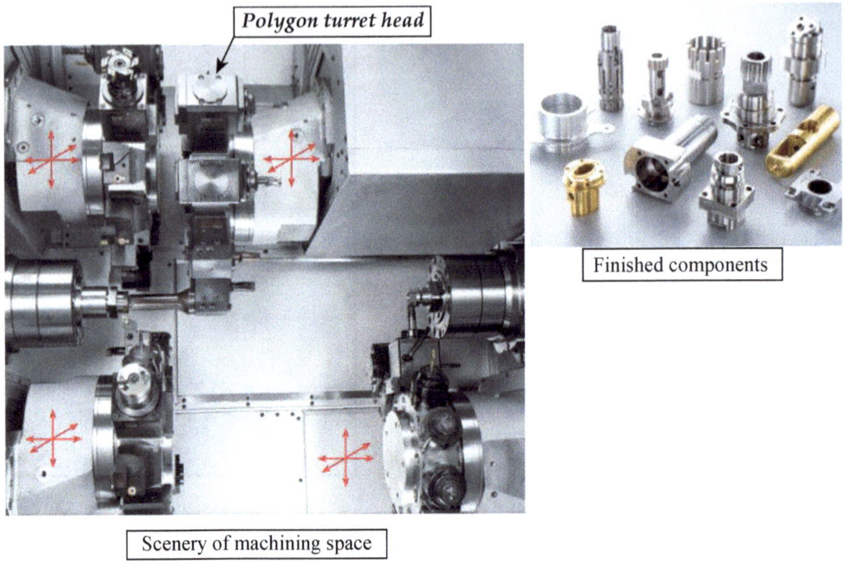

Fig. 2.20 Machining space of Mill-Turn consisting of three same turret heads (Type TNX65/42, by courtesy of Traub 2009)

Within the choice of machine tool context, furthermore, we must pay special attention to the economic problem. For example, Fig. 2.20 reproduces the layout of the turret head in the machining space of "*Mill-Turn*", i.e., allocation of the same three turret heads with built-in-rotary tool function. It is worth suggesting that Traub, one of the leading German machine tool manufacturers, asserts the large economic benefits by responding much more requirements of the user, although not facilitating the costly tilting head. In this context, we must recall the necessity of establishing an economic evaluation for the modular design. From the past, we have endeavoured it; however, up to now, there are no reliable and quantitative evaluation methods as yet, although Kersten et al. (2009) reported a trial.

2.3 Trial to Establish Choosing Guide for Kind of Machine Tool and Machining Method

As already shown in Fig. 2.2, the facing problem may be called the "*Choice Problem of Rational Processing Methods*" and is closely concerned with process planning. It is however regrettable that there are no effective and convenient guidelines widely acceptable for such a choice as yet. This can be attributed to the difficulty in decision-making in full consideration of the technological, economic and social prerequisites related to each machining requirement. In addition, we have a myriad of machining methods possible, once the part drawing is released.

In contrast, there are growing demands and importance for such a choice methodology or technology in accordance with the enhancement of the component quality with keen price. In fact, a root cause of difficulties lies in the determination of *one-to-one relationship* between the part drawing and process plan. More specifically, we must convert the geometrical information on the part drawing into the machining-related information, i.e., information conversion into differing properties. For example the *cylindrical profile (geometrical information)* is to be in reality by various *turn tops (machining information)* as already suggested in Figs. 1.6 and 1.7.

As will be clear from such difficulties in information conversion, we cannot guarantee the one-to-one relation, and thus the process design engineer determines, in general, process planning on the basis of her or his long-standing experiences and also with the trial-and-error method. As can be easily imagined, there have been no generalised and objectified guidelines for choosing the utmost suitable machining method when a part drawing is given.

In process planning, this is thus called "*One-to-one Paring Problem*" between the part drawing and machining methods available, and to give a clue for solving such a crucial problem, a trial will be discussed in the following.

In short, an idea is to use the machining symbol, which is, in general, enacted by the National Standards, e.g., JIS B0122 (Symbols of Metal Working Processes), and by which the process plan can be described. It is furthermore convenient that the machining symbol can be converted into the combination of several alphabetical letters in the order of the predetermined rule, i.e., machining function description, as similar as the functional description of the machine tool.

Admitting that we have still some uncertainties in the conversion, it is possible to estimate some pairs among the kinds of the machine tool and machining methods, by which the finished component can be produced correctly according to the part drawing.

Figure 2.21 shows a challenging trial to correlate the kind of the machine tool and the work to be machined. In the top middle and bottom of Fig. 2.21, the sample work and its process plan are shown, and furthermore, the machining symbols described in the process plan can be converted into the machining function description as shown in the top right of Fig. 2.21. In short, the movement of "$C/(X + Z) +/C_a (X + Z)$" is necessary to finish the sample work, whereas TC can move "$C/(X + Z) +/(A_a + C_a)(X + Z)$" as can be estimated from the top left of Fig. 2.21. As compared the machining function description with the functional description of TC, we can judge that TC is capable of machining the sample work.

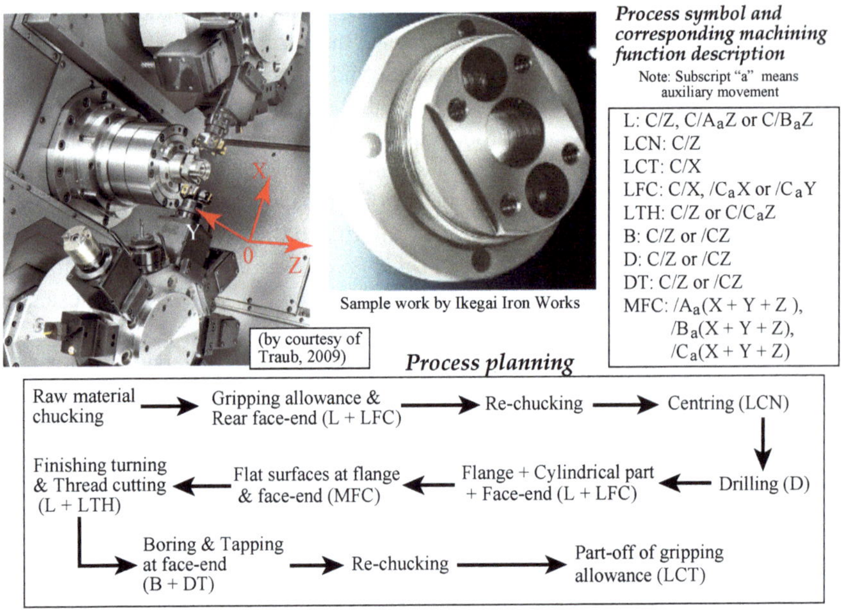

Fig. 2.21 Sample work with its process planning and conversion of "Process Symbol" into "Machining Method Description"

2.4 Linkage Diagramme of Structural Configuration Entities

As can be readily seen, the *flow of force* and functional description can be considered the powerful tools in the choice of the suitable pair of the machining method and machine kind, and also in the improvement for the machining technology hereafter. It is however suggested that the *flow of force* does not contain satisfactory information for evaluating the synergy effect of an innovative contrivance on the machining technology, which is obtainable by integrating the functionality and performance of the structural configuration entities and concern.

For example, Fig. 2.22 shows first rear face parting-off, and Fig. 2.23 shows furthermore complex turning, i.e., integration of reverse-feed turn top, grooving and parting-off. In fact, such the turning methods were impossible in the past, and we needed re-chucking of the work. In contrast, a combination of TC of twin-spindle type and the novel cutting tool can facilitate these turning methods without re-chucking the work at present. It emphasises that TC of twin-spindle changes the machining scenery by its hand-off function to large and various contexts.

It is thus necessary to detail the "Vertex" in the *flow of force* by giving the properties, i.e., identities of corresponding entity. Of special interest, we can thus propose a concept shown in Fig. 2.24, which is called "Linkage Diagramme for

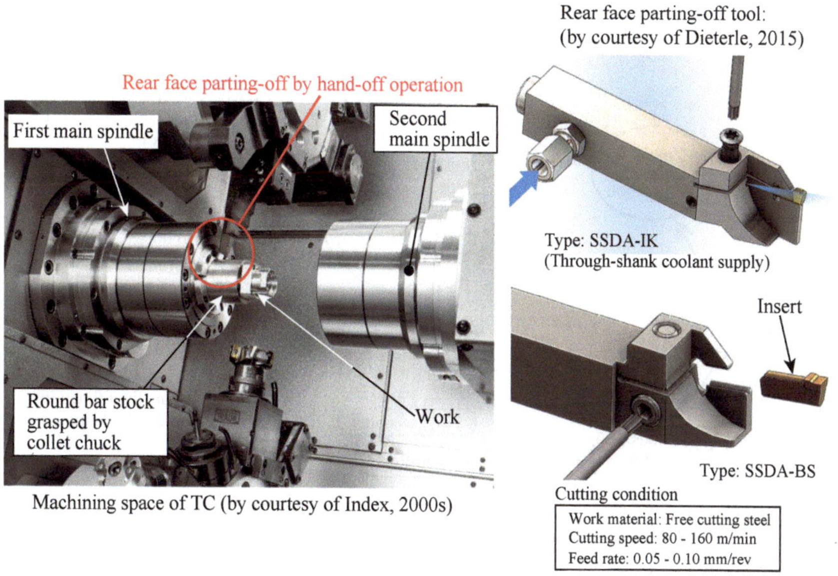

Fig. 2.22 Innovative turning by combination of twin-spindle and new cutting tool—rear face parting-off

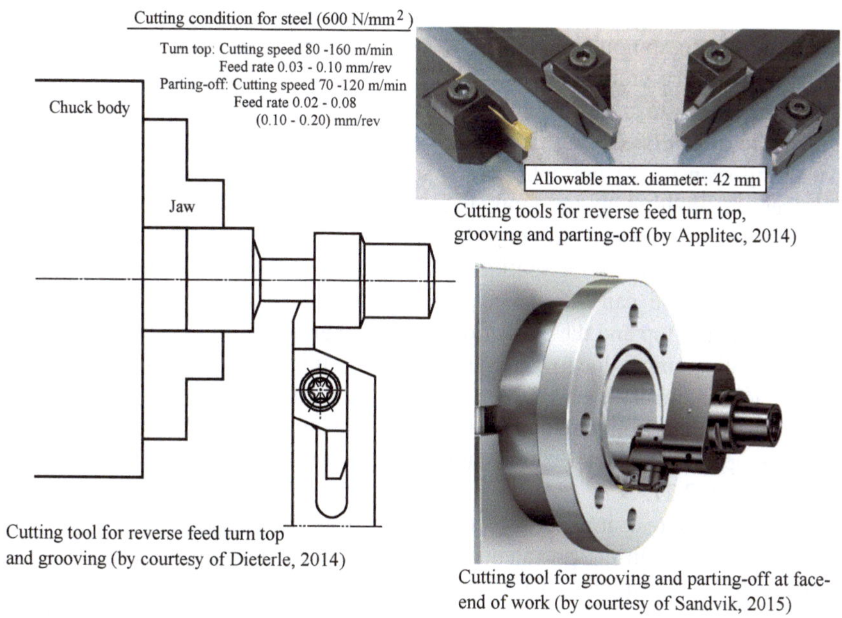

Fig. 2.23 New horizon in process planning by innovative cutting tools

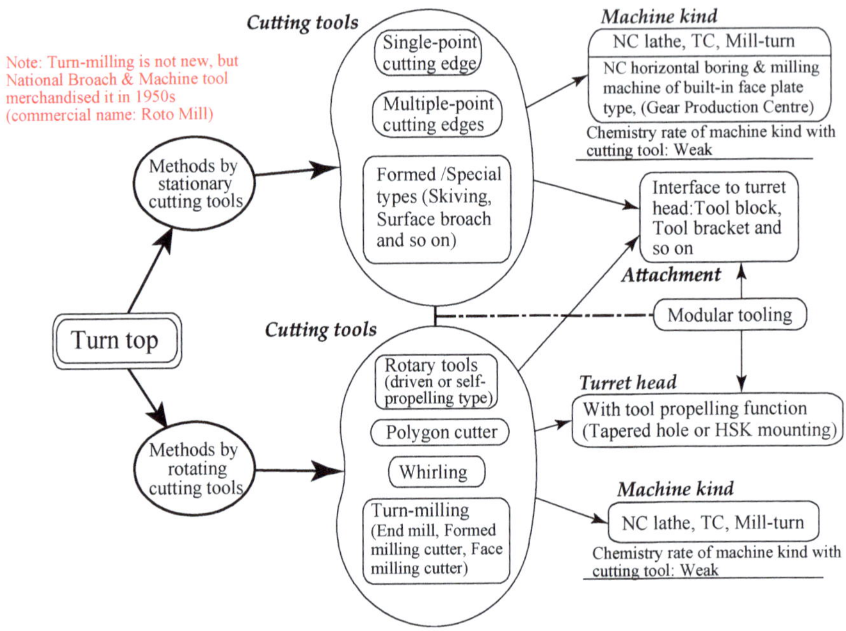

Fig. 2.24 Linkage diagramme ranging from machining method to structural configuration entities—turn top

Structural Configuration Entities and Concerns". In the case of Fig. 2.24, we can grasp at glance the preferable linkage ranging from the machining method, through the cutting tools, to the structural configuration entities. Finally, we can choose a suitable combination among the related entities, i.e., not optimum, but preferable machining scenery. At present, there is something to be seen for such a concept; however, in addition to turn top, Figs. 2.25 and 2.26 show such linkage diagrammes for drilling and grinding. In Fig. 2.25, the automatic back spotfacer of Hahn and Kolb-brand is for spotfacing, counterboring, deburring and chamfering.

Importantly, we can see the apparent difference in the linkage diagrammes between the turn top and the cylindrical or surface grinding. In fact, grinding shows the simple diagramme due to the simplicity in grinding tools. More importantly, we can see the chemistry rate between the kind of available machine tools and the machining method from the linkage diagramme. In general, the chemistry rate is not so strong apart from the special machining method like gun (deep hole) drilling as shown in Fig. 2.27.

Of special interest, it appears that such the "Linkage Diagramme" increases its importance duly with the prevalence of the machining function-integrated kind of various types. Apparently, we face much more difficulty in the choice of the suitable combination between the machine method and the machine kind with the

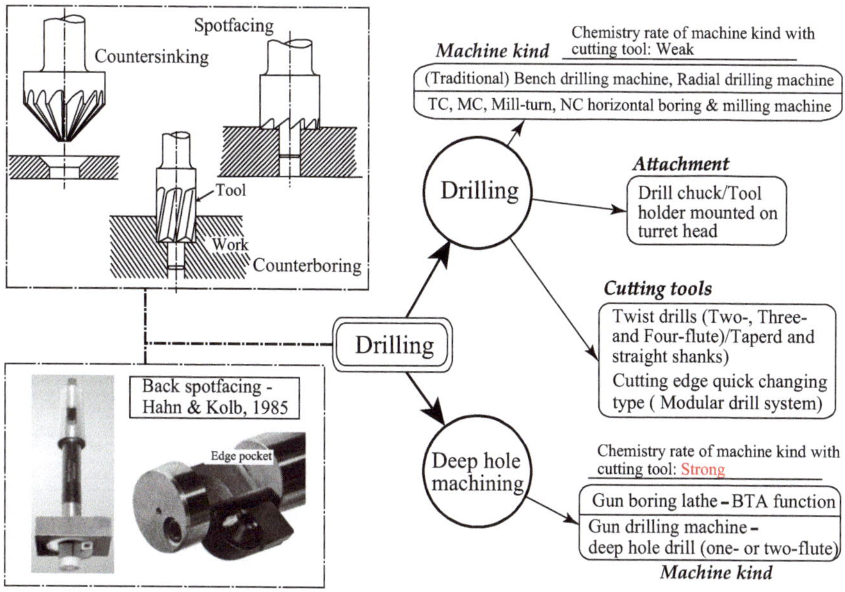

Fig. 2.25 Linkage diagramme ranging from machining method to structural configuration entities—drilling

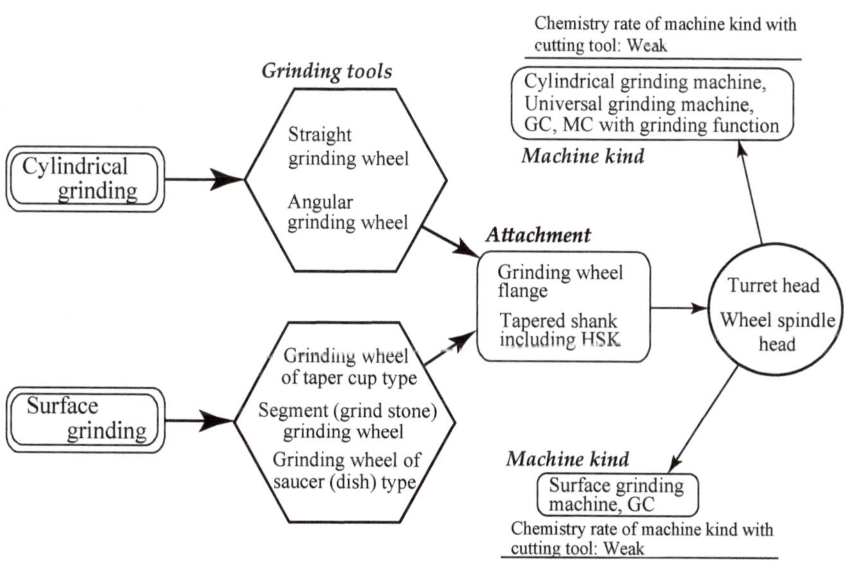

Fig. 2.26 Linkage diagramme ranging from machining method to structural configuration entities—grinding

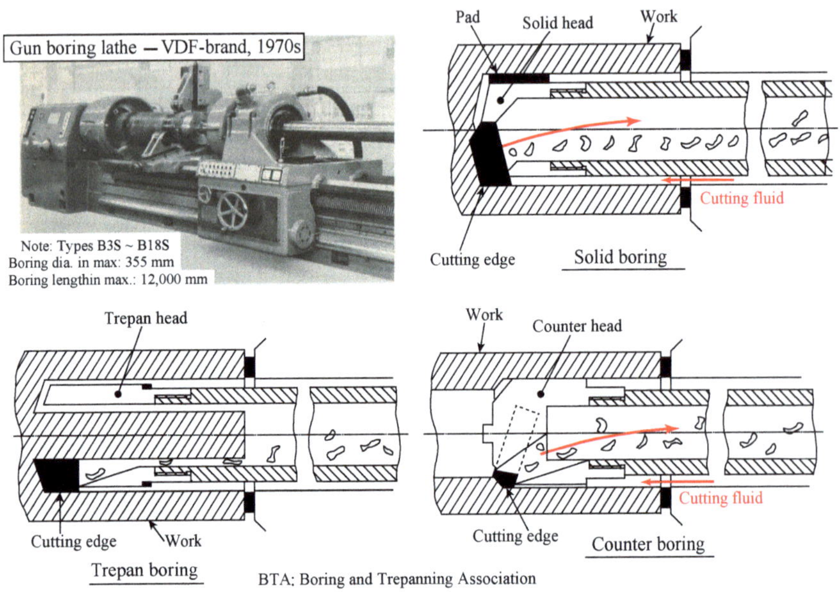

Fig. 2.27 Deep hole drilling and boring of BTA type by gun boring lathe

increase of the machining function-integrated kind, a classification of which is, for further reference, shown in Fig. 2.28. Importantly, we must eye the possible direction in near future of the machining function-integrated kind to the "One-machine Factory" as will be discussed in Chap. 7 (see Fig. 7.13).

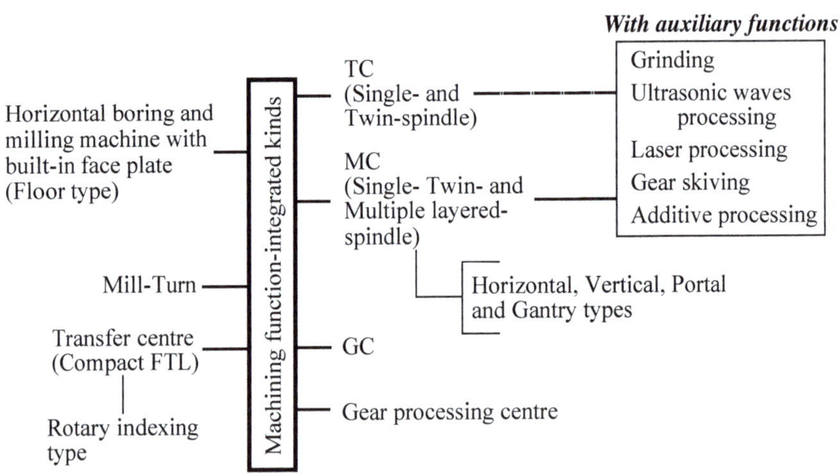

Fig. 2.28 A classification of machining function-integrated kinds in 2010s

References

Ito Y (2008) Modular design for machine tools. McGraw-Hill

Ito Y (2015) Applicability of "Platform Concept" to machining function-integrated machine tools—a new "Raison d'être" for modular design. Bull JSME Mech Eng Rev 2–1:1–12

Kersten W et al (2009) Kostenorientierte analyse der modularisierung. ZwF 104–12:1136–1141

Schmidt J, Bechle A (2003) Wälzschälen als neues Hochleistungs-Bearbeitungsverfahren. ZwF 98–11:589–593

Spur G, Stöferle Th (1980) Handbuch der Fertigungstechnik Band 3/2 Spanen. Carl Hanser Verlag, S. 415–417

Thomas GG (1972) Production technology. Oxford University Press

Weck M (1974) Rechnerunterstützter Entwurf und Maßnahmen zur Ausführung flexibler Fertigungssysteme. Industrie-Anzeiger 96–74:1683–1689

Part II
Determinants for Functional and Performance Specifications of Machining Space

Abstract

The machining space consists of, in principle, a linkage of structural configuration entities, e.g., nose of main spindle and tool seat in turret head, as already described in Chap. 2, and together with the attachment and tool, these carry out some functional and performance activities necessary to produce the component in cooperation with one another. In consideration of both the material and information flows in the machining space, i.e., one of the mathematical models for machining space like FMC, Part II will first describe the supply of the raw material and discharge of the swarf. Then, the structural configuration entities and due attachments will be detailed. Importantly, Part II will provide us with the suitable combination of the structural configuration entities with the attachment and tool. In addition, the present perspective of the in-process measurement will be quickly noted.

Paraphrasing, Part II may sublimate the concept of a linkage of structural configuration entities to the concrete choosing guide for the kind of machine tool, machining method, and concerns, provided that the part drawing is given. In this context, the present and near future perspectives for the platform concept will furthermore describe placing the stress on its user-oriented characteristic features. The platform concept is one of the advanced variants in the modular design and grows its importance almost daily together with the modular tooling system.

Headnotes

In discussing matters of machining space concerns, it is better to consider the material and information flows as already shown in Fig. 2.1. From such an interrelation in flows, we may identify the engineering problems to be solved, and as exemplified by process planning and operational planning (refer to Chap. 2), a facing issue is the skillful and dexterous fusion of both the hardware and software. In fact, process planning can be carried out, provided that we have the information for the machining facilities, work preparation, processing tools for work, allowable cost, delivery time, and so on.

Of special interest from the hardware aspect, the machining space consists of, in principle, a linkage of structural configuration entities as already suggested in Chap. 2. Importantly, together with the attachment and tool, these entities can function and carry out various manufacturing activities with close cooperation with one another. Actually, primary concern is machining, and of course, we must discuss the preparatory work, e.g., turret indexing, tool changing, jig, and fixture preparation, and furthermore the raw material and cutting fluid supplies, and discharge of finished component and swarf. In addition, we must discuss in-process and on-the-machine measurement for the finished component and also for the healthy condition of the machine-attachment-tool-work system.

More specifically, we must place the stress on the function and performance of the spindle nose rather than the structural configuration and supporting bearings of the main spindle system. In this context, the necessity is the accommodating method of the chuck on the spindle nose. As can be readily seen, such a viewpoint differs from that so far prevailed in the machine tool engineering.

Having in mind the importance of eying the structural configuration entities and their functional and performance activities in the machining space, in Part II, Chap. 3 will describe first the supply of the raw material and semifinished component in, and also discharge of the swarf from the machining space. Then, each structural configuration entity will be detailed together with reviewing and discussing the attachment, which is accommodated on the structural configuration entities, in Chaps. 4 and 5. In addition, the present perspective in the in-process measurement will be quickly noted in Chap. 6.

Within the utilisation of machine tool context, furthermore, a novel structural design, i.e., "*platform concept*" is being developed. Importantly, the *platform concept* is of advanced type of the modular design and also one of the user-oriented concepts as that for transfer line (TL) in the past. In fact, the *platform concept* may elevate its effectiveness by providing the structural body component closely allocated around the machining space with the larger interchangeability as will be discussed in Chap. 7. More importantly, the modular tooling system may accelerate the percussion effects of the *platform concept*.

Chapter 3
Raw Materials for Work, Engineering Materials and Swarf Discharge

3.1 Supply of Raw Materials to Machining Space and Material Efficiency

Although depending upon the product kind, we supply various raw and semi-finished materials to the machining space, which are, in general, processed beforehand by metal non-removal processing as shown in Fig. 3.1.

For example, casting and welded component are, in general, supplied to the machining space as a whole raw material and semi-finished condition, respectively, and then they are machined in part as shown in Figs. 3.2a, b. In contrast, the formed raw materials are removed unnecessary allowance by machining across the whole configuration, and in the case of cold rolling and extrusion, we may obtain the round bar, angle steel, I-beam and so on. In general, we cut a part of such raw materials to the size necessary by the sawing machine at the work preparation area and carried on machining according to the requirement. Obviously, we must consider the material efficiency in all the work preparation procedures.

As will be described in Sect. 3.2, we can obtain the information of the work material symbolised by the combination of alphabetical and numerical letters, i.e., material symbol, from the part drawing, and finish the work as per the indication of the part drawing. In this context, the production engineer should know, for example, the differing features of the austenitic stainless from the martensitic stainless steels.

As widely known from the past, it is preferable to superimpose the information for raw material processing shown already in Fig. 3.1 together with the material symbols described on the drawing.

In this context, we need to conduct the documentation for the grass root-like knowledge in machining the raw and semi-finish materials, which are derived from their processing methods by nature as follows.

(1) Hot forged material: It is common sense that the raw material hot-forged has the scale (oxide layer) of certain thickness at its surface layer. Obviously, such a

Y. Ito and T. Matsumura, *Theory and Practice in Machining Systems*,
DOI 10.1007/978-3-319-53901-0_3

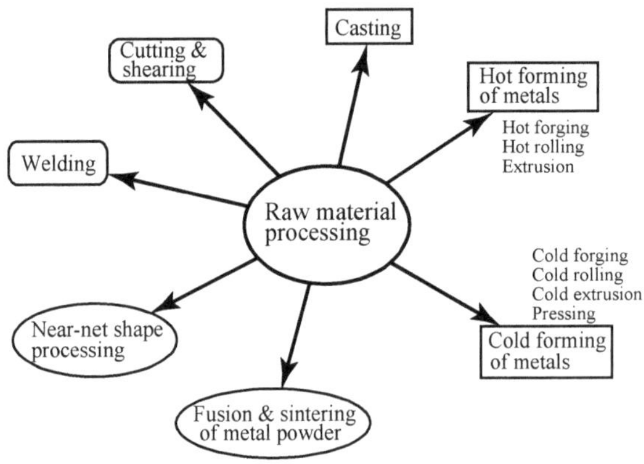

Fig. 3.1 Processing of raw and semi-finished materials for serving machining

scale damages severely the cutting tool, and in the worst cast the cutting tool is broken.

(2) Cold working material: In general, the raw material includes the residual stress and strain caused by the crystal distortion, and thus we must conduct tempering and normalising. Even when tempering and normalising, for example, the unfinished top jaw for the jaw chuck should be cut off from the raw steel material, so that the radial direction of the finished jaw in the chuck body coincides with the drawing direction of the raw steel material. By such a work preparation, we can prevent the breakage of the top jaw in higher-rotational speed by the centrifugal force. In contrast, the cold-drawn bar can save the machining cost by not processing the outer surface in certain machine elements.

(3) Cast iron: Although the important surface is cast by placing it underneath in the mould, we often observe the inclusion of foreign debits and particle like casting sand. Such foreign micro-material damages the cutting tool, and thus we must eliminate it at the work preparatory area by the manual operation.

(4) Heat treatment: Especially grinding for carbon steel and alloy steel heat-treated, even the engineer for machining concern must learn the differing features in direct quenching (hardening) and case hardening, and also the "*Isothermal Transformation Diagramme*". The latter is concerned with the "*Austemper*", "*Martemper*" and "*Marquench*", and for example, the hardened steel by "*Austemper*" is not necessary annealing, but tougher than that ordinarily quenching and annealing. In grinding, the hardened steel shows better machinability than other materials; however, loading (pores being stuffed by chip) is at issue in case of austempered steel.

(5) Hardening of gear teeth: In gear grinding after hardening and annealing, the undue proportion in tooth profile changes by either flame or induction hardenings.

In addition, we must be aware of the correlation between the material symbol and the machinability as will be exemplified in the following. In short, the material composition implies the machinability to some extent, although such data are not documented systematically (see Sect. 3.2).

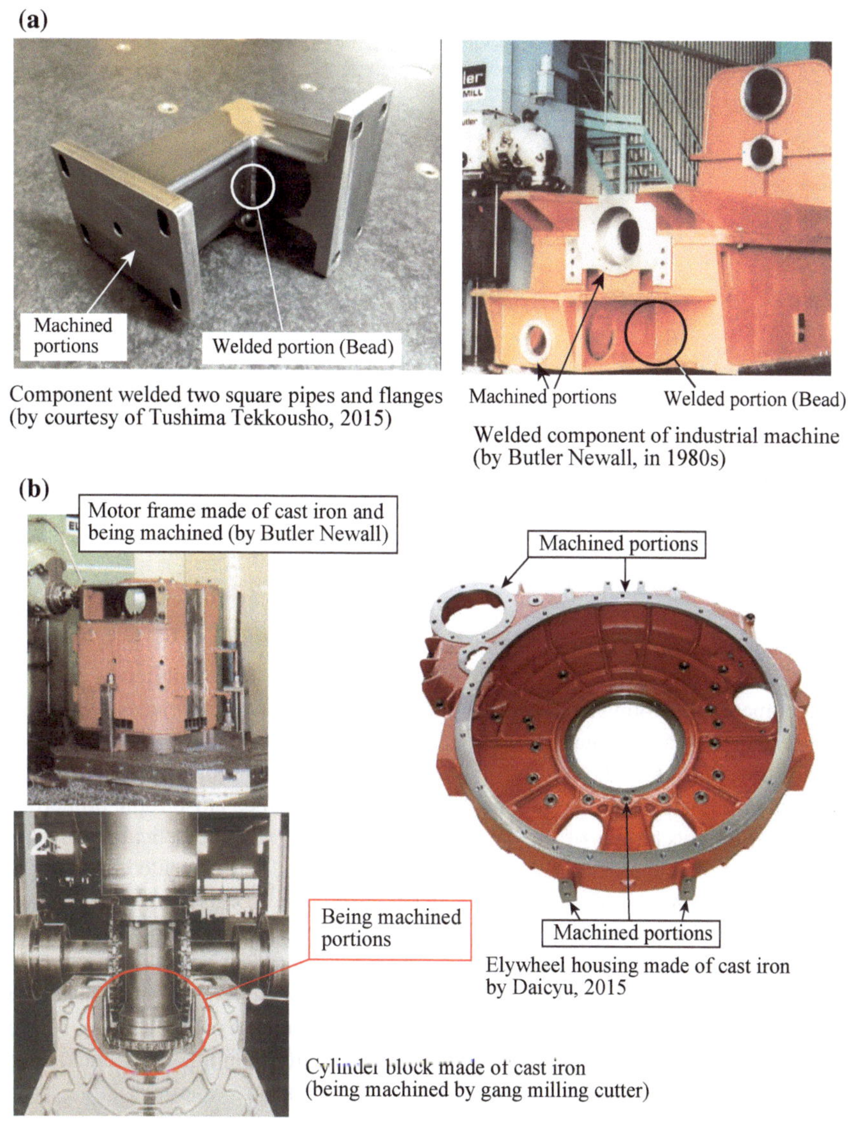

Fig. 3.2 Machining portions in raw and semi-finished materials: **a** Welded component and structure. **b** Casting

(1) The free cutting brass of Pb type, i.e., C3600 series as per JIS, shows the higher machining efficiency together with better surface finish and swarf dischargeability by generating the chip.
(2) The calcium-deoxidised steel (killed steel) shows the better tool life together with the reduction of the cutting force.

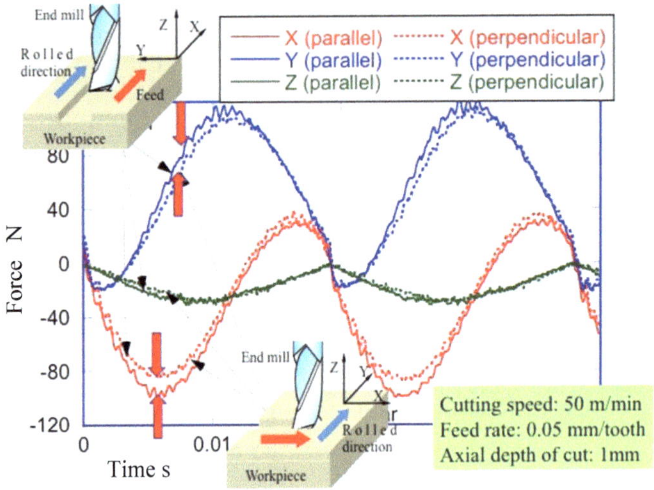

Fig. 3.3 Effects of anisotropic properties in Ti-alloy on cutting force (by courtesy of Matsumura 2016)

Although growing the importance of such grass root-like knowledge with the advance of the automatisation, it is regrettable that there are no publicised database or documentation noting it. In general, the engineer gets such knowledge by the oral tradition from senior to junior.

To this end, Fig. 3.3 shows the differing magnitude in the cutting force depending upon the rolling direction of the raw material, i.e., anisotropic properties, while groove machining of the Ti alloy (Ti–6Al–4V) with the helical square end mill of 10 mm in diameter, 30° in helix angle and made of cemented carbide. As can be readily seen, we can observe the considerable effects of the anisotropic properties on the cutting force, which becomes large when the cutting direction is parallel to the rolling direction (Matsumura and Tamura 2012). More specifically, the relative angle between the rotating cutting edge and the crystal distortion direction changes from nil to 90°, for example, in cutting parallel to the rolling direction. In fact, we can observe a considerable difference in the swarf thicknesses between the parallel and the perpendicular cutting modes.

3.2 Engineering Materials in General

From the part drawing, surely we can get the material symbol, composition and physical properties of the material. The engineer for machining concern must, at least, understand the correlation among the material symbol, its properties and applicable product. Figure 3.4 shows thus some materials for the jet engine of the

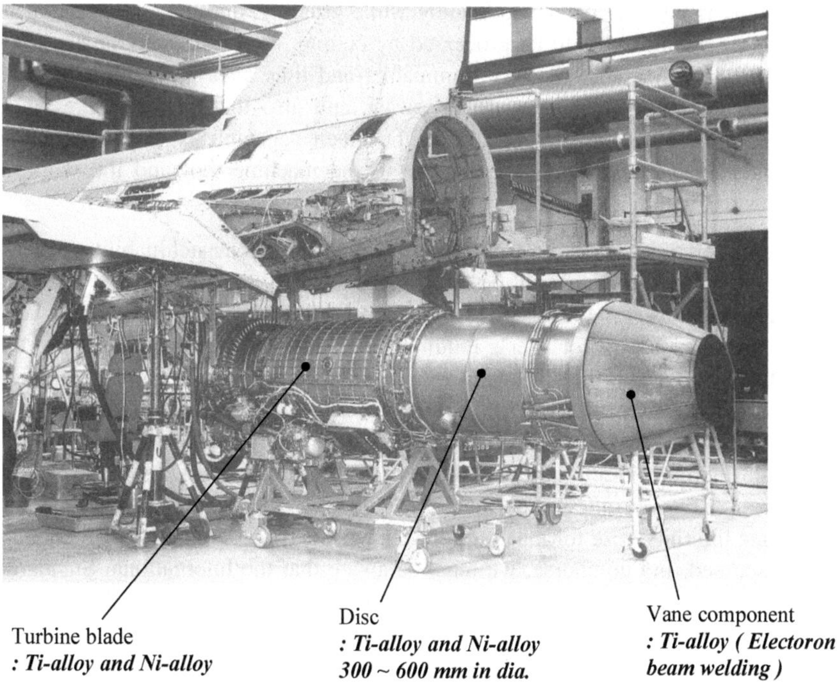

Fig. 3.4 Turbofan for "Gripen" in late 1980s (by Saab)

combat plane, i.e., "Turbofan" of General Electric/Volvo Flygmotor-brand (Type RM 12) for Gripen, which is rated at 80.5 kN with afterburning.

Importantly, the engineer must be aware that the material symbol does not indicate any information necessary to machine. In fact, it is rare cases to get the information for machining, i.e., machinability, from the material symbol. In general, the machinability may be represented by the magnitude of cutting or grinding force, anti-wear and -damage ability of cutting tool, possibility of loading in grinding wheel, flammability of swarf and so on.

From one aspect, the machinability may interpret a part of grass root-like knowledge mentioned in Sect. 3.1. In contrast, it is very convenient if the data sheet for the material symbol describes such information in the form of remarks as follow.

(1) In Brass turning, the cutting edge penetrates often into the work, because of acting the negative back component of the cutting force. It is thus better to choose the tighter backlash eliminator for feeding mechanism in the machine tool.
(2) In 4–6 brass turning, the swarf is chip-like, whereas 7–3 brass generates long ribbon-like swarf. In short, the machinability of 4–6 brass is better than that of 7–3 brass.

(3) Al-alloy is liable to occur the loading while grinding, and thus the component made of Al-alloy should be processed by cutting, but not by grinding.
(4) Ti-alloy and Mg-alloy is very flammable, and thus always we must pay the special attention to the disposal of swarf, and also the danger of fire while machining. When occurring the fire, the green vegetable is very effective to suppress the fire without any damage to the machine tool and the work as compared with the use of fire extinguish agent.

More importantly, we must accumulate the data sheet for machinability, at least the specific cutting force, as will be discussed in Part III and Appendix I. Such the data are mandatory for the machine tool manufacturer in any occasions of conducting the consultancy for user's requirements and receiving the order from the client. In short, we call the "difficult-to-machine" material, which can be classified two-pronged way. One is the engineering material with tough and soft, and the other is that with brittle and hard. Apart from such information for machinability, we will furthermore discuss the material symbol in the following.

Figure 3.5 shows some representative engineering materials for machining, and even when limiting those for machining, we have a myriad of engineering materials to be discussed. In due course, we must be aware that the International Standards (IS) has not covered all the engineering materials even now. In fact, the material symbol is standardised in part by IS, whereas nearly all National Standards have been, with wider scope, enacted the material symbol.

As a result, we have relied on the comparison table for the material symbol since 1950s, and for example, the German engineers once produced such a table, in which they compared those of France, German, Japan, Spain, Sweden, UK, US America, USSR and so on one another, as reported elsewhere.

With advance of the globalisation and furthermore localised globalisation, such a comparison table increases its importance beyond our expectation. Typically, the cutting tool manufacturer, e.g., Tungaloy, has recently publicised such a table.

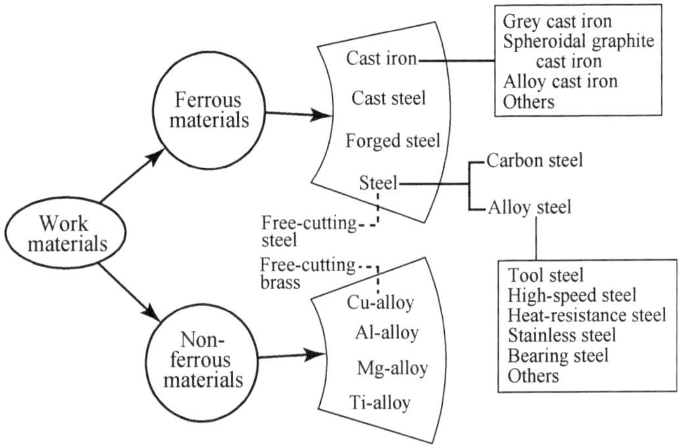

Fig. 3.5 First-hand view for representative engineering materials

Table 3.1 A comparison table for material symbols (by courtesy of Tungaloy)

Material kinds	Standards					
	IS	(Japan) JIS	(USA) AISI SAE	(UK) BS BS/EN	(FRG) DIN DIN/EN	(France) NF NF/EN
Carbon steel	C45 C45E4 C45M2	S45C	1045 1046	C45 C45E C45R		
Ni–Cr–Mo steel	20NiCrMo2 20NiCrMoS2	SNCM220	8615 8617(H) 8620(H) 8622(H)	20NiCrMo2-2 20NiCrMoS2-2		
	41CrNiMo2	SNCM24 0	8637	–	–	–
	41CrNiMoS2		8640			
Stainless steel	X12CrMnNiN 18-9-5	SUS202	202	284S16	–	–
	X10CrNil8-8	SUS301	301	301S21	X12CrNil7-7	Z11CN17-08

According to it, we can get, in general, the same or similar material across some industrial nations; however, in certain cases, it is difficult to get the same material, although IS standardised the material symbol as exemplified in Table 3.1.

To this end, Table 3.2 reproduces the cutting data for "Inconel", and as will be clear from it, the cutting condition depends upon the machining method, cutting pattern, tool material and supply of cutting fluid. Paraphrasing, the machining symbol itself does not provide us with the satisfactory information for the machining technology, and we must adjust properly the machining condition in consideration of especially the machining method and tool material (Breuer 1997). Of note, the ordnance is, in certain cases, made of hard-to-machine material, and thus often its cutting conditions are in military confidentiality.

Table 3.2 Machining conditions for Inconel 718 (by Breuer)

Cutting patterns		Tool materials	Cutting conditions		
			Cutting speed (m/min)	Feed rate (mm/rev)	Depth of cut (mm)
Turning	Rough and dry	Whisker-reinforced ceramics	250	0.2	0.5 ~ 0.9
	Finish	Carbide K10	30	0.2	0.5
Boring	Diameter 32 mm Internal cooling (10% emulsion)	Carbide P20-P40 TiC/TiN coating	40	0.05	–
Milling	Rough Cone face cutter D = 80 mm, Number of teeth 5	Whisker-reinforced oxide ceramics	950	–	2 ~ 4.8
	Finish and 10% emulsion Cone face cutter D = 80 mm, Number of teeth 5	Carbide K10	75	–	0.5

3.3 Swarf and Its Discharge from Machining Space

Figure 3.6 is a flow of swarf processing ranging from the swarf generation, through discharge from the machining space, to the swarf disposal including its recycling. In matters of machining technology concerns, the swarf generation at and discharge from the machining space are at issue. When considering the growing importance of the environmental problems in the globe, however, it is furthermore necessary to generate the swarf for ease of recycling. Figure 3.7 shows thus the briquetting press for recycling the swarf as a raw material.

In swarf generation at and disposal from the machining space, there have been a considerable number of the engineering problems from the past as follows. As reported elsewhere, these problems are fatal, but not satisfactorily solved as yet.

(1) Reduction of generated volume of swarf together with regulating the shape and size of swarf suitable for the discharge.
(2) A less pill-up volume of swarf in machining space to reduce the unfavourable heat sources.
(3) Design of structural body component suitable for swarf discharge.
(4) Total enclosure design to prevent swarf scattering, leakage of cutting fluid, and crushing-through of projectile.

In discussing the swarf problem, furthermore, it is better to establish a reliable classification system of swarf; however, even now, we cannot have it as exemplified the research of Klocke et al. (2010). In fact, they have relied on the classification system of INFOS as shown in Fig. 3.8, which was one of the proposals for the swarf classification in the past.[1]

In this context, we have a further confusion with the advent of the new material, and a root cause of difficulties in the establishment of the classification system lies in the variety of the machining method and each method generates also various swarfs. In addition, even in the same machining method, the shape and size of the swarf depends upon the work material and its heat treatment, combination of work material with the tool material, either dry or wet cutting, cutting conditions, and so on.

For the sake of further understanding, Fig. 3.9 illustrates various patterns in the swarf generation in the case of the orthogonal cutting model.

3.3.1 *Swarf Generation and Cutting Fluid*

In general, we can obtain the finished surface of high quality when using the cutting condition, by which the ribbon- and tangle-like swarfs are generated. Needles to

[1]INFOS classification system is German-originated, and thus to understand neatly the feeing for the shape of swarfs, German terms are described together with English terms in Fig. 3.8.

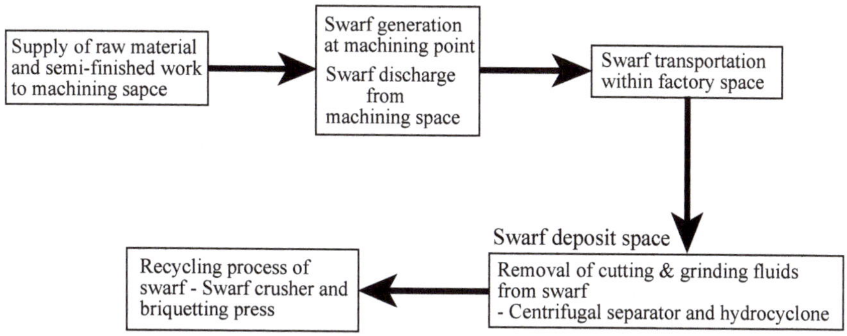

Fig. 3.6 Processing flow of swarf

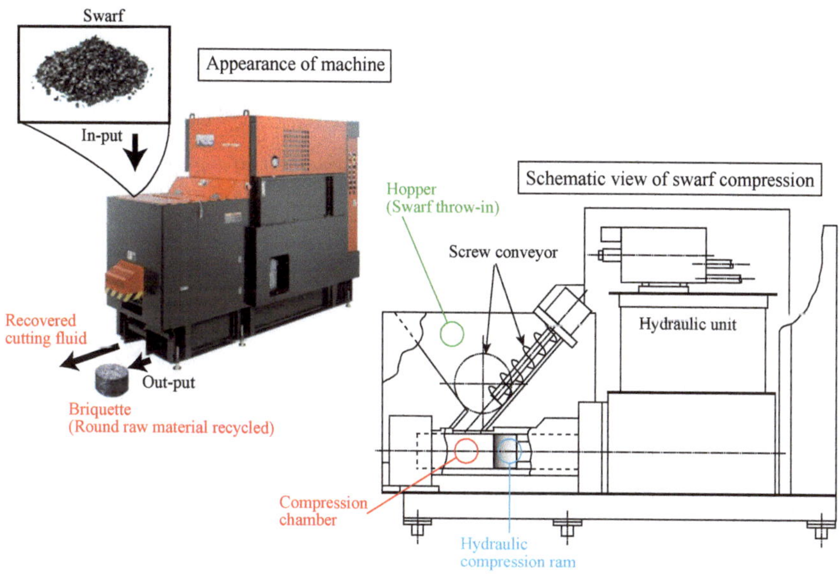

Fig. 3.7 Briquetting press (by courtesy of Nicotech 2012)

say, such swarfs are difficult to processing something necessary, and thus we need to contrive the suitable processing technology for the ribbon- and tangle-like swarfs.

There are three representative ways to generate the swarf for ease of processing, i.e., use of cutting tool with chip breaker, supply of higher-pressurised fluid to cutting point, and control for form-generating movement as exemplified by the "Woodpecker-like feed of drill" and also by the change of NC information for tool path.

In both the single-point cutting tool and the milling cutter, nowadays, the tool with throw-away tip including the insert is very popular, because of its handy management of tip in choice of material, tool life control, re-grinding, disposal and

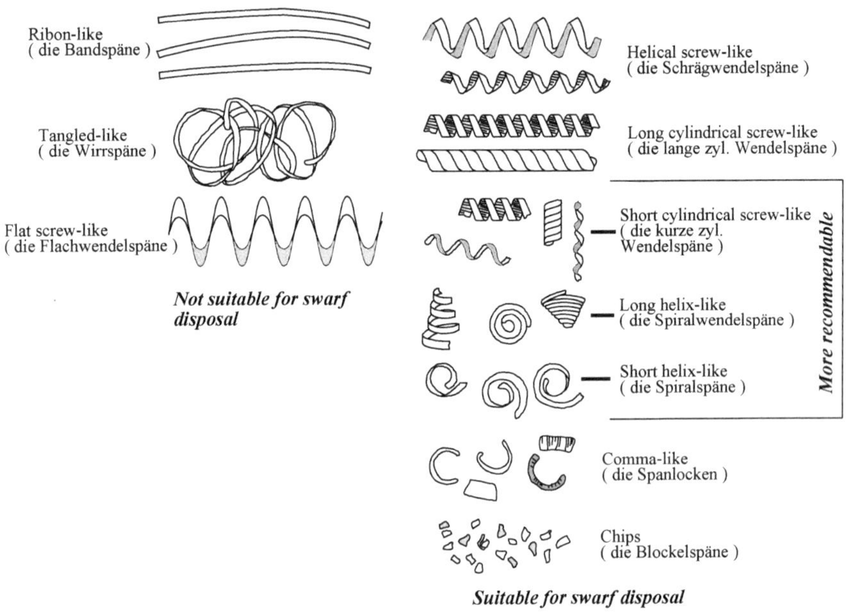

Fig. 3.8 A classification system of swarf by INFOS

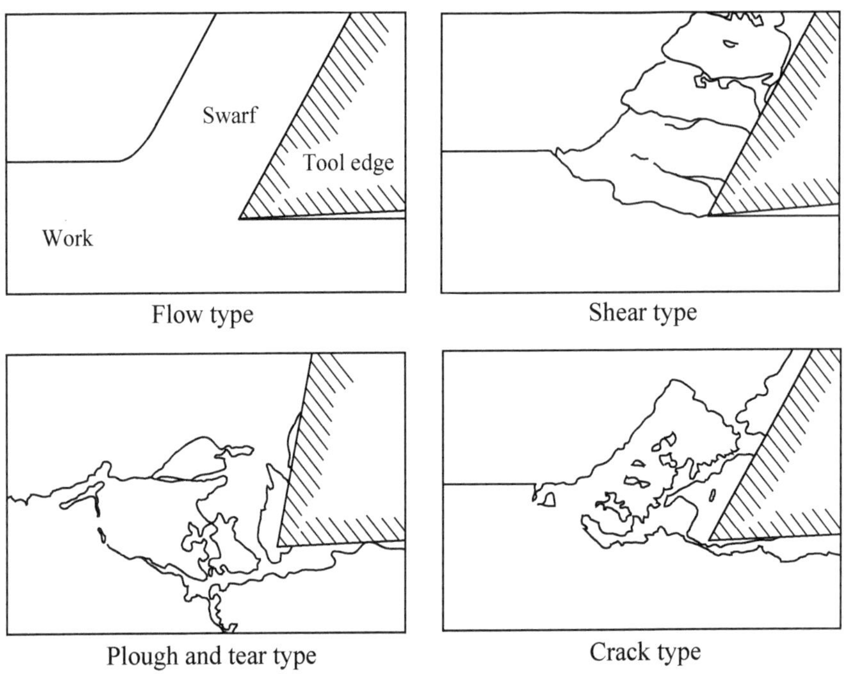

Fig. 3.9 Various patterns in swarf formation

Fig. 3.10 Scenery of turn top by throw-away tip with chip breaker (by courtesy of Tungaloy, Type TUS)

so on. Importantly, the throw-away tip with chip breaker gives us various advantages to the ease of swarf generation control, although the tip shape becomes very complicated by placing the small step, channel, dimple and so on at the tip surface. Figure 3.10 is scenery demonstrating the effectiveness of the tip with chip breaker.

Figure 3.11 displays furthermore various throw-away tip together with boring tool with internal supply of the cutting fluid with high pressure (see catalogues of cutting tool manufacturers). It is very interesting that the throw-away tip has been recently integrated with the cutting fluid supply to enhance the cutting performance. As well known, we can benefit directly by supplying the cutting fluid, e.g., reduction of the cutting force, increase of tool life and improvement of surface quality, and indirectly the ease of swarf discharge.[2]

In fact, the cutting fluid with high pressure can assist the swarf breakage. For example, Technical University [Technische Hochschule (TH)] Aachen has reported the effect of the higher pressurised cutting fluid on the swarf breakage in parting-off the work made of Inconel 718 (Klocke et al. 2010). In fact, the swarfs are comma-like and chip in Fig. 3.8, when supplying the cooling fluid (emulsion type) of 80 bar and 300 bar, respectively. The cooling fluid is supplied through the

[2]The cutting fluid is one of the crucial issues in machining, and can be classified into the "oil-based" and "water-based" cooling lubricants or coolants. At present, the former is for Ti-alloy and Inconel machining, and a synergy of the vegetable oil, solid lubricant and extreme-pressure sulfur additive instead of the old-fashioned one, i.e., that including the chlorine compound. In contrast, the latter is being prevailed; however, the water quality including the incidence of bacterial degradation is very important. For example, the soft waters found in Scotland and hard waters of limestone and chalk areas, i.e., those found in Kent, can cause foaming and form a scum of hard water soaps, respectively. Thus, there have been a considerable number of quick notes.

King N (Dec. 1991/Jan. 1992) Ah, the smell of it. Manufacturing Engineer: 7–9.

(2011) Water-soluble high-performance cooling lubricant—The bacteria hunter. Swiss Quality Production: 52–53.

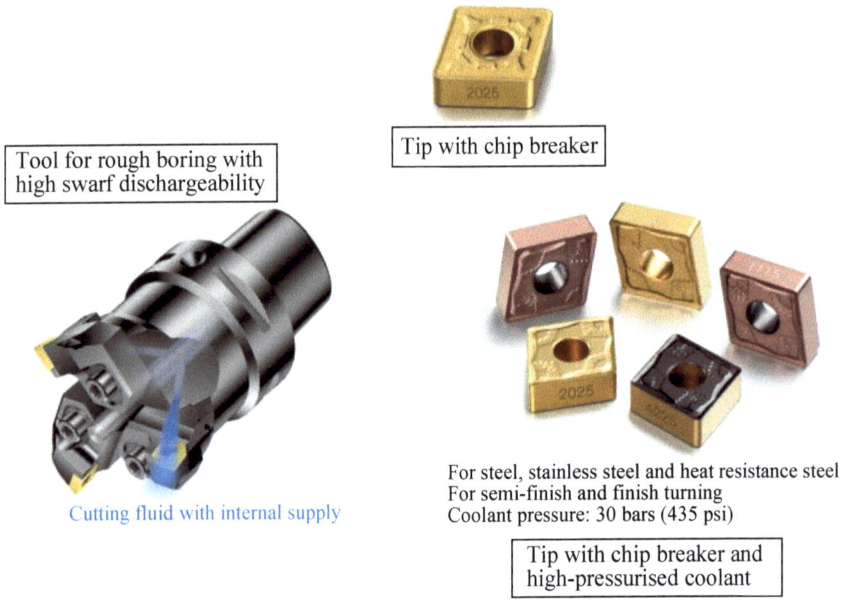

Fig. 3.11 Internal cutting fluid supply and tip with high-pressurised coolant (by courtesy of Sandvik 2015)

hollow clamp plate for the cBN insert (cutting speed: 250 m/min, feed rate: 0.1 mm/rev, cutting width: 6 mm).

In the case of conventional twist drill, the swarf discharge is, as widely recognised, more difficult than that in other machining methods, because the swarf is discharged through the narrow flutes. In addition, the margin of the drill rubs the inside of hole, although the drill has the body clearance along its longitudinal direction.

It is thus common that drill feeding is woodpecker-like movement, and in addition, the drill has the internal fluid supply function to reduce the friction within drilling mechanism and also to ease the swarf discharge as shown in Fig. 3.12. In short, the cutting fluid can be supplied from the face-end of drill body.

At present, the drill manufacturer contrives furthermore various remedies, as follows.

(1) The flute is polish-finished to reduce the friction as that of Iscar-brand.
(2) In the deep-hole (gun) drilling, Gühring employed the wider and mirror-finished flute to ease the swarf discharge around 2005. This gun drill is made of sintered carbide of super-fine particle, and for machining the motorcar component made of cast iron and Al-alloy.
(3) Walter merchandises the twist drill made of sintered carbide, which has the shallow groove with short pitch around the edge portion of the margin. As can be imagined, the cutting fluid becomes the vortex flow, and plays something definite.

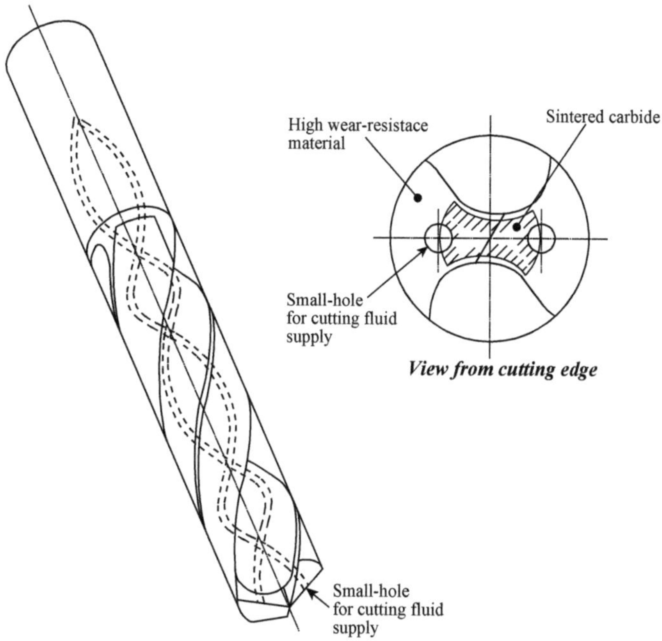

Fig. 3.12 Twist drill with supply hole for cutting fluid (by Sandvik 1998)

In the twist drill, thinning is very important to narrow the chisel point and duly to reduce the axial component (thrust) in the drilling force. As can be expected, thinning may be also indirectly effective to the swarf discharge, because the narrower dull chisel point results in the change of swarf formation. Figure 3.13 illustrates the twist drill placing the stress on the thinning patterns, and each thinning can be characterised as follows.

(1) Point-thinning is the very popular with ease of making, and for steel, cast iron and non-metal material drilling.
(2) X-thinning is applicable to the relatively thick chisel, to be superior in the drill engagement to the work than XR-thinning, and extremely for drilling of the deep-hole and difficult-to-machine material.
(3) XR-thinning is to improve the drill engagement to the work, to increase the drill life, and easier in thinning as compared with X-thinning.

Within the swarf disposal context, another remedy is to reduce the generation volume of the swarf. A representative machining method is "*Trepan Boring*", one of the variants of the Boring and Trepanning Association (BTA) method, as shown in Fig. 3.14. As can be readily seen, the core may be used as a raw material of other components. It is also recommended to employ the raw material of near-net shape.

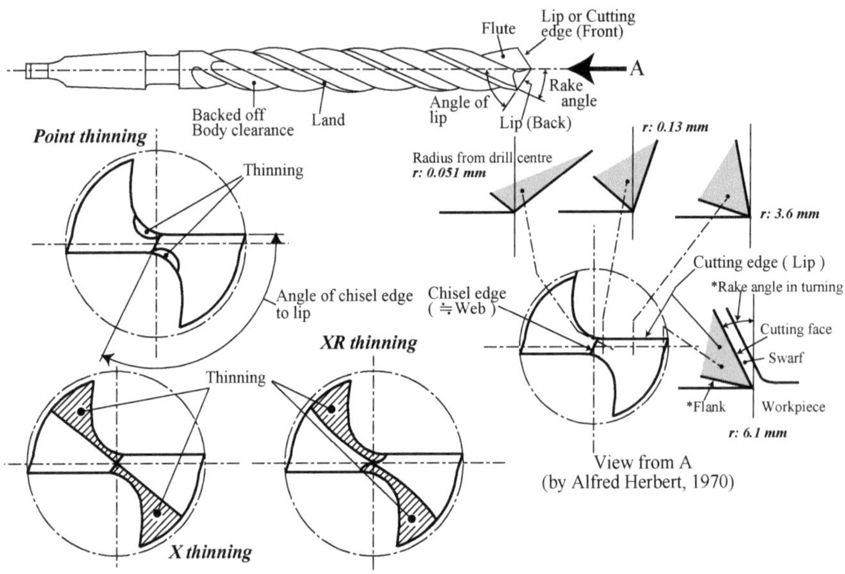

Fig. 3.13 Twist drill in general and various thinning methods

In general, the generated volume per hour V of the swarf can be written as

$$V = A \times u \times n \times v \tag{3.1}$$

where,

A cutting area,
u depth of cut,
n number of works machined per hour,
v specific swarf volume, i.e., that generated from unit volume of raw material.

Table 3.3 gives us the specific swarf volumes for several raw materials while milling and drilling. Importantly, we must be aware of the difficulty to find such the data apart from those reported by The MTIRA (Gough 1970).

To this end, Figs. 3.15 and 3.16 show some grinding fluid supply systems, which may assist indirectly to improve the loading condition of the grinding wheel. In these systems, a facing issue is to penetrate the fluid through the entrained flow of air around the grinding wheel. Actually, Nakagawa Seiki contrived the system shown in Fig. 3.15 for the centreless grinder in 1960s, where the grinding fluid is supplied from the inside of the wheel cover with a less influence of the entrained flow of air. In contrast, Fig. 3.16 uses positively the porosity of the grinding wheel (see Fig. 1.15, Graham and Whiston 1978).

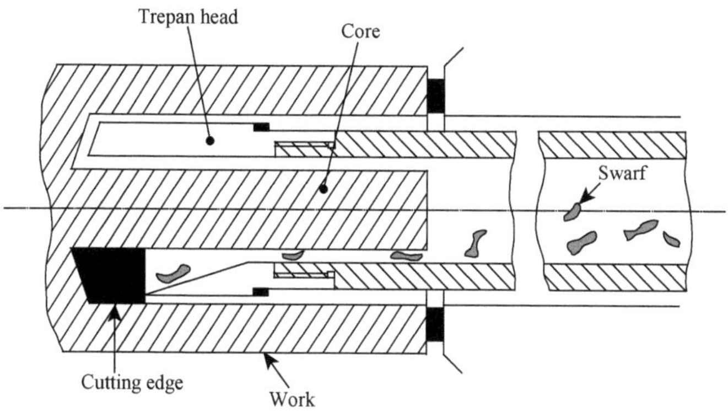

Fig. 3.14 Schematic view of "Trepan Boring"

Table 3.3 Swarf volumes from specific machining operations (by The MTIRA)

Work material	Type of machining operation	Volume of swarf from 16 cm^3 (1 in^3) of solid metal
Mg-alloy	Face milling	2500 ~ 5000
Al-alloy	Spar milling	800
Steel	Face milling	500 ~ 700
Cast iron	Slab milling	250
Cast iron	Face milling	115 ~ 160
Cast iron	Drilling sizes 25–75 mm	160
Medium-carbon steel	Drilling sizes 25–75 mm	650 ~ 1300

3.3.2 *Swarf Discharge from Machining Space*

Following the swarf generation, we must quickly discharge the swarf from the machining space to avoid unfavourable influences, e.g., heat accumulation by generated swarf, invasion of chips into the guideway, degradation of positioning accuracy of the work by sticking the thin flake-like swarf to the jig and fixture.

In this context, the "*Slant Bed* (bed of chip-flow type)" is the utmost famous remedy in the structural design as shown in Fig. 3.17. In the preferable case, the swarf falls directly in the mobile swarf skip as shown also in Fig. 3.17. More specifically, the structural body component related to swarf discharge should be provided with the slope like the inclined partition and cross slide, on which the swarf rolls downward, and then wide aperture at its underneath, from which the swarf falls down to the transportation equipment placed at or under the factory floor as shown in Fig. 3.18. Importantly, the wider the aperture, the more deterioration is the rigidity of the structural body component. The designer should manage the

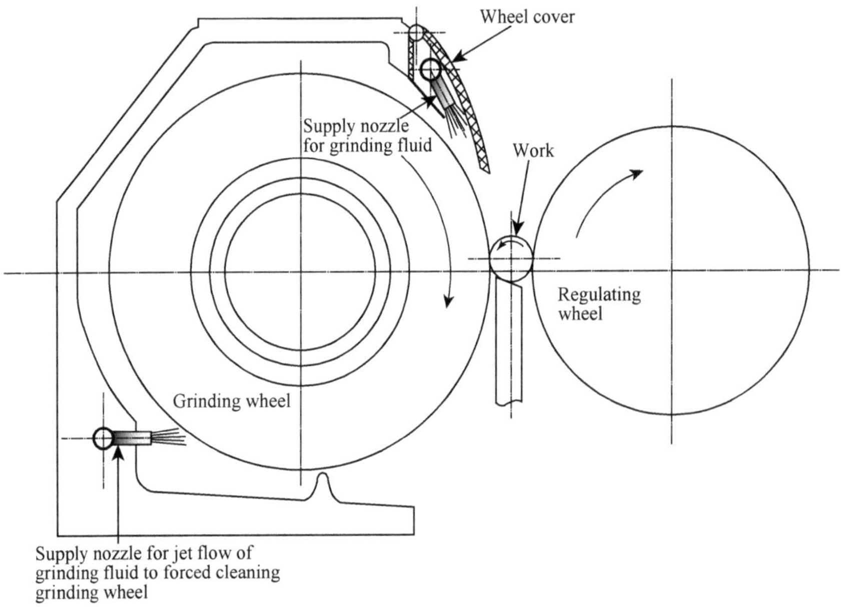

Fig. 3.15 Grinding fluid supply system in consideration of entrained flow of air—In case of centreless grinder (by Nakagawa Seiki 1960s)

suitable allocation of the stiffening rib, partition, cell, double-wall and so on to leverage between the swarf dischargeability and the rigidity of the machine body.

It is thus necessary to accumulate the data regarding the slope angle, on which the swarf rolls freely, and also the bulk density of the swarf, by which we may estimate the ease of the transportation of the swarf. The MTIRA publicised such data in the past as shown in Fig. 3.19 and Table 3.4 (Gough 1970), and to our regret, after then we have certain difficulties to get such data. In addition, we must be aware that the data shown in Fig. 3.19 and Table 3.4 are only for reference, and that the thin-flake like swarf of Al-alloy is stuck to even the vertical wall.

In the meanwhile, the total enclosure has been employed in NC machine tool to enhance the marketability from the viewpoint of the industrial design. In addition, the total enclosure is very effective to assist the swarf collection, to prevent the swarf scattering and to guarantee the safety of the operator. In contrast, the total enclosure deteriorates the thermal stability of the machine body in certain cases.

In relation to swarf processing, a facing issue in the total enclosure is also to leverage between its swarf dischargeability and structure design without any "*Drum Effects*", i.e., membrane vibration. In general, the total enclosure is made of steel sheet and plate, and thus liable to occur the drum effect. To prevent it, the reinforcing members, e.g., rib and spar, should be allocated in the inside of the total enclosure in consideration that the industrial design is in well fruition. As a result, the reinforcing members become obstacles for the swarf discharge.

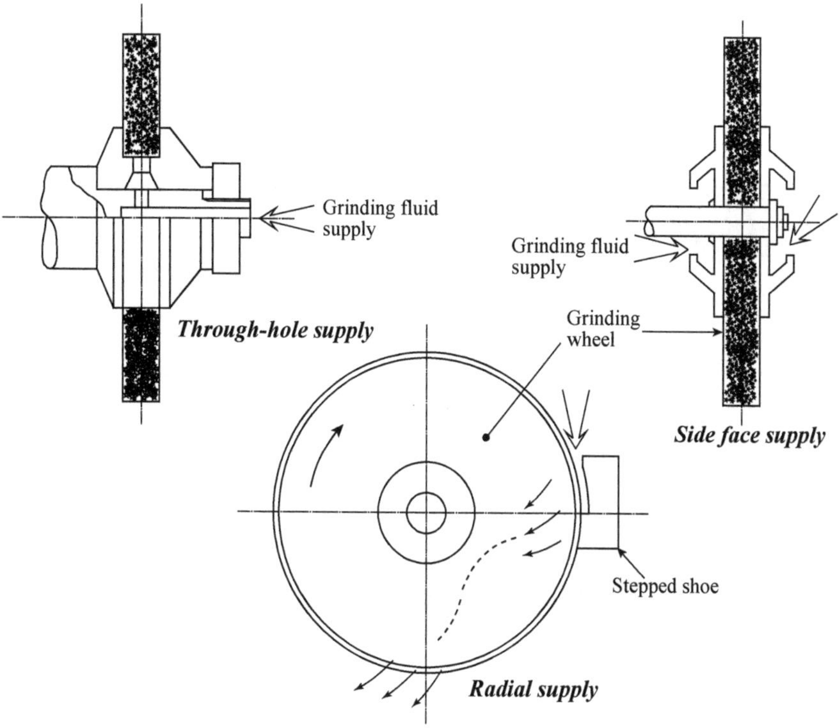

Fig. 3.16 Grinding fluid supply system of wheel impregnation type (by Graham and Whiston 1978)

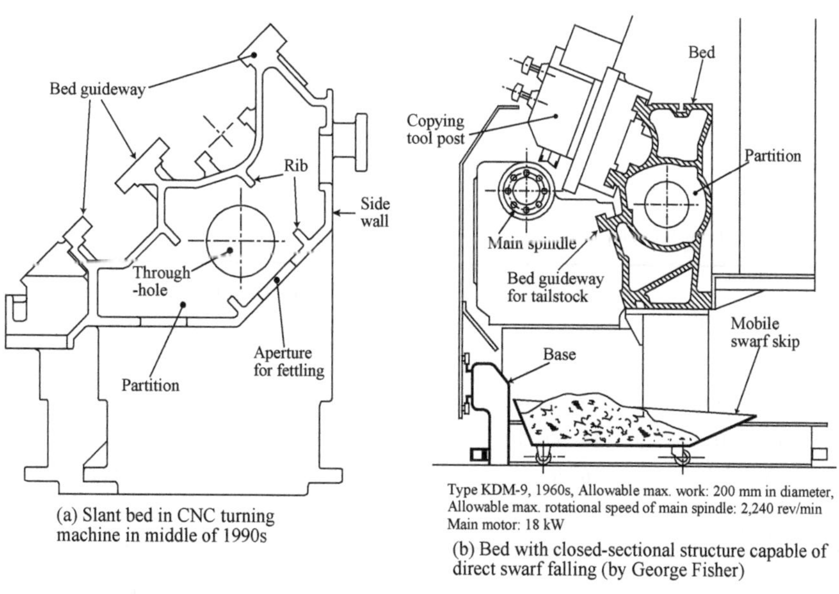

Fig. 3.17 Bed structures suitable for swarf discharge

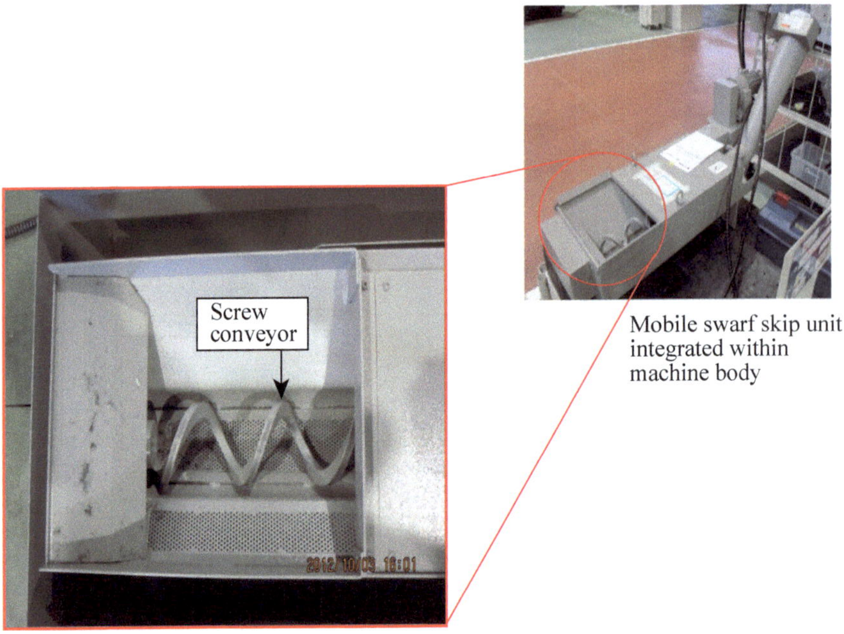

Fig. 3.18 Swarf transfer from machine tool (by courtesy of Amada Machine Tools 2013)

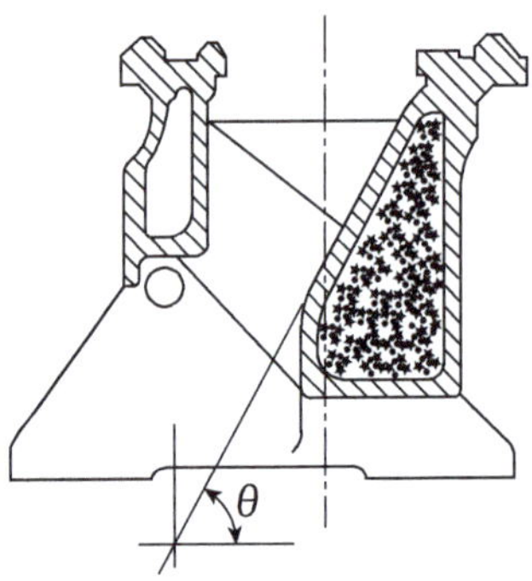

Average of angles θ (deg.)

Swarf material	Type of finish on slope surface of structural body component		
	Smooth (2–20 μm CLA)	Fine-machined or painted (20–150 μm CLA)	Rough-machined (150–300 μm CLA)
Cast iron	13–26 (34)	19–30 (43)	28–45 (58)
Mild steel	14–60 (78)	24–45 (77)	27–72 (79)
Al-alloy	16–46 (50)	30–32 (74)	31–72 (90)
Phosphor bronze	15–24 (32)	22–25 (40)	——

Fig. 3.19 Slope angle of structural body component for rolling freely swarf particle (by The MTIRA)

Table 3.4 Experimental data on bulk density of swarf (by The MTIRA)

Type of swarf	Bulk density (kg/m^3)	
	From actual sample used	From published data
Steel		
Long bulky	525	240
Short bulky	360 809	560
Chips	Mean 1474 (1022 ~ 2000)	1040
Phosphor bronze		
Light	1538 1938	1121
Heavy	–	2242
Al-alloy		
Long	Mean 243 (56 ~ 390)	160
Short	Mean 459 (229 ~ 961)	320
Cast iron		
Short bulky	Mean 1329 (503 ~ 1746)	–
Fine chips	Mean 1682 (1025 ~ 2114)	–

References

Breuer E (1997) Securely and swiftly in Inconel. EPE. Carl Hanser, Munich, p 10–14

Gough PJC (ed) (1970) Swarf and machine tools. Hutchison of London, London

Graham W, Whiston MG (1978) Some observations of through-wheel coolant application in grinding. Int J Mach Tool Des Res 18:9–18

Klocke F et al (2010) Automatisierte Produktion –ohne Sapnbruch undenkbar. ZwF 105 (1/2):21–25

Matsumura T, Tamura S (2012) Force prediction in milling of titanium alloy. In: Proceedings of ASME 2012 international symposium on flexible automation, St. Louis USA, June 2012

Chapter 4
Structural Body Components Closely Allocated Around Machining Space—Core for User-oriented Platform System

As will be clear from Chap. 2, the leading structural body components are the headstock, tailstock and turret head in the turning centre (TC), and the spindle head and table in the machining centre (MC), when discussing the machining space. More specifically, the function and performance specifications of the machining space can be determined by the (1) spindle nose and tapered hole at the face-end of spindle, (2) barrel and its tapered hole in the tailstock, and also (3) tool seat in the turret head, where the attachment and tool are mounted on through the tapered hole, bay-type flange, tenon-slot joint and so on. In addition, we must understand the (4) *allowable maximum cutting torque* for the main spindle and table, and if possible, the (5) on-site calculation method of the cutting force.

Of course, such determinants change to some extent depending upon the kind of the machine tool. For example, the grinding spindle noses in the grinding centres (GCs) of horizontal and vertical type are similar to TC and MC, respectively.

Importantly, the matters of items (1)–(3) concern should be guaranteed the interchangeability with the attachment and tool in large and various extents, so that the user-oriented platform may demonstrate its capability satisfactorily. In certain cases, we can expect some percussion effects more than our expectation by user's own contrivances.

Figure 4.1a, b show the main spindles for the numerical control (NC) lathe and MC placing the stress on the spindle nose and tapered hole at front face-end in the main spindle. In NC lathe, the spindle nose is short tapered and in the most case enables the chuck to be accommodated. The tapered hole can facilitate the accurate placement of the centre by *Morse taper*; however, with prevailing NC lathe and TC, the chuck work is dominant in turning and thus centre renders generally useless at present. As a result, the straight bore is often used instead of the tapered hole. In retrospect, the straight bore was common style in the automatic turning machine with turret head, which is one of the original kinds of NC turning machine at present. In addition, the through-hole in the main spindle is for draw bar placement to operate the jaw in the chuck.

Y. Ito and T. Matsumura, *Theory and Practice in Machining Systems*,
DOI 10.1007/978-3-319-53901-0_4

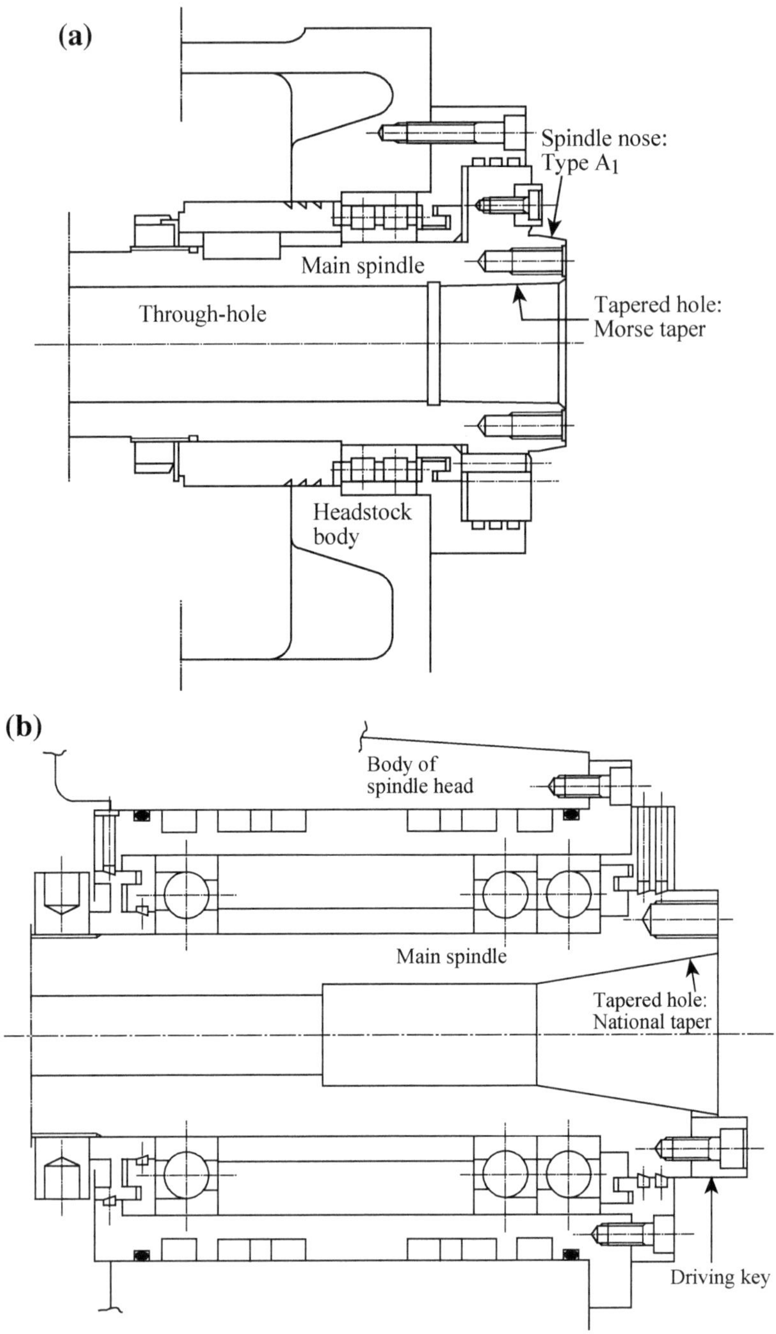

Fig. 4.1 Spindle nose and front face-end configurations: **a** NC lathe. **b** Small- and medium-sized MC

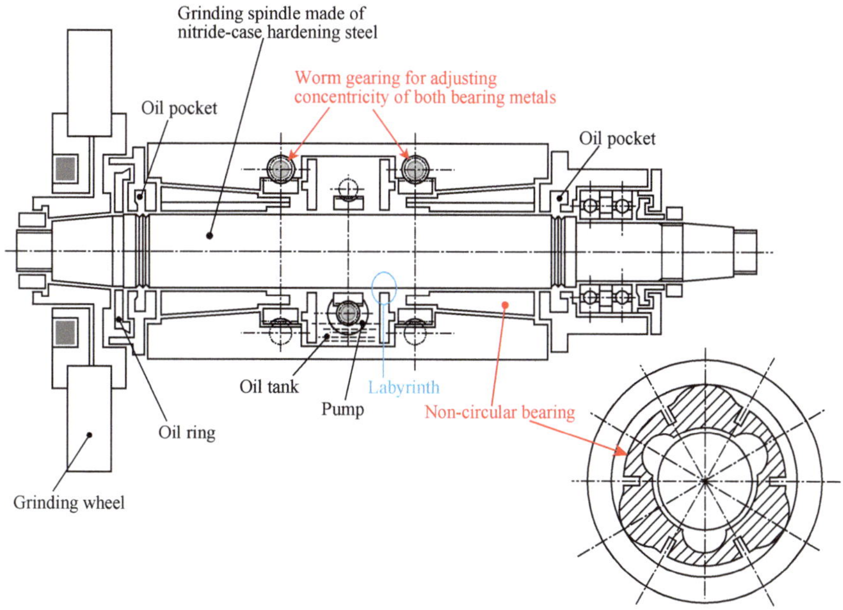

Fig. 4.2 Grinding spindle of bedway grinding machine (Schamltz-brand 1961)

In MC, the tapered hole is, in general, of "*National Taper*" and can facilitate to mount various tools from the automatic tool changer (ATC). Importantly, the "*National Taper*" has been replaced by the hollow-shank taper (HSK; der HohlSchaftKegel) system as will be discussed in Sect. 4.1. In addition, the face-end of spindle nose consists of the driving key with the threaded hole to mount the face milling cutter for heavy-duty cutting. Of note, in the era of traditional milling machine and planomiller, the outer surface of spindle nose was finished to the fitting tolerance to accommodate accurately the adapter.

In short, the taper profile is primary concern to be the interchangeability in reality. For further understanding, Fig. 4.2 shows a grinding spindle, and as can be readily seen, the grinding wheel can be mounted on the spindle through the wheel flange with tapered hole (refer to Sect. 4.2).[1] In addition, Fig. 4.3 shows the tail-stock of the engine lathe, and we can observe similar configuration as that of spindle nose. Even in the tailstock, the outer surface of barrel is finished with fitting tolerance to accommodate the adapter, if necessary.

[1]It appears that the grinding spindle shown in Fig. 4.2 is old-fashioned; however, the spindle nose is the same as that being in practice. In many respects, this grinding spindle is worth reproducing because of its outstanding main bearing of non-circular type, which is to be in reality by using the elastic deformation of the bearing metal (pad).

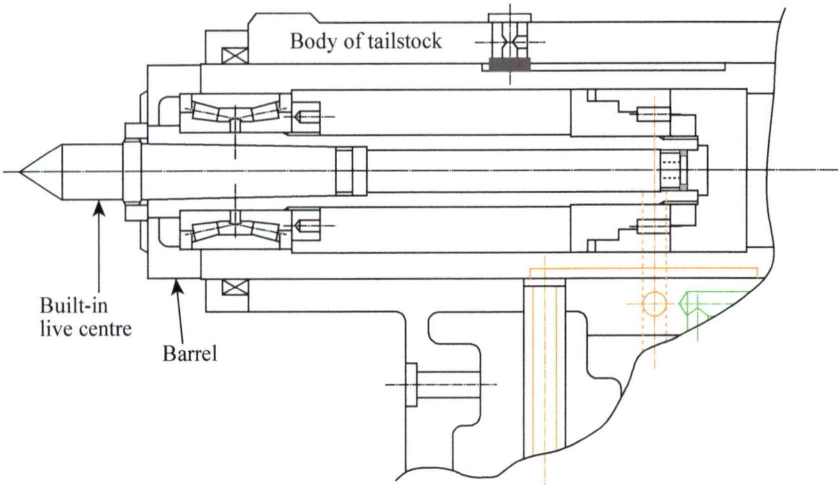

Fig. 4.3 Tailstock and barrel with built-in live centre and of oil hydraulic travelling type

Summarising, the spindle nose and tapered hole at front face-end of the main spindle are very important to guarantee the interchangeability, and thus some quick notes will be given in the following.

4.1 Tapers in General and Facing Engineering Problems

As will be clear from the above, the taper profile is the utmost important entity across the whole spindle nose, barrel of tailstock and turret head. This is because the taper connection is very simple, but can realise the accurate positioning between two bodies-in-contact together with ensuring the acceptable joint stiffness. More specifically, the taper can be classified into the (1) self-holding and (2) self-releasing types by the threshold regarding the frictional angle, which regulates the self-locking phenomenon. In the former, the holding force should be reinforced by the tongue or key way driving when the frictional holding force is not enough. Obviously, the latter can ease holding and releasing, for example, the tool shank to be connected; however, it necessitates the positive locking device, which induces, in certain cases, the deterioration of the positioning accuracy.

Figure 4.4 summarises thus nearly all kinds of the taper including those of old-fashion, such as Jarno taper and Reed taper, and also indicates some kinds of the machine tool available. At present, the "*Metric, Morse and National tapers*" are widely employed, and needles to say, we can secure the interchangeability between different sizes of the same taper kind with the taper reducing sleeve, and also between the different kinds of the tapers with the adapter.

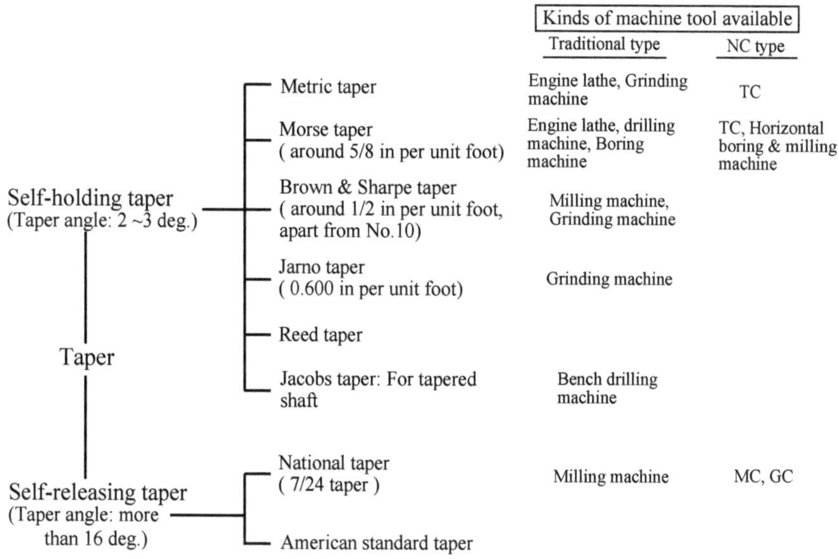

Fig. 4.4 Kinds of taper

Of note, it is very interesting that the Jacobs taper is for the tapered shaft, and that has been and being widely employed in the bench drilling machine. Importantly, the International Organization for Standardization (ISO) has enacted the eight types in Jacobs taper, and the taper is around 1/19 in the case of around 20 mm in nose diameter, which is equivalent to the diameter of the main spindle of the bench drilling machine.

Although the taper connection is very simple and handy jointing method, there is a fatal shortage by nature in the self-holding taper, i.e., "*Linking Phenomenon and Fretting Corrosion*". Paraphrasing, the more accurate in contact of two bodies at taper, the more dangerous is their linking, and in contrast, the more loose in contact at taper, the easier is occurrence of fretting corrosion.

More specifically, the two bodies-in-contact at the taper, e.g., cutting tool mounted on the tapered hole, should be in the best positioning accuracy by improving the contact condition across the whole mating surface. In contrast, the fully perfect contact condition results in "*Linking*", and even in the worse case, it is impossible to disintegrate the two bodies-in-contact. In the case of non-perfect contact, the micro-slip occurs often, and produces the oxide debris, which colors a taper red or dark red, i.e., "*Fretting Corrosion*". Of course, the taper is worn and becomes bad profile.

Thus, we used to modify the contact condition by the stepped taper, taper with recess, and giving the slight difference between both the taper angles in mating bodies.

4.2 Spindle Nose and Tapered Hole

The taper connection is one of the machine tool joints, and generally speaking, shows the weakest rigidity as compared with other joints. When the taper connection includes within the *flow of force* of a system, thus, its total rigidity is liable to deteriorate. Importantly, the static and dynamic behaviour of the machine tool joint depends considerable upon the relative stiffness of the two bodies-in-contact (Ito 2008).

In addition to the joint stiffness, we have further another problems at the spindle nose and tapered hole at front face-end of main spindle, when accommodating, for example, the chuck and cutting tool. As widely known, the spindle nose in the main spindle of turning machine has been standardised by, for example, Japanese Industrial Standards (JIS) as shown in Fig. 4.5. More specifically, the chuck and adapter are accommodated to the short taper at the spindle nose by simultaneously contacting with the flange-end or nose-end face (front face-end of main spindle), depending upon the independent four-jaw chuck or self-centring three-jaw chuck.

In this context, we must be aware of the following.

(1) Although the short taper is standardised, we cannot guarantee neatly the interchangeability between those enacted by JIS and American Standards Association (ASA). The former and latter are based on metric and inch systems, respectively, and thus we have the very small conversion errors.

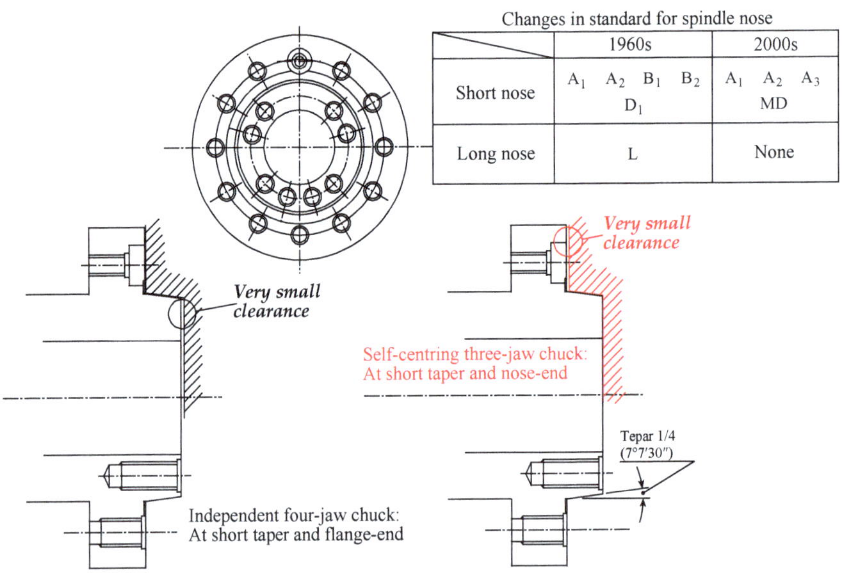

Changes in standard for spindle nose

	1960s	2000s
Short nose	A_1 A_2 B_1 B_2 D_1	A_1 A_2 A_3 MD
Long nose	L	None

Fig. 4.5 Chuck mounting on spindle nose of Type A_1—Difference between independent four-jaw and self-centring three-jaw chucks in 1960s

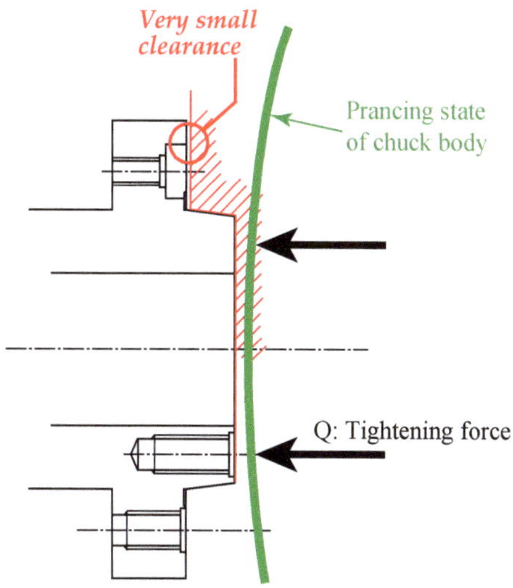

Fig. 4.6 Prancing of chuck body by tightening force

(2) As also shown in Fig. 4.5, the standard changes with the advance of the related technology.
(3) In the accommodation of self-centring three-jaw chuck, we have still an essential problem to be solved, i.e., cocking (not smooth and not even) motion of jaw within T-slot. This cocking is caused by the prancing deformation of chuck body as shown in Fig. 4.6, which is similar to the elastic deformation of the circular plate concentrated-loading at its centre. Fortunately, with the advance of chuck technology, this cocking motion disappears gradually (see Chap. 5).

In relation to the spindle nose, furthermore, we must pay a special attention to the tapered hole at the nose-end face. In fact, the spindle nose is liable to elongate dynamically while milling with higher-speed as shown in Fig. 4.7 (Rall and Findeklee 1998), and also expand due to the centrifugal force while rotating in the higher-speed as shown in Fig. 4.8. Of special note, the most serious problem is that the tool shank shows the considerable bedding-in into the main spindle. This behaviour is extremely relevant in the case of "*National Taper* (NT)" connection (7/24 Taper).

In retrospect, there have been a considerable number of the contrivances to overcome the substantial problems mentioned above, e.g., flange mount tooling system proposed by Mori Seiki and Kuroda in 1993,[2] and also a patent in USA

[2]In the flange mount tooling system, the two surfaces-in-contact can be realised by using the "Short taper of 20 deg" and "Front face" in the spindle nose like the placement of the four-jaw independent chuck to the main spindle of the engine lathe. This system can guarantee the reproducibility of positioning accuracy better than 3 μm when using ATC.

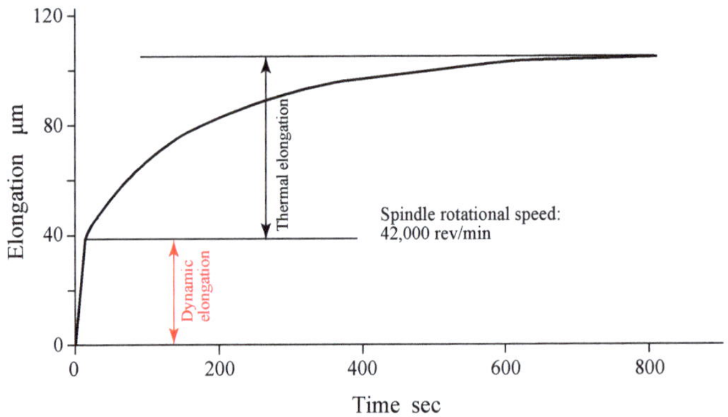

Fig. 4.7 Dynamic elongation of main spindle while milling by higher-speed

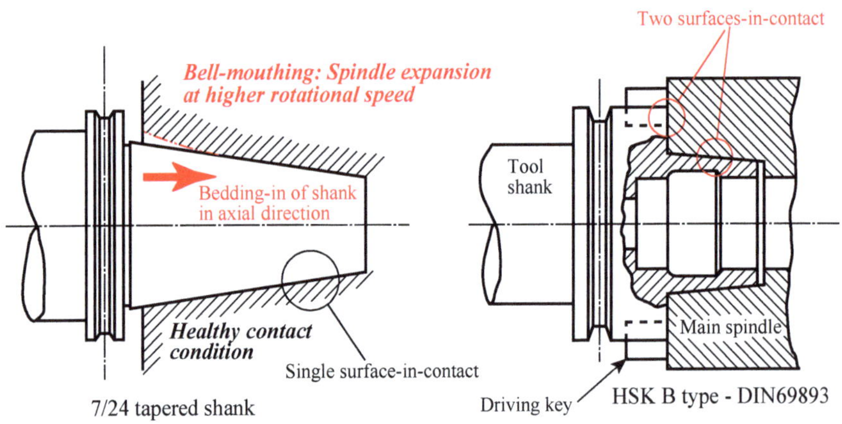

Fig. 4.8 Shortcomings of 7/24 taper at higher rotational speed due to "Bell-mouthing" and its remedy by employing HSK

(No. 5,322,304, June 21, 1994), in which the two surfaces-in-contact becomes to be reality using the "Taper with Saw-like Configuration" and "Front Face Contact".

In short, the preferable remedy for the bedding-in problem is to employ the two surfaces-in-contact joint, and the most representative joint is HSK, which is compatible simultaneously with the higher-speed and heavy-duty cutting. In this case, a primary concern is thus to enhance both the joint stiffness and the positioning accuracy, and in due course, HSK improves especially the positioning accuracy of cutting tool. Figure 4.9 reproduces the two types of HSK enacted within the German Industrial Standards (DIN; Deutsche Industrie Normen). As can be readily seen, HSK can be characterised by its configuration of two surfaces-in-contact.

To understand how far the two surfaces-in-contact benefits, Fig. 4.10 shows some comparisons of the joint stiffness among the flat, i.e., two surfaces-in-contact,

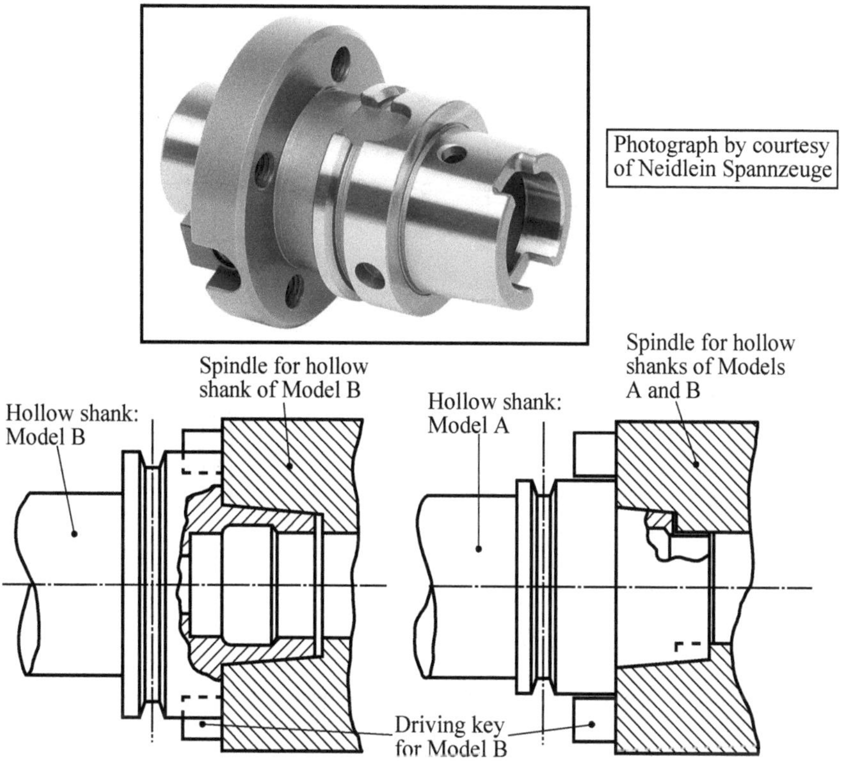

Fig. 4.9 Hollow shanks—Models A and B according to DIN 69893

Curvic coupling and NT types (Hazem et al. 1986). As can be readily seen, the rigidity deterioration is considerable at the joint as compared with that of equivalent solid, i.e., spindle-tool shank connection without joint (monolithic tool shank), and furthermore the flat joint shows larger rigidity than those of other joints. Although the taper connection is very important in attachment and tool mounting as mentioned above, the research into the taper connection has not been so active.

In addition, a new horizon emerges in the taper connection by merchandising a new product. In short, on the strength of the beneficial features of the HSK, Sandwik has employed the taper connection with polygon shape as shown in Fig. 4.11.

In grinding, the grinding wheel is the most popular tool, and mounted on the grinding spindle through the wheel flange. Figure 4.12 shows the typical wheel flanges: one is as per DIN, and the other has been on market by the Japanese manufacturers, where each grinding machine manufacturer has employed its own standard as shown together in right below of Fig. 4.12. In fact, apart from DIN the

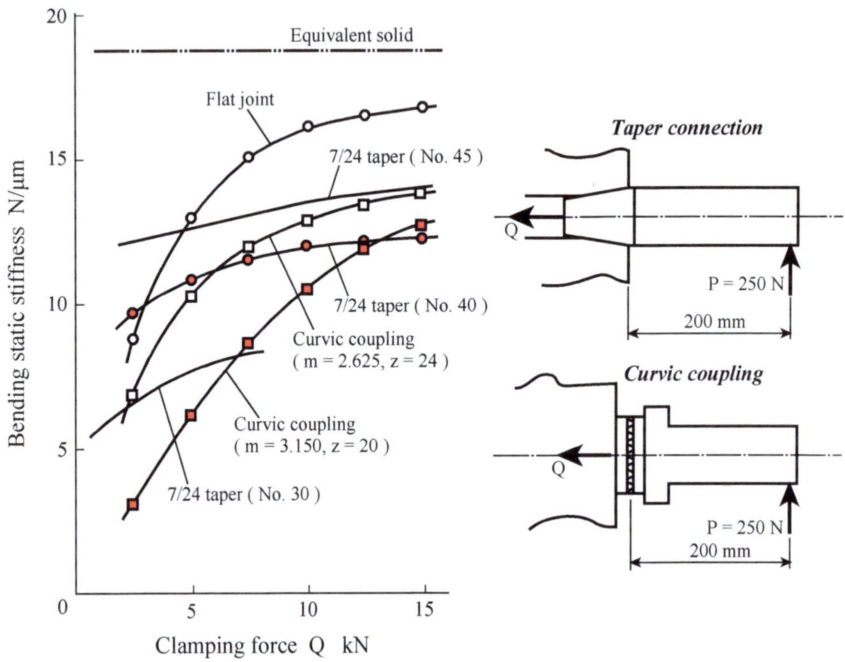

Fig. 4.10 Comparison in bending static stiffness for various tool shank connections—In dry contact condition

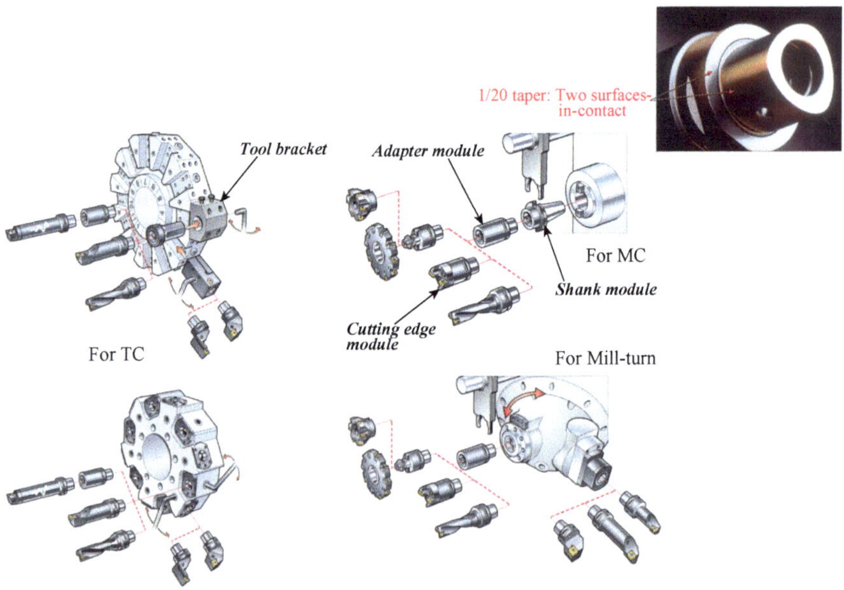

Fig. 4.11 Tool layout and necessary attachments in machine tools (by courtesy of Sandvik 2007)

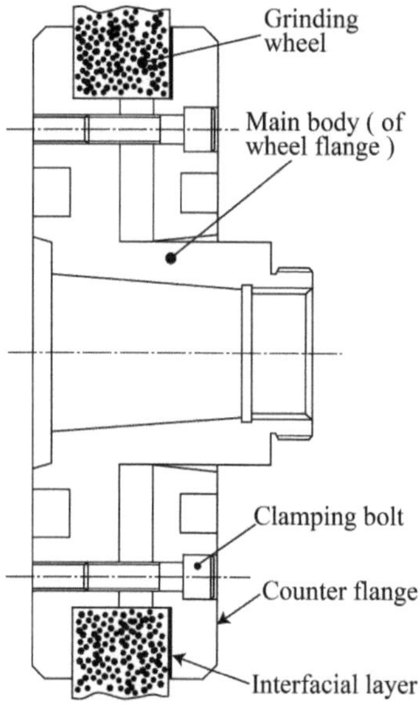

Wheel flange as per DIN 6375

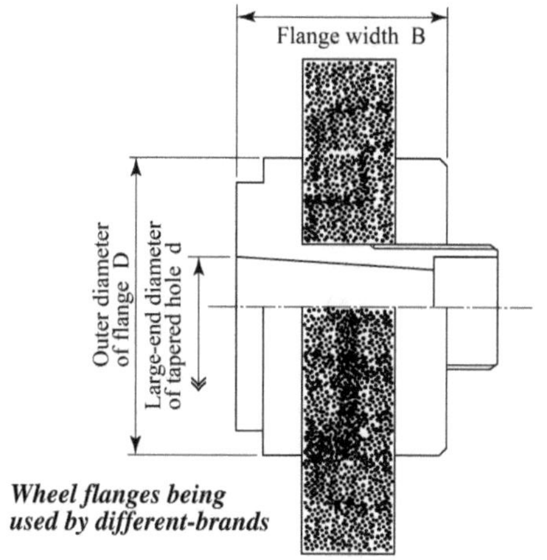

Manufacturers \ Dimensions	D mm	d mm	B mm	Taper
Jung	104	33	65	
Mitsui high-tech	79	26	45	1/5
Studer	178	44	62	
Toshiba Machine	170	65	65	1/5
Tsugami	168	40	71	3 deg.

Fig. 4.12 Some examples of wheel flange

wheel flange is not standardised in general as yet.[3] In retrospect, Brown and Sharpe proposed its own taper system as can be readily seen from Fig. 4.4.

Although employing widely, the wheel flange as per DIN involves some problems as follows.

(1) Difficulty in maintaining relative positioning accuracy between work spindle and grinding wheel, because of deflection of paper ring (interfacial layer) and counter flange.
(2) Ease of acting bending stress on grinding wheel, resulting in wheel breakage in certain cases.
(3) Difficulty in realising uniform tightening across all the clamping bolts.

To solve such problems, for example, Saljé of Technical University of Braunschweig (TH Braunschweig) once proposed a new wheel flange in the 1980s as shown in Fig. 4.13, where the following remedies were employed.

[3]The wheel flange appears as to be the milling chuck and tool holder in MC and TC, and thus should be discussed in Chap. 5. In consideration of a less umber of families as compared with the jaw chuck, however, we discuss it herein. Of note, we will discuss also the tool holder including the tool bracket and tool block in Sect. 4.3, because of the same reason.

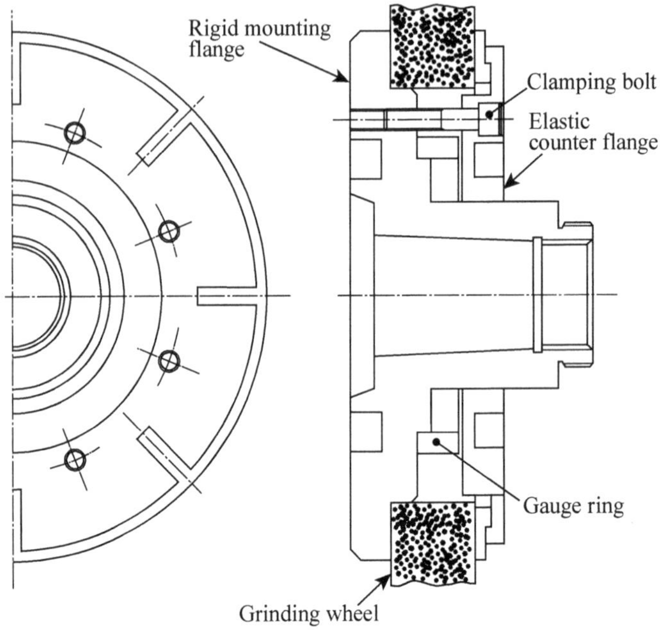

Fig. 4.13 A New proposal for wheel flange (by Saljé 1980s)

(1) Fruition of better positioning of grinding wheel than ever before by pressing grinding wheel to rigid mounting flange.
(2) Non-sensitivity for unevenness in side face of grinding wheel by using elastic counter flange.
(3) Realisation of uniform tightening by pressing counter flange on gauge ring.

In principle, even now the problems mentioned above are not completely solved as yet.

Accidentally, the two surfaces-in-contact was at issue even in grinding on that occasion as same as that for cutting. Figure 4.14 reproduces a trial, and in due course, we benefited from it to large context. Obviously, HSK is being prevailed in the grinding sphere and a noteworthy improvement is carried out as shown in Fig. 4.15. In this improvement, the gripper is allocated far from the face-end of main spindle, and thus the chatter vibration with higher frequency may be suppressed because the large damping source is closer to the grinding point. This suggestion is based on our long-standing experience on the factory floor.

Importantly, Weck (1993) suggested the beneficial features of HSK in grinding as follows.

(1) Higher-speed grinding with hard grain is to be in reality.
(2) Increase of damping in grinding system.

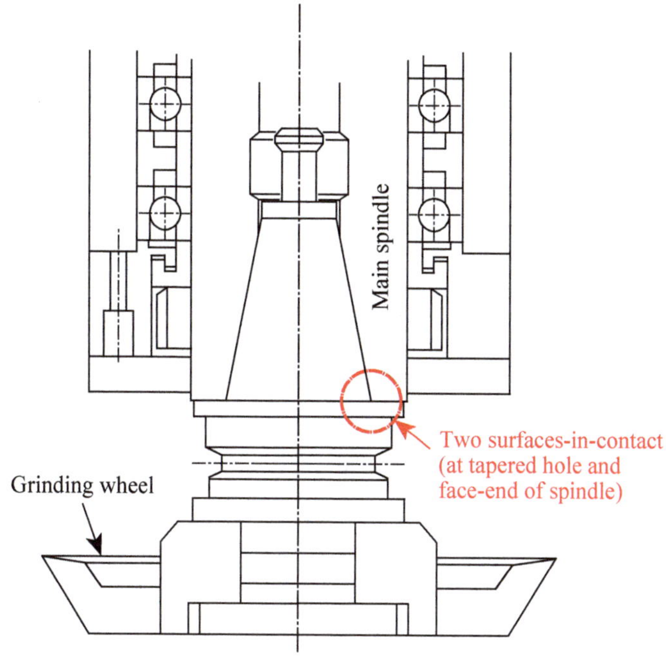

Fig. 4.14 Two surfaces-in-contact mounting of grinding wheel on main spindle with National taper in late 1980s (by courtesy of Okamoto Machine Tool Works)

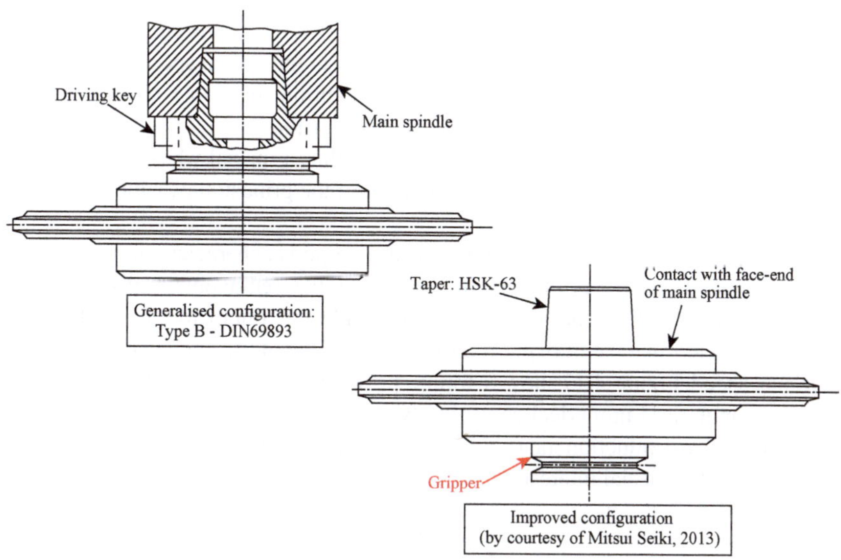

Fig. 4.15 Wheel flange of HSK type

Fig. 4.16 Segment grinding wheel for vertical heavy-duty surface grinding machine (Type BRV of Kovosvit-brand, in the 1960s)

(3) Reduction of dressing force and ease of wheel setting.
(4) Enhancement of variation of grinding by quick changing the wheel, using the formed wheel, and also using cBN and diamond grain.

Within the grinding wheel context, a further important variant is the segment wheel for the heavy-duty grinding as shown in Fig. 4.16. In this case, the grinding wheel is 750 mm in diameter, rotates 590 mm/min in maximum speed and is driven by the motor of 30 kW in output power. In addition, the table of the grinding machine can accommodate the work of 2100 kg in maximum. As can be readily seen, the grinding wheel is liable to wear out, and thus a special wheel flange should be used to accommodate a considerable number of the grindstones, which are easily exchangeable.

To this end, it is worth suggesting that the torque-power characteristics of the main motor is one of the determinants in discussing the removable capacity of the machining space. More specifically, the allowable maximum torque of the main spindle can determine the maximum removable volume of the undue portion at each rotational speed. Figure 4.17 shows a torque-power characteristics of AC servo motor for the computerised numerical control (CNC) turning machine, which is similar to that of three-phase induction motor, i.e., leading main driving motor for the traditional machine tool. Of special interest, the torque of the induction motor can be written as

$$T = 71620\,(PS/N)$$

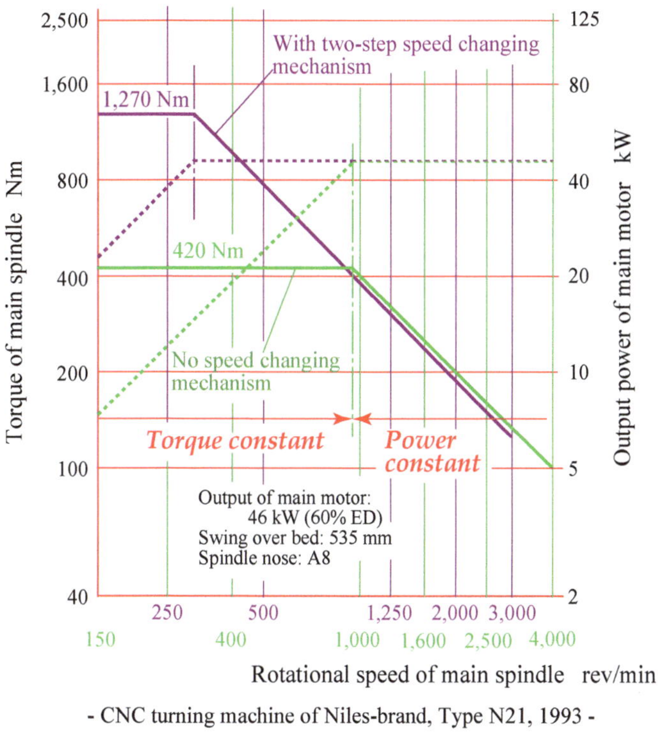

Fig. 4.17 Characteristics of main spindle driving by AC servo motor

where,

T = torque kgf-cm, PS = Motor power, N = Rotational speed of main spindle rev/min.

In accordance with the design principle, for example, the size of spindle nose should be accommodated the chuck, which can be burdened the maximum cutting torque. Importantly, other structural configuration entities are also designed the same principle.

In short, the machine tool user must pay the special attention to the kind of the torque-power characteristics of the main motor, i.e., implicit structural configuration entity, in the investigation into the suitable machining space. In this context, at least, we must differentiate the following characteristics.

(1) AC servo motor with permanent magnet field: Constant torque characteristics with rotational speed.
(2) Brushless DC servo motor: Constant torque characteristics within the lower rotational speed and over such a speed, torque decreases with rotational speed.

4.3 Turret Head

The turret head is one of the most important structural configuration entities surrounding the machining space especially in CNC turning machine, TC and *Mill-turn*. In the case of *Mill-turn* based on MC, furthermore, the universal milling head, i.e., milling head of tilting type, can replace the turret head. There are thus a handful of kinds in the turret head depending upon the structural configuration as shown in Fig. 4.18. For the ease of understanding, Fig. 4.18 indicates also some examples already shown elsewhere in this book, and Figs. 4.19 and 4.20 reproduce other turret heads, i.e., those of disc and twin types, which can be characterised as follows.

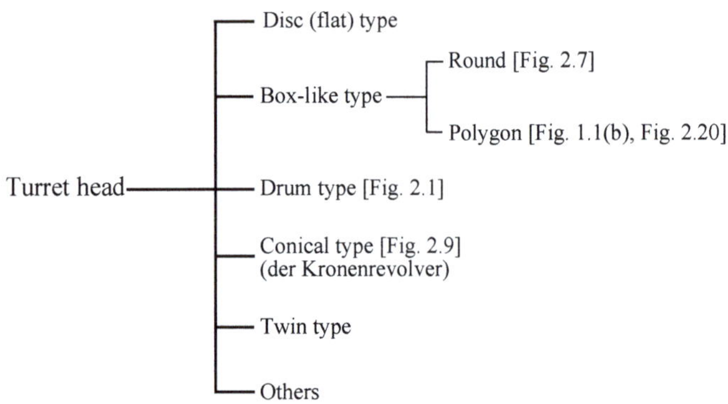

Fig. 4.18 Classification of turret head

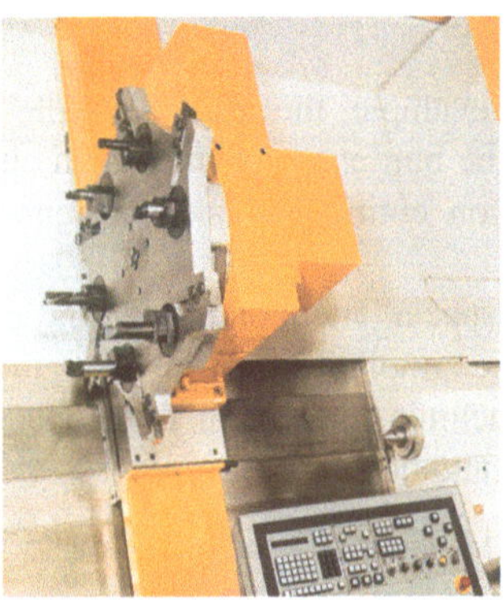

In CNC production lathe
- Type PNE, VDF Boehringer-brand
(On-the-spot investigation, October 1979)

In turret lathe of drum type
- Type PIREXA, Pittler-brand, in 1960s

Fig. 4.19 Examples of disc turret in 1960s and 1970s

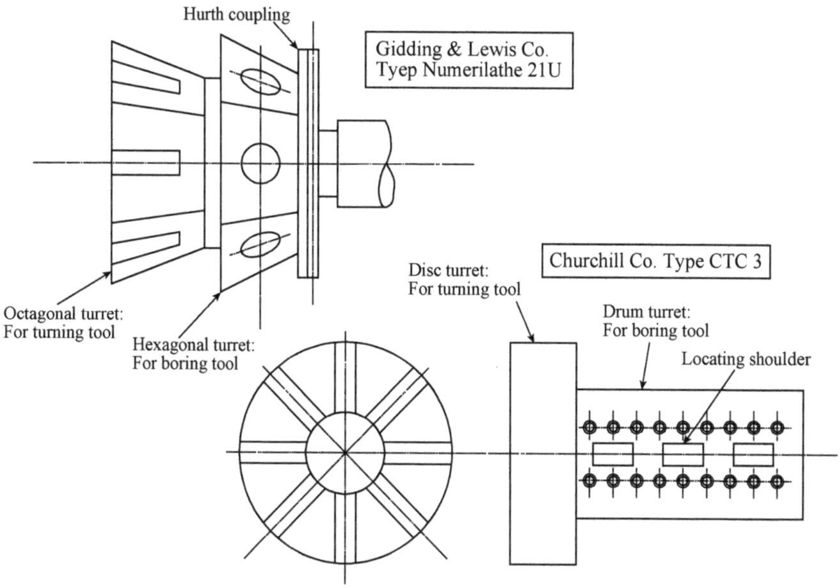

Fig. 4.20 Twin-Turret head (Based on sketch by Ito, at IMTS Chicago 1982)

(1) The disc type is originated the traditional automatic and automatic turret lathe to aim at the ease of use together with light-weighted and compact structure with satisfactory rigidity. Of note, those of VDF and Pittler are based on the turret heads in the traditional turret lathes of Scheu-brand and Pittler-brand, respectively.
(2) Although involving considerable difficulties in the structural design, the twin type aims at the larger flexibility in the tool layout, provided that the satisfactory rigidity with light-weighted structure is to be in reality.

Obviously, primary concern in the turret head design is the joint stiffness between the tool seat in the turret head and tool holder including the tool bracket. Geminately, another root cause of difficulties lies in the fruition of the light-weighted turret head with acceptable rigidity under the spatial constraint. Paraphrasing, the more rigidity a turret head with multiple-function, the larger shape and heavy structure are its precondition resulting in the narrower machining space, although the user does need the wider machining space. From the past, this is the utmost hindrance in the design of turret head, i.e., less spatial allowance for turret accommodation in a machine tool as a whole.

More specifically, the turret head becomes unwieldy by proving, at least, the following functions to enhance the high operability and flexibility in tooling.

(1) Functions facilitating the quick indexing, quick changing the tool holder, and driving the rotating tool.
(2) Structural configuration for increasing the number of tool (holder) seats and wider allowable space for each tool seat.

As can be readily seen, these design requirements impede the realisation of the rigid turret head with compact and light-weighted structure.

Of various turret heads shown in Fig. 4.18, the polygon type has been prevailed, and to enhance the flexibility for tooling layout, the octagonal, decagon and dodecagon configurations are common. Importantly, the turret head of conical type has an interesting feature. In fact, the turret seat is declined to the spindle axis, and thus by indexing the turret, the axial direction of the tool can be positioned either along or perpendicular to the spindle axis.

As will be clear from the above, we may apply the design technology for the turret head in the era of the traditional turret lathe. For example, Fig. 4.21 shows a comparison between the old- and the new-fashioned tooling layouts in the single-spindle automatic of Index-brand in the 1970s and Traub-brand in the late 1990s. As can be seen, there are no essential changes between both the layouts in the single-spindle automatic; however, it can be considered that a new aspect is the enhancement of another functions, e.g., "*Hand-off*" of work being machined from the primary to secondary machining spaces. More specifically, the work is first machined at the primary machining space, in turn transferred to the secondary machining space by the mandrel of shaft tolerance-basis type after completing the necessary processes, and finally finished in the secondary machining space.

Paraphrasing, we must consider furthermore the resurrection of tool layout technology of NC lathe in 1980s, and from such a viewpoint, Fig. 4.22 shows the old-fashioned locating mechanism of the tool bracket to the tool seat and an example of the quick tool changing system. In the past, the locating and fixing functions for the tool bracket were arranged separately to maintain both the accurate

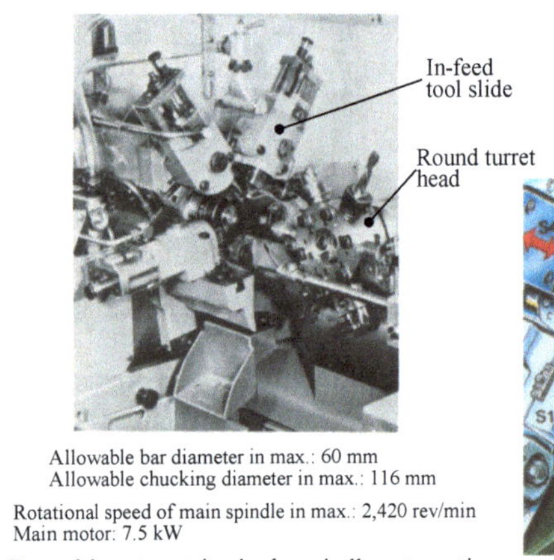

Round-box turret in single-spindle automatic of programmable control (Type ER60, Index-brand, in 1970s)

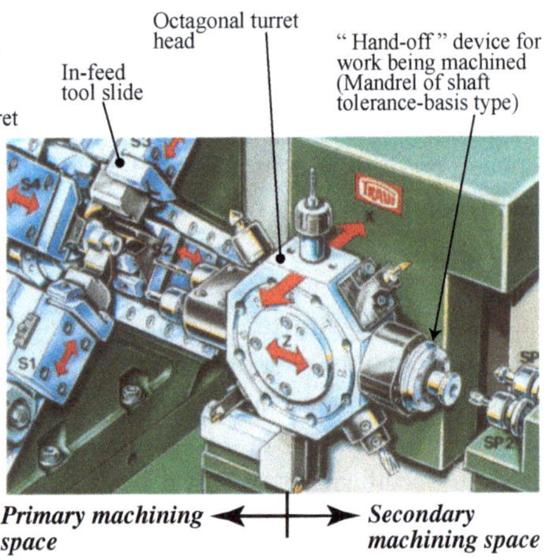

Turret in CNC single-spindle automatic with " Hand-off " function (Traub-brand, in late 1990s)

Fig. 4.21 Changing trends in turret heads of single-spindle automatics

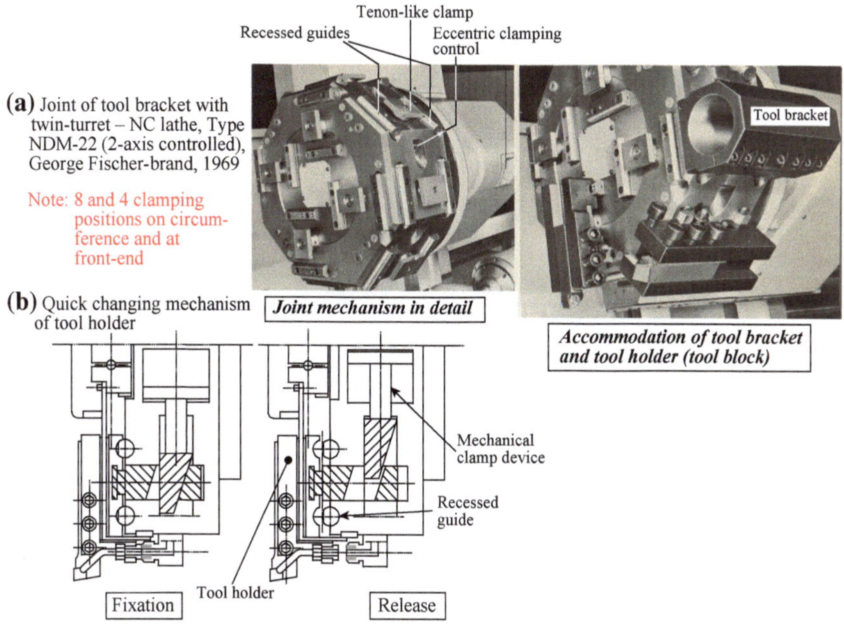

Fig. 4.22 Joint of tool holder with turret head

positioning and the rigid clamping; however, these two functions are integrated into one at present as can be readily seen from Fig. 4.11 in accordance with the advance of the corresponding technology. In that of Sandvik shown in Fig. 4.11, for example, the reproducibility in positioning accuracy is better than ±2 μm.

At present, the cutting tool is mounted either directly or indirectly, e.g., through the tool holder, on the tool seat in the turret head as already shown elsewhere. In the former, the taper connection, one of the machine tool joints, is at issue, for which we discussed already in Sect. 4.1. In the latter, a crucial issue is the jointing problem of the tool holder including the tool block and tool bracket to the tool seat as already exemplified in Fig. 4.22.

For mounting the tool holder on the turret, in fact, a less number of structural configurations for jointing have been only in practices, and as will be clear from our long-standing experience, there are the two-pronged engineering problems as follows.

(1) To secure the rigidity between the tool holder and the turret body, i.e., realisation of high joint stiffness.
(2) Together with maintaining the rigidity, to enhance the reproducibility of positioning accuracy of tool.

Importantly, the tool holder, especially tool bracket, may reinforce the versatility in the tool layout under the constraint number of the tool seat. In contrast, each tool holder with the tool should adjust its weight to be nearly equal across the whole tool

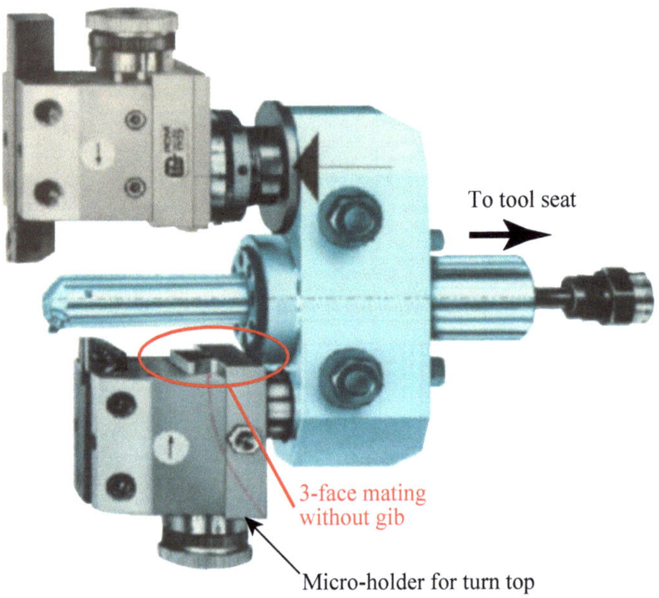

Fig. 4.23 Tool bracket for precision turn top and boring (by PCM Willen 1970s)

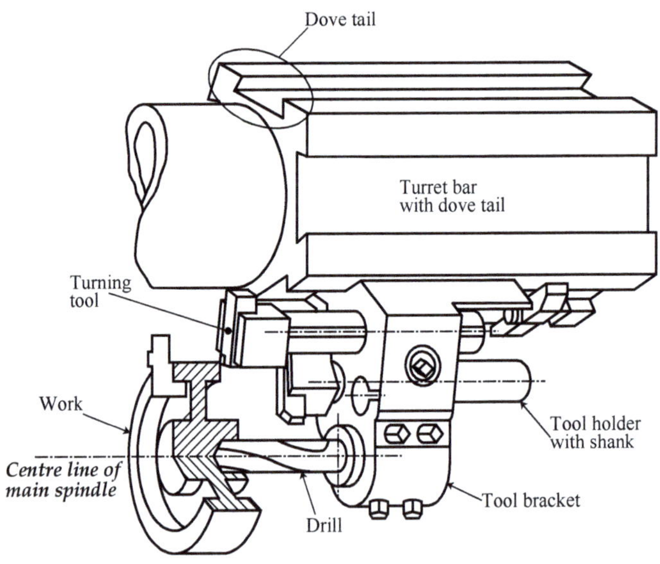

Fig. 4.24 Plain cutter holder for drilling, rough turning, facing and chamfering in traditional automatic turret lathe

holders, so that the indexing accuracy is not influenced by the acceleration and deceleration of the turret head. In the past, thus, the massive tool bracket was employed; however, with the increase of indexing speed, the simple and light-weighted tool bracket has been used, e.g., two-axis drilling tool cassette. Of note, SU-matic has merchandised such modernised tool holders, and nearly all TC manufacturers purchase them.

In retrospect, we used to mount the turning and boring tool bracket as shown in Fig. 4.23, which was contrived by PCM Willen in Sweden in the 1970s. This tool holder can be characterised by providing the dovetail without gib to have better rigidity and to be light-weighted tool holder in reality. Figure 4.24 reproduces the massive tool bracket for complex machining in the automatic turret lathe of Cleveland type. This multiple turning head travelling along the turret bar is for drilling, rough turning, facing and chamfering, and can benefit to machine the component, which requires of the longitudinal movement with short range. Reportedly, the vertical twin allocation of the main spindle and turret bar may solve the "*Cold turret—Hot spindle Problem*",[4] and thus we may conduct machining with better accuracy.

Supposedly, the polygon turret head has enough number of the tool seat at present, and NC itself has enough machining flexibility. There are thus no necessities to contrive the tool bracket for complex machining; however, it emphasises again that we must eye the accumulated knowledge for the connecting methods between the turret head body and tool holder in the traditional turret lathe, automatic lathe and automatic, if necessary.

Within the matters of tool holder concerns, furthermore, we can observe the growing importance in the quick changing and modular tooling systems. The tool bracket has, more or less, such the system function; however, these two have apparently been mushroomed either in separate or often in synergic way within the tool holder and modular tooling system as will be discussed in Chap. 7.

4.4 Linear Travelling Table with Built-in-Rotary Table and Auxiliary Indexing Round Table

The table is one of the structural configuration entities especially in MC. In fact, the work is accommodated on the table and machined by the cutting and grinding tools mounted on the main spindle. In general, the table with T-slots has been widely used, and in certain cases, the table has the bolt- and threaded-holes to set and fix the work. Figure 4.25 is a classification system of the table, and both the linear travelling and the trunnion types have been prevailed in MC. In contrast, the rotary

[4]In the conventional structural configuration, the turret head is allocated far from the main spindle, and thus we observe the considerable temperature difference between the turret head and the main spindle. In general, the main spindle is in higher temperature than the turret head, i.e., a pair of hot spindle-to-cold turret head. In the case of the vertical twin allocation, such a temperature difference appears not to be significant.

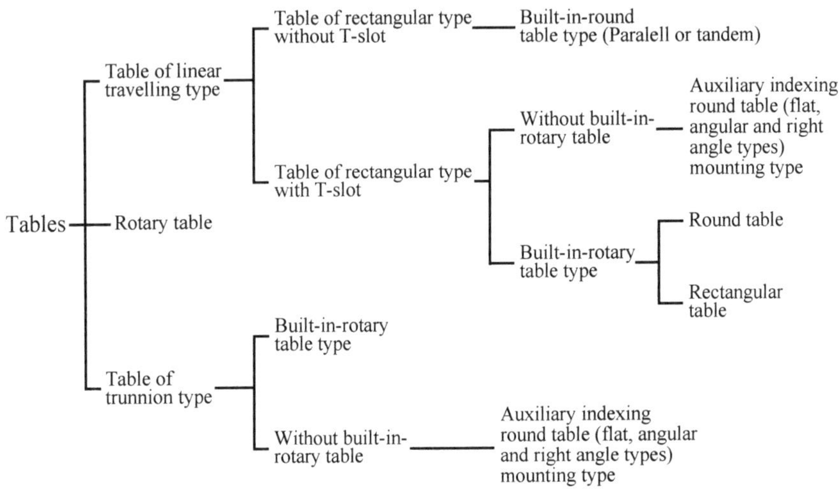

Fig. 4.25 Classification of tables

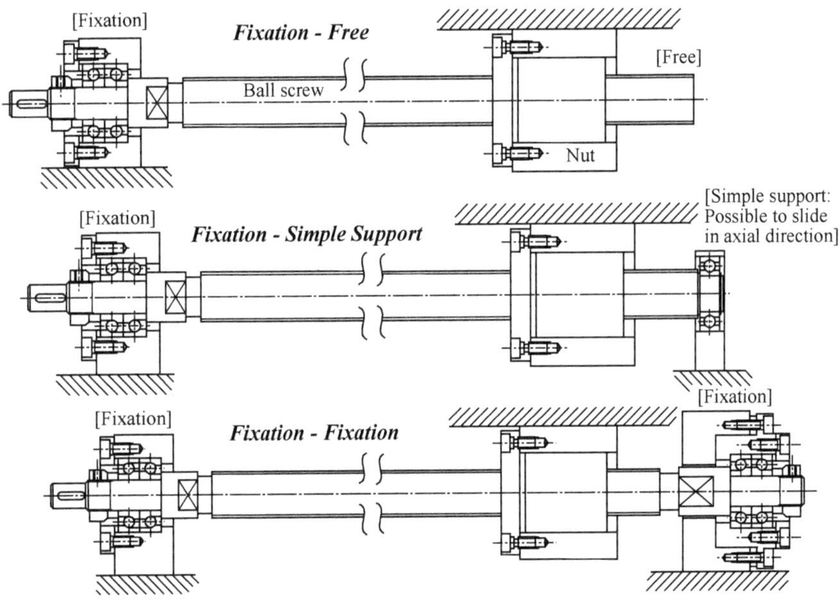

Fig. 4.26 Three representative supporting systems for ball screw

round table has been employed mostly to NC vertical turning machine as already shown in Fig. 2.10.

In general, the tables for linear travelling and rotary movement are driven by the ball screw and pinion-gear mechanism, respectively, and it is necessary to move

Fig. 4.27 Double-pinion driving system for table of large-sized machine tool (by Pfauter, Type PE 1990s)

smoothly the table with the best guiding accuracy without any backlash (see Appendix B). In the ball screw driving, thus, there are the three representative supporting systems as shown in Fig. 4.26, and in due course, the double-not mechanism can facilitate the elimination of the backlash. In contrast, the double-pinion mechanism has often been used in no-backlash driving for the large-sized machine tool either for the lower travelling speed or for the higher rotational speed like NC vertical turning machine as shown in Fig. 4.27.

In retrospect, MC has been developed on the basis of the traditional milling machine, horizontal boring and milling machine, planomiller and so on, and thus even now, there remain some structural configurations in the past. Figure 4.28 shows the rectangular and round built-in-rotary tables in MC. In the case of the built-in-round table (1500 mm in diameter), the table is driven by the helical double-pinion engaging with the gear ring, which is fixed underneath of the table, and also driven by the motor through worm gearing. Figure 4.29 shows furthermore a special built-in-rotary table system in linear travelling table without any T-slots, i.e., double-decker like table configuration. In principle, these rotary tables are for indexing and also for the lower speed rotation when accommodating the heavy and unwieldy work.

Importantly, in the case of NC vertical turning machine, the table should be rotated with the higher speed under heavy loading as like as the main spindle of TC. More importantly, the table within the quinaxial controlled MC and "*Mill-turn*" should be also rotated under heavy-duty cutting for heavy work like NC vertical turning machine.

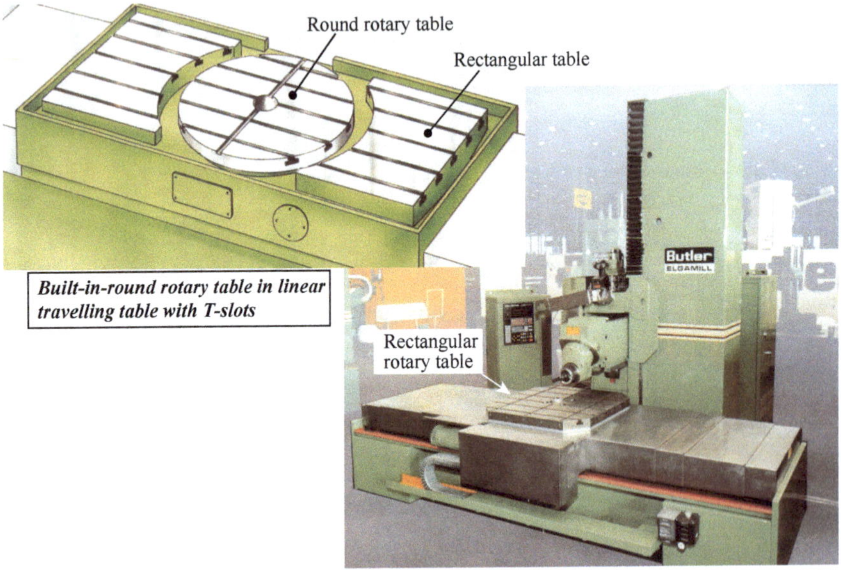

Fig. 4.28 Built-in-rotary table in MC (by Butler Newall, Type ELGAMILL HE 1985)

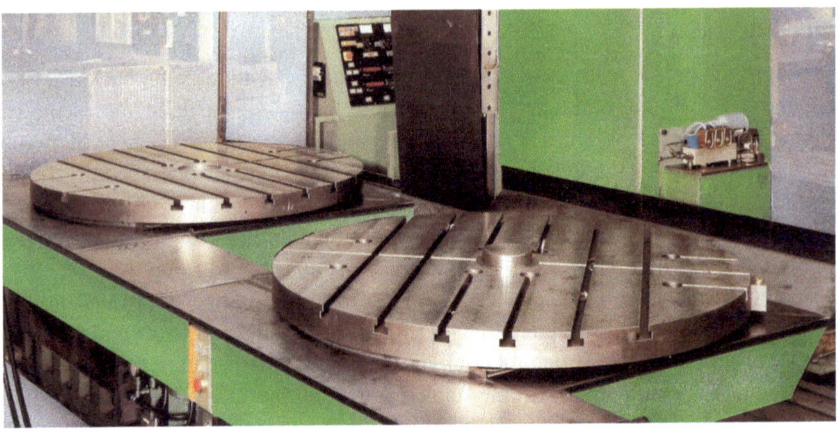

Fig. 4.29 Built-in-round rotary table in MC (by Butler Newall, Type ELGAMILL HE 1985)

Obviously, the core technologies are to rotate the table under the higher speed and heavy loading, and to integrate such driving mechanism within the limited space as implicated by the built-in-rotary table in the trunnion of quinaxial controlled MC.

In this context, it is worth noting that the cross-cylindrical roller or-tapered roller bearing enables the advanced built-in-rotary table to be enhanced to a large extent as will be clear from Fig. 4.30. More specifically, Timken contrived once the cross-tapered roller bearing in the 1960s together with showing its advantageous

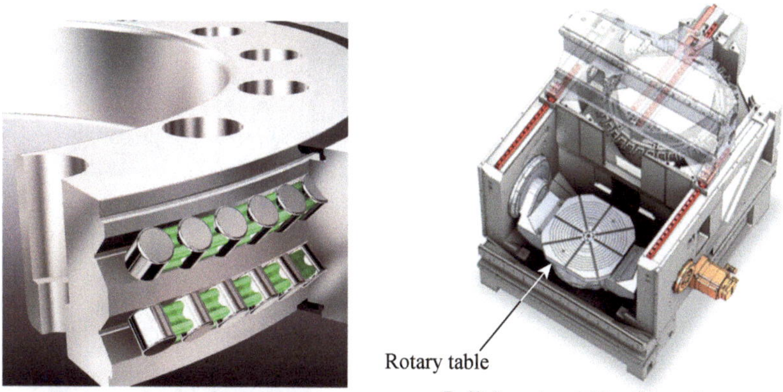

Rotary table supporting bearing of cross-cylindrical roller type (by courtesy of THK, 2013)

Built-in-rotary table of trunnion type (by courtesy of Hermle, Type B300, 2008)

Fig. 4.30 Built-in-rotary table of trunnion type and its supporting bearing of cross-roller type

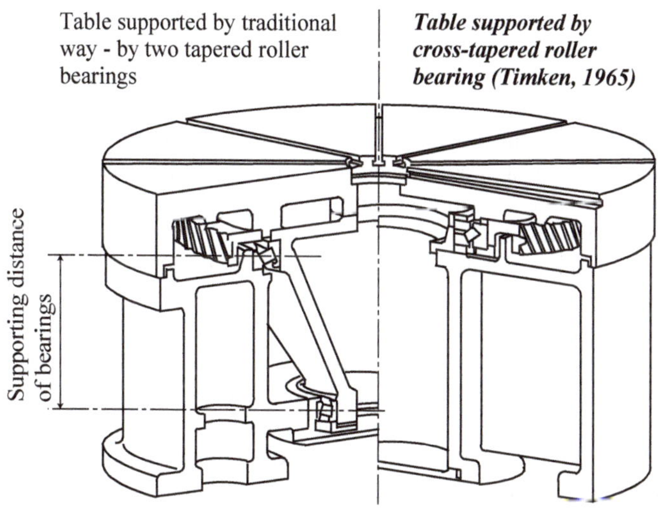

Fig. 4.31 Comparison of table supporting by cross-tapered roller bearing to traditional bearing arrangements

features. In general, the table is supported by a couple of bearings, which necessitates the bearing distance of certain length; however, the cross-roller bearing needs not such a bearing distance, i.e., bearing distance being nill.

Paraphrasing, as shown in Fig. 4.31, the cross-roller bearing can be in fruition of the rigid and compact table, which can place within the narrower space and can rotate higher speed under heavy loading. In the 1960s, however, there were no acceptable markets, whereas with advance of the quinaxial controlled MC, the cross-roller bearing can be resurrected as reported widely elsewhere.

Fig. 4.32 Double-worm driving mechanism for work table of CNC hobbing machine (by Lorenz 1960)

Of note, the work table in hobbing machine is rotated by double-worm gearing as shown in Fig. 4.32 to eliminate the backlash. In worm-worm rack driving, one of the variants of worm gearing, for the large-sized table in planomiller, it is common to heat-treat the worm rack by the sulfidisation to reduce the frictional energy loss.

In the mean while, the auxiliary indexing round table is, in principle, one of the attachments; however, it is convenient to discuss it herein, because of its close tie with the primary table system as already shown in Fig. 4.25.

In the era of the traditional machine tool, the auxiliary indexing round table was often employed in the milling machine by mounting it on the linear travelling table with T-slot. At present, we must use, for example, the auxiliary NC rotary table and NC rotary table of trunnion type, which are advanced types of the auxiliary indexing table in the past, if necessary. In short, the linear travelling table can play the role of that with built-in-rotary table by mounting the auxiliary round table, although the work holding rigidity is, in general, low.

The core technologies in the auxiliary indexing round table are thus as similar as those in the built-in-rotary table, i.e., indexing and rotary mechanism, and also backlash eliminator, apart from the lower rotational and indexing speed.

As already shown in Fig. 4.25, thus, worm gearing has been used in the auxiliary indexing round table, and at present, the roller gear cam and direct drive with the motor becomes popular. More specifically, Fig. 4.33 shows the worm gearing of double-lead type and roller gear cam, which are extremely suitable to eliminate the backlash in meshing. As can be readily seen from Fig. 4.33, worm gearing of double-lead type can be characterised by the continuously changing teeth thicknesses, which are generated by slightly varying the modules for both the tooth flanks. Thus, we can eliminate the backlash by shifting the worm in axial direction.

In contrast, the roller gear cam appears as one of the variants of worm gearing, where screw-like roller gear and roller follower is engaged without any backlash by

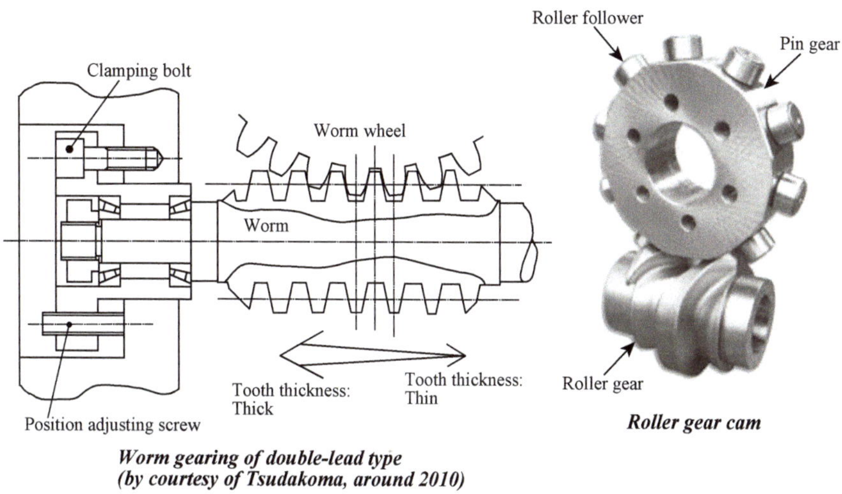

Fig. 4.33 Driving mechanisms for NC rotary table

giving the pre-load to the roller follower. For the sake of further understanding, Fig. 4.34 illustrates another mechanisms in the variants of the ball screw without any backlash.

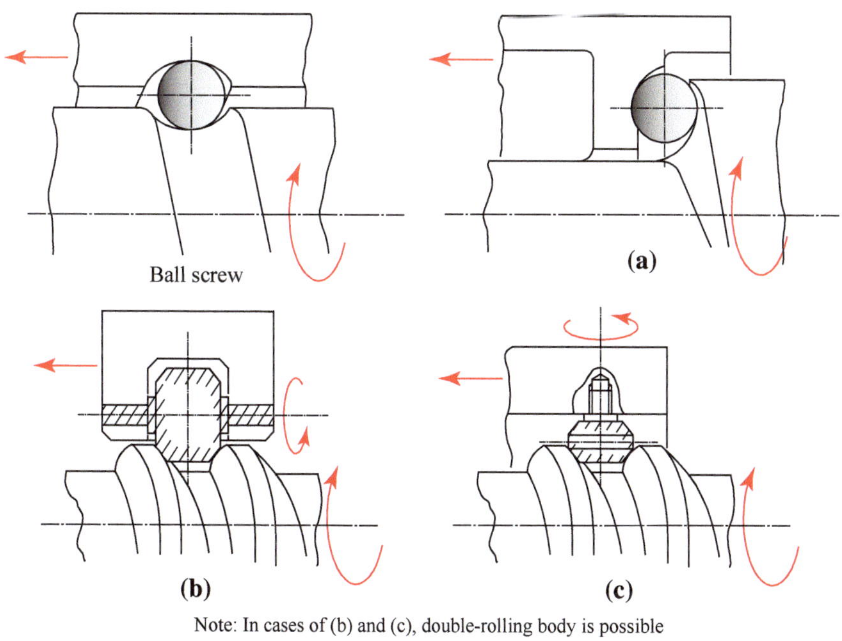

Fig. 4.34 Variants of ball screw for linear motion without backlash (by Bankmann)

Importantly, a facing problem for worm gearing is to reduce the friction in meshing, whereas we must aware that worm gearing is for large speed reduction, and also for the self-locking by giving the smaller helical angle less than the frictional angle. In fact, the self-locking is a characteristic function of worm gearing, because other fixing devices render useless. Thus, we have so far endevoured to improve the tooth profile and material in worm gearing, and in general, worm gearing consists of the worm made of hardened alloy steel and worm wheel made of brass alloy.

References

Hazem S et al (1986) A new modular tooling system of curvic coupling type. In: Davis BJ (ed) Proceedings of 26th MTDR conference, pp 261–267

Ito Y (2008) Modular design for machine tools. McGraw-Hill, New York

Rall K, Findeklee J (1998) Kompensation axialer Spindelverlagerungen beim Hochgeschwindigkeitsfräsen. ZwF 93–11:558–561

Weck M (1993) Trends of manufacturing technology looking towards the 21st century. Industrial Technical Seminar, Kobe

Chapter 5
Work and Tool Holding Devices—Chuck, Centre, Mandrel and Tool Holder

As widely known, the better machining accuracy, higher machining speed and heavy-duty cutting capability can be facilitated with secure grasping and accurate holding of both the work and the tool. In due course, there are a considerable number of the devices called the attachment for such purposes, and in general the attachment can be roughly classified into those for work and tool holding. In this context, we must be aware that some devices may change their roles from work to tool holding depending upon the machining requirement.

Figures 5.1 and 5.2 show the classification for work and tool holding devices, respectively. As compared with the work holding devices, we have a less number of the tool holding devices, because nearly all (cutting and grinding) tools are mounted directly on the main spindle, turret head, and tool spindle. Importantly, these devices can be classified and used depending upon the holding accuracy of the work and tool. More importantly, the holding accuracy is subjected to the relative stiffness of the holding device to the work and tool.

Generally speaking, the chuck and tool holder including tool bracket are in the leading position in the case of the axi-symmetrical work, and the mandrel and jig are for the more accurate holding than the chuck. As a result, the mandrel is used in the various forms depending upon the machining requirements. In certain cases, drilling is carried out by using the guide bush, i.e., a kind of jig, and furthermore the tool shank is converted into the small-sized work holder as exemplified in Fig. 5.3, which is of course capable of containing within the automatic tool changer (ATC) of the machining centre (MC).

As widely accepted, the work holding technology should be furthermore dealt with in accordance with the work configuration, i.e., either the axi-symmetrical rotational work or the box-like work including the plate-like shape. In the case of the box-like work, it is however worth suggesting that the work holding device is

Y. Ito and T. Matsumura, *Theory and Practice in Machining Systems*,
DOI 10.1007/978-3-319-53901-0_5

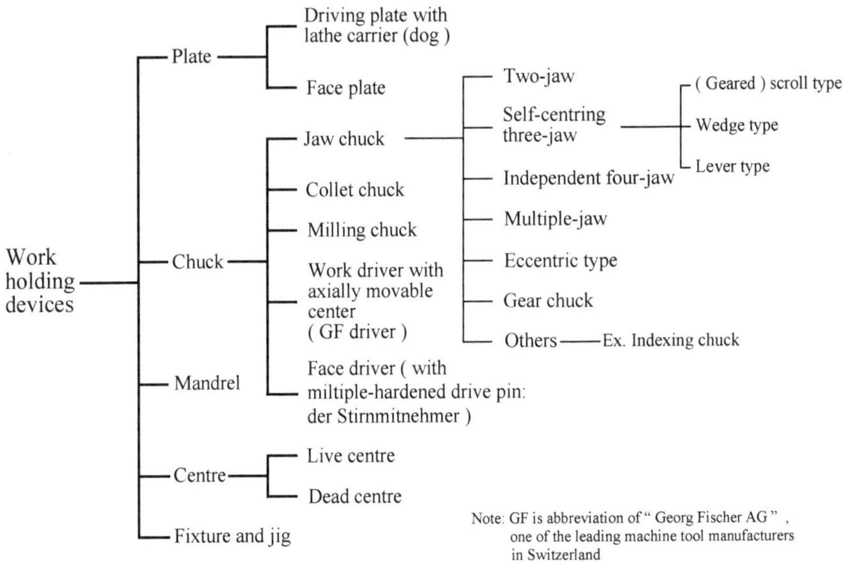

Fig. 5.1 Classification of work holding devices

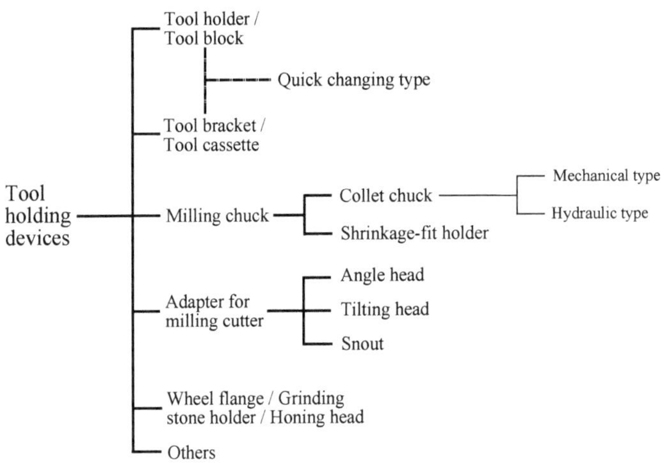

Fig. 5.2 Classification of tool holding devices

generally called fixture and jig, and they are very complicated and difficult to the systematisation and generalisation. In short, we may show literally such a situation as "*each component having its own fixture or jig*" as similar as the mandrel.

Figure 5.4 shows a typical fixture for a casting component, where we can see clearly the roles of the locator and clamping plate (see also Fig. 1.4a). Such the state

Fig. 5.3 A special holder for small-sized work capable of containing within ATC of MC (ROBOMATIC of SIGMA-brand, 1987)

of the fixture is the same even now as shown in Fig. 5.5, although the related technology is duly modernised. Intuitively, from these examples we may understand the difficulty in the systematisation for the fixture and jig. In fact, the fixture consists of a combination of the base plate (surface plate), angle plate, adjustable plate, vice, clamp, chuck, small jack and so on, ranging from the large to small parts. In certain case, we use the round table, indexing table and jaw unit.[1]

Against this context, some manufacturers have merchandised the fixture and jig of modular type. For example, Nabeya has merchandised a modular fixture system, where the basic modules are the base element, supplementary base, locating element, clamp unit and clamping parts, e.g., jack, bolt and nut.

Having in mind the present and near future perspectives in the work and tool holding devices, some quick notes for the leading devices will be given in the following.

[1]The fixture and jig is an assembly of a considerable number of parts, and thus the joint of both the parts determines its performance to large extent. From the viewpoint of the machine tool joint, there are various interesting research subjects extremely considering the differing magnitude of the joint deformation at each clamping point in the fixture and jig. In retrospect, there have been no noteworthy researches since 1960s apart from those of Shawki and Abdel-Aal (Inter. J. Mach. Des, Res. 1965; 5:183 and 6:31).

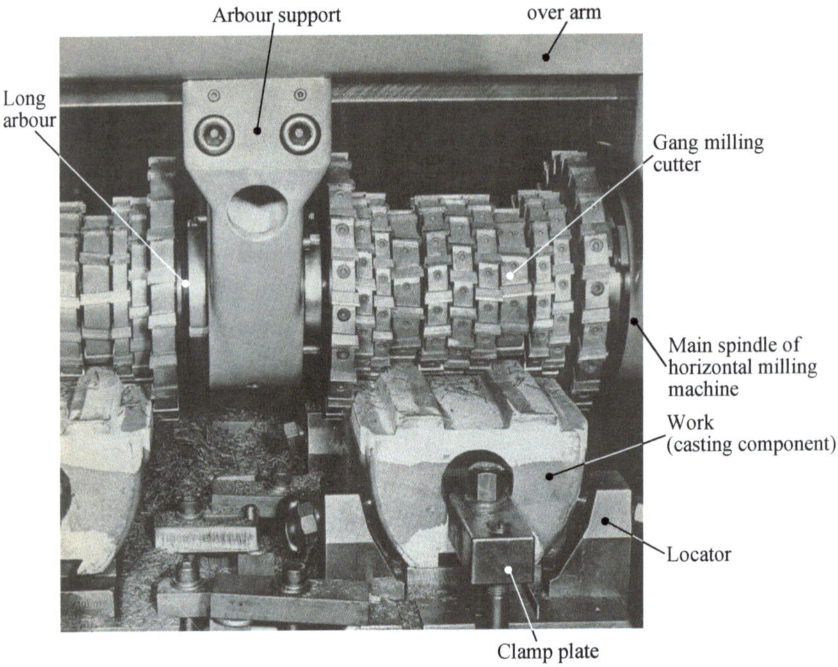

Fig. 5.4 Fixture and jig for heavy-duty cutting of casting component by gang milling cutter (by Heller 1960s)

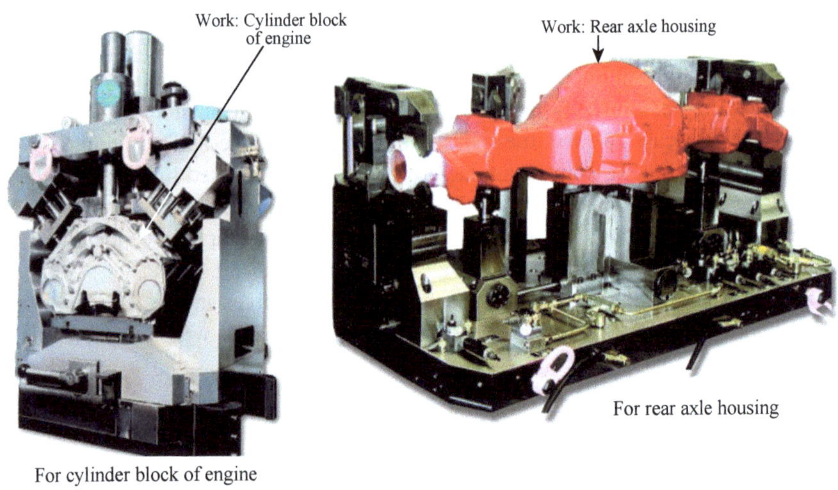

Fig. 5.5 Hydraulic clamping devices (by courtesy of Röhm 2015)

5.1 Chuck and Concerns

There are multifarious kinds and types in the chuck as already shown in Fig. 5.1, and of these, the jaw and collet chucks are the utmost representative kinds. The jaw chuck can facilitate to grasp only the work, whereas the collet chuck is used either to grasp the bar work or the tool with shank, e.g., milling cutter, drill and tap. Importantly, the jaw chuck has been investigated to some extent in the academia, although other devices have been contrived and developed on the basis of the factory floor experiences.

5.1.1 Jaw Chuck

The jaw chuck can be classified into several types depending upon first the number of jaws and then jaw travelling mechanism, and we have now two-, three-, four- and multiple-jaw chuck. Of these, we have often used the self-centering three- and independent four-jaw chuck, which are shown in Figs. 5.6 and 5.7, respectively, and for example, the two-jaw chuck is widely used in the car wheel lathe.

With the prevalence of the numerical controlled (NC) machine tools, however, the independent four-jaw chuck is disappearing gradually. As widely known, the independent four-jaw chuck is very convenient to grasp the work with complicated shape and size, which necessitates turning in part. More specifically, each jaw can be adjusted independently, and in addition to it, the chuck body has, in general, a considerable number of the slots and in certain cases four T-slots. Obviously, these

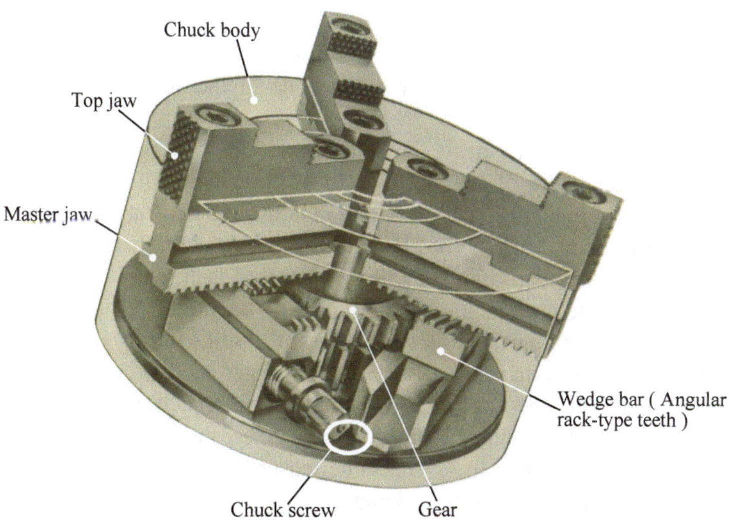

Fig. 5.6 Self-centring three-jaw chuck of wedge type (by Howa-Reishauer, in 1960s)

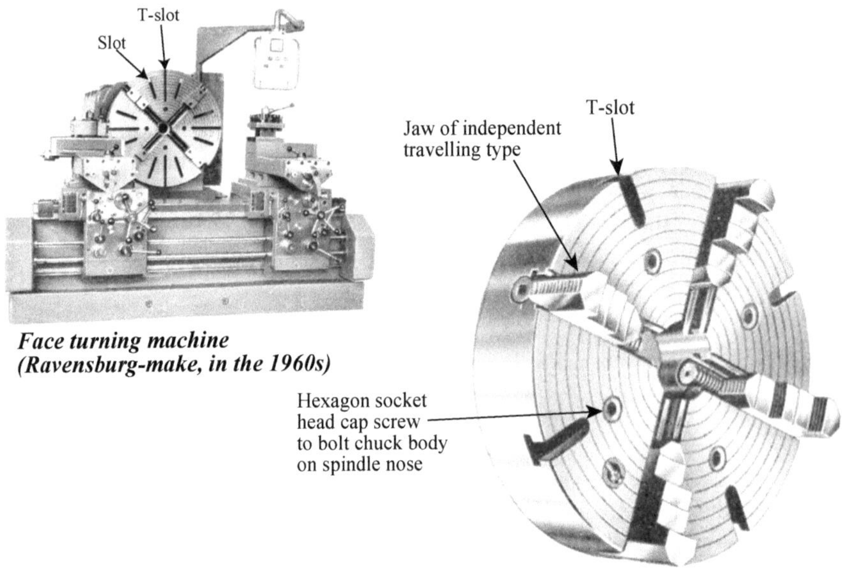

Fig. 5.7 Independent four-jaw chuck and its mounting on face turning machine (by Stau)

can assist the secure fixation of the unwieldy work, i.e., difficult-to-mounting work, by using also the additional jaw, jig and fixture as similar as the accommodation of the box-like work on the rotary table in the vertical turning machine.

In short, the independent four-jaw chuck is, in principle, suitable for the manual operation by the skilled technician, but not compatible with the automatisation for work holding, and for the sake of further understanding, its mounting state on the face turning machine is also shown in Fig. 5.7 (Stau 1963). Of special interest, Fig. 5.8 shows the modernised special purpose-oriented chuck, which is suitable for the automatisation, and which appears to be one of the successors of the independent four-jaw chuck.

Admitting that the family of the jaw chuck is widespread, we must discuss the dominant kind in the era of NC machine tools. An important precondition is to automatise work holding and releasing, and thus the self-centring three-jaw chuck with scroll, lever or wedge type has been in leading position, because of its compatibility with the automatisation as will be discussed later.

Needles to say, the jaw can, in general, travel by the power of pressurised oil or air; however, its travelling range is short. Thus, the self-centring three-jaw chuck consists of the combination jaw, i.e., a couple of master and top jaws, to be capable of adjusting preferably the work grasping range at present. In due course, the top jaw can, in general, be changed, and in certain cases, the automatic jaw changer (AJC) can facilitate it, while using the same master jaw. As can be expected, the chuck itself is furthermore of quick changing type as will be discussed later, provided that the chuck is small-sized. Of note, the jaw with manual operation can

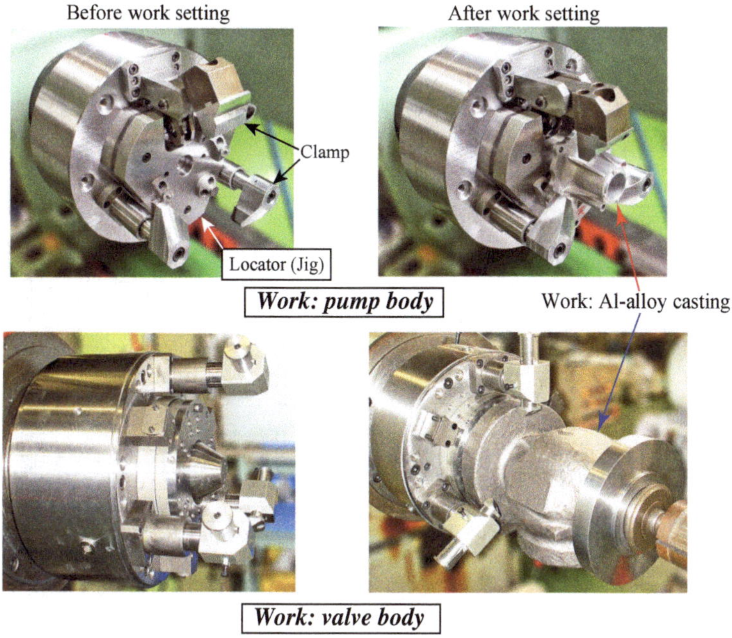

Fig. 5.8 Special purpose-oriented chuck in 2010s (by courtesy of Howa Machinery)

travels the longer range than the powered chuck as shown already in Fig. 5.6, and in the past the jaw was monolithic.

Within this context, there are four major engineering problems as will be clear from Fig. 5.6, i.e. (1) jaw travelling mechanism, (2) jointing method of top jaw with master jaw, (3) work holding accuracy and gripping force, and (4) allowable maximum rotational speed. In addition, it is vital in the chuck design that the chuck body can contain compactly the jaw travelling mechanism, and that the enough gripping force can be guaranteed even in the higher-rotational speed.

Jaw Travelling Mechanism

In Fig. 5.6, the master jaw is guided by the T-slot and can move to radial direction by engaging its rack with the wedge bar, which slides to the circumferential direction through the gear rotated manually by the chuck screw. As can be readily seen, three jaws can move simultaneously, and its core is the wedge bar. In general, there are, as already mentioned above, three representative jaw travelling mechanisms, i.e., scroll, wedge and lever types, and Fig. 5.9 shows the jaw travelling mechanism in both the lever and wedge types of old-fashion. Importantly, each type has several variants as exemplified by the wedge type shown in Figs. 5.6 and 5.9.

In retrospect, the scroll chuck was very common in the era of traditional engine lathe, because of its high manual operability by the turner. In the scroll chuck, the master jaw with scroll engages with the scroll plate, which can rotate by the chuck screw through the pinion. In short, the scroll-scroll plate meshing is equivalent to

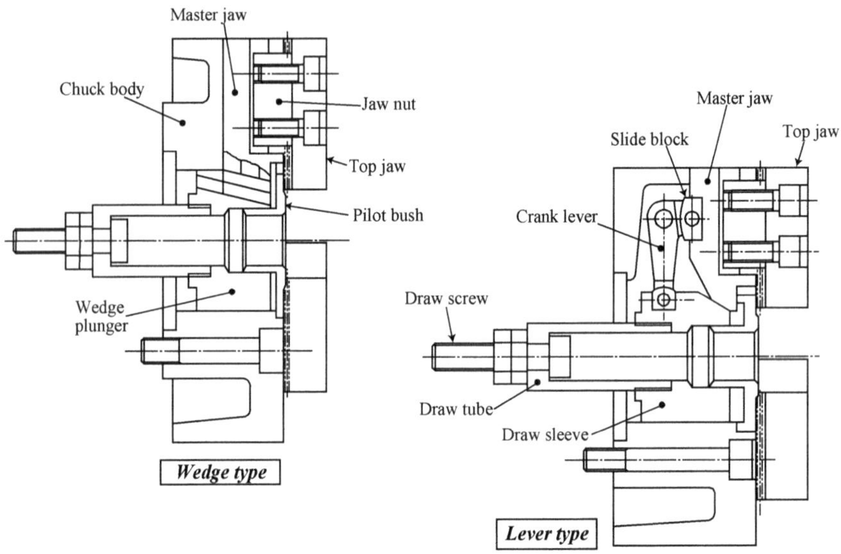

Fig. 5.9 Schematic views for three-jaw self-centring chucks of wedge and lever types

the master jaw with rack—wedge bar—gear meshing in Fig. 5.6. Of note, the self-centring chuck of scroll type is called the "Geared scroll chuck" by American English.

In these chucks, a fatal defect is first to guide the master jaw by the T-slot, and then another problem is, as already mentioned, the jointing method of the top jaw with the master jaw. As can be easily imagined, the T-slot is liable to occur the stress concentration as shown in Fig. 5.10, and in addition, the jaw cannot travel smoothly by the deformation of the chuck body as already mentioned in Chap. 4 (see Fig. 4.6). Thus, a tentative remedy was to improve the profile of the T-slot like the "Wedge-hook" for the wedge plunger as shown together in Fig. 5.10 (Mueller 1977). In fact, the wedge hook is a target of the patent in the chuck of wedge type, because the shape and size of the hook can furthermore determine the conversion efficiency of the draw force for the chuck actuator to the gripping force.

Within this context, one of the innovative remedies is not used the T-slot, but to employ the "*Fir Tree-like Joint*" patented by Forkardt or the spherical joint as shown in Figs. 5.11 and 5.12. By these mechanisms, the jaw travels radially together with dragging-in the work to the face-end of the chuck body, although the jaw stroke is short. Importantly, such a dragging-in effect is preferable to solve *bell-mouthing of the jaw* (gripping faces to depart from squareness with chuck face), when holding the work, and in the case of Fig. 5.12, furthermore, it is said that the gripping force can recover from its reduction state by the rocker block, while rotating at the higher-speed. As widely recognised, the centrifugal force may cause the reduction of the gripping force, and up to now, we have still unsolved problems as will be discussed in detail later.

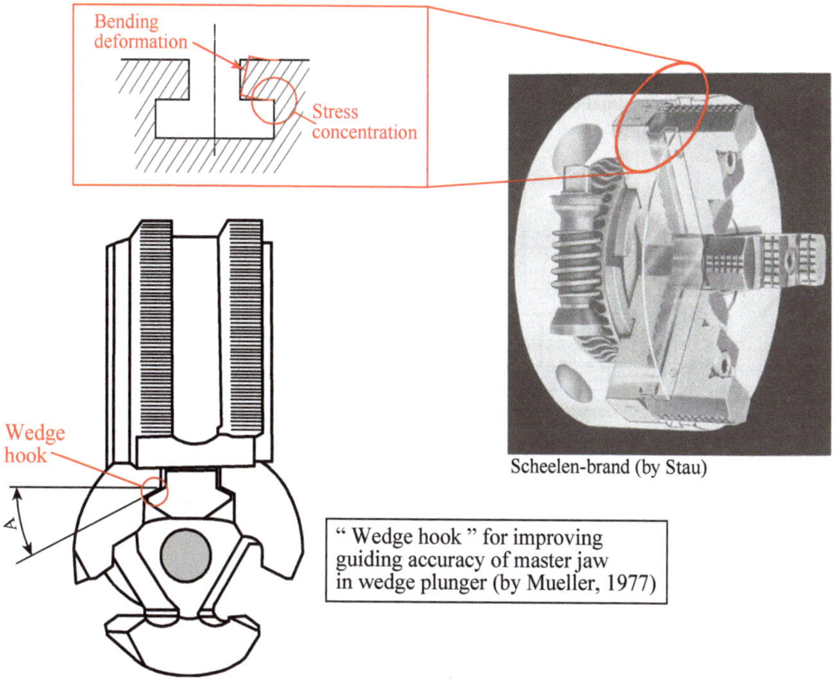

Fig. 5.10 Deterioration of guiding accuracy by deformation of T-slot and its improvement

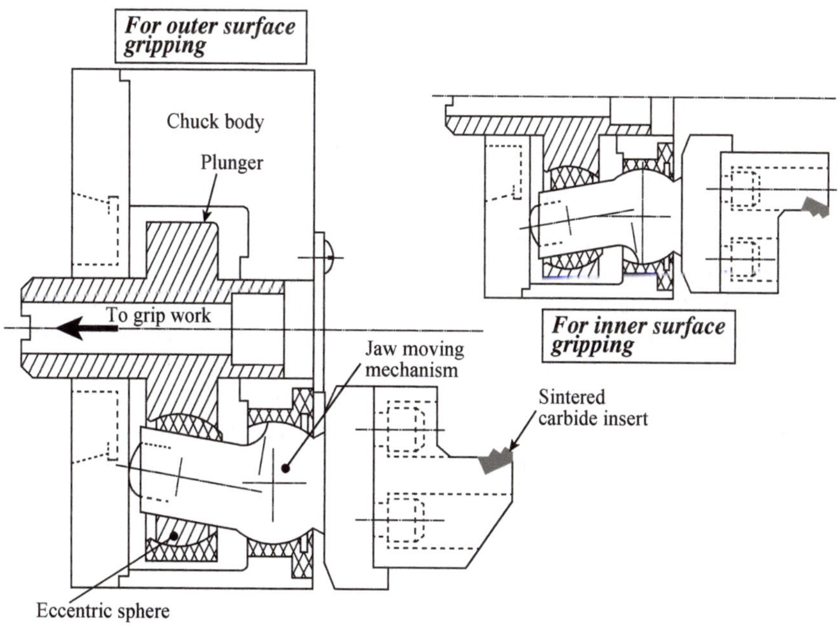

Fig. 5.11 Ball-rock chuck with jaw quick changing type—From outer to inner surfaces gripping (by Teikoku Chuck 2011)

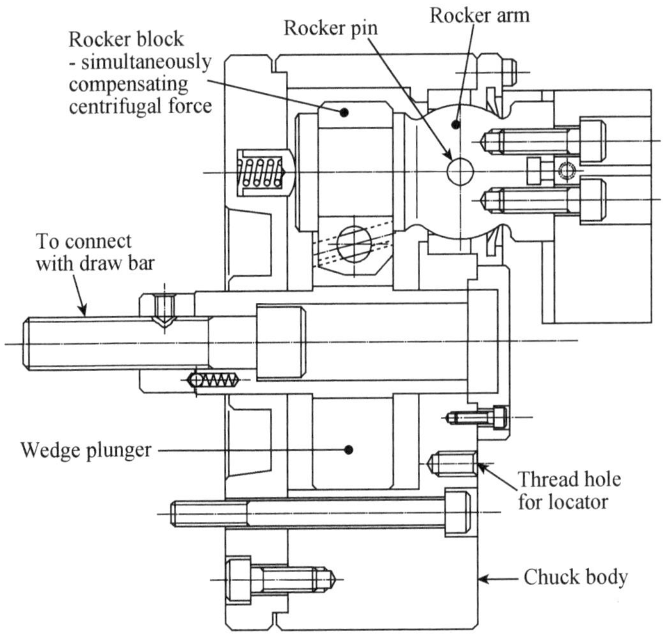

Fig. 5.12 Self-centring 3-jaw chuck of rocker type using wedge plunger (Type HO25M, Howa Machinery 1980s)

For the jaw travelling mechanism, we have often a new patent even now. For example, Buck Logansport Co. proposed an "*External Wedge Mechanism*" in 1995 to reduce gripping-force loss drastically and to shortening the cycle time. This mechanism can be characterised by a couple of the fast-stroking rotating wedges, which are placed on the face-end of plate and actuated by the hydraulics (Owen 1995), and in due course, the productivity is improved considerably in the higher-speed and larger-volume operations like Al-alloy wheel turning.

Top Jaw Changing Method and AJC

In the jaw chuck, the top jaw is, in general, for gripping the outer surface of the cylindrical work, whereas for the inner surface gripping, either the same top jaw should be re-bolted in reverse way or another inner surface-oriented jaw should be bolted onto the master jaw. In this context, the jaw chuck of ball rock type eases such a preparatory work by only loosening four clamping bolts, rotating the top jaw at a half around, and tightening the bolt again as already shown in Fig. 5.11.

In the automated jaw changing method, another issue is, as already mentioned, the short stroke of jaw travelling, and we must often change the jaw in accordance with the machining requirement. In consequence, the joint between top and master jaws is one of the crucial points, and thus we can duly find a considerable number

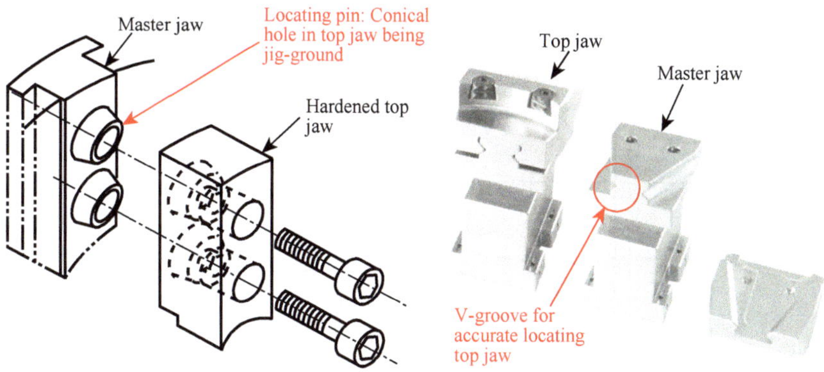

Contrivance by Micro Centric, around 2000

Contrivance by Röhm, 2010 (by courtesy of Röhm)

Fig. 5.13 Top jaw changing methods with better positioning accuracy

of the patents and noteworthy contrivance as shown in Fig. 5.13.[2] In general, we must fulfill the better positioning accuracy of the top jaw with better reproducibility, the higher rigidity, and also the quick changing mechanism. For example, that of Röhm is employed the V-groove to ensure the positioning accuracy better than 0.02 mm.

In the top jaw changing issue, we must discuss AJC extremely relating to the flexible machining system for the axi-symmetrical component. More specifically, to reinforce the flexibility, TC placed within the system should have the wider accommodation capability to the diameter variation of the work by changing the top jaw. In AJC, as well as the modular tooling and hollow shank (HSK; Hollow Schaft Kegel), the top jaw should be in the two surfaces-in-contact with the master jaw. In addition, the allowable fitting tolerance between both the jaws is to be wider and thus we used to employ some adjustment mechanisms, e.g., adjusting bolt shown in Fig. 5.14. In fact, AJC shown in Fig. 5.14 is patented and the belleville washer in it can adjust the uneven clearances at the contact surfaces A and B, by which the simultaneous contacts at the both surfaces is to be in reality. In short, the top jaw can be locked by means of the eccentric cam, which is capable of rotating by the pinion-rack bar mechanism. Although not showing in Fig. 5.14, the pinion is placed at the opposite side of the camshaft, the pinion is engaged with the rack and the rack bar is placed inside of the hollow draw nut (Ito et al. 1989).

On the strength of AJC, the idea of the automatic chuck changer (ACC) was duly proposed in the 1980s with the advance of flexible manufacturing. On that occasion, we faced the necessity for increasing the flexibility of the work holding devices beyond AJC. As a result, ACC has been in practical use in certain cases, where the

[2]In the independent four-jaw chuck, in general, the jaw is monolithic, and thus should be once released from and again engaged to the screw for travelling the jaw after reversing the gripping surface for either the outer or inner surface of the work.

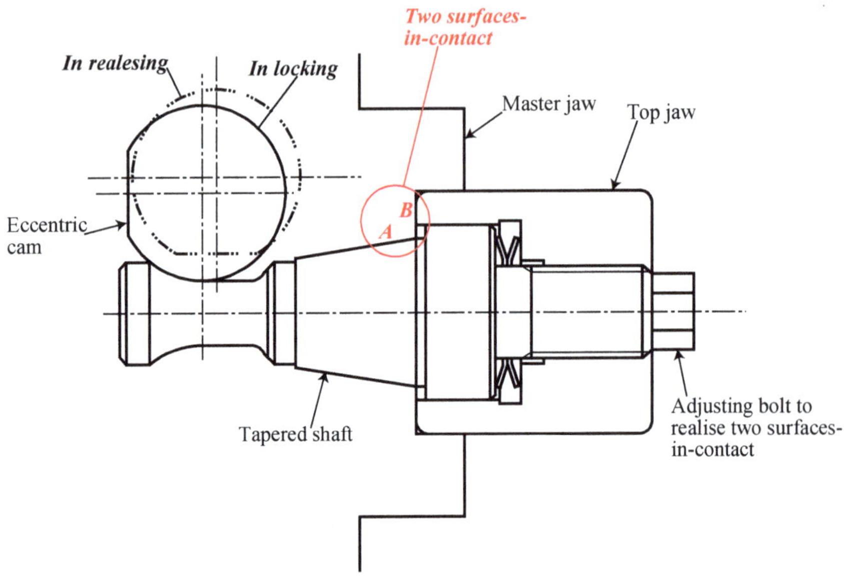

Fig. 5.14 An example of AJC—Japan Patent Tok Kai-Hei 1-321104

adapter is placed between the spindle nose and the chuck module. Importantly, the "*Hurth Coupling*" and "*Curvic Coupling*" can often join the adapter and the chuck module.

Because of having a large family, even when discussing the jaw chuck with narrow scope, there is another classification based on the power sources for work holding. In fact, there are three representative types, i.e., manual, air including diaphragm and hydraulic types, and of these, the chuck of diaphragm type has been used in gear holding.

Having in mind that chucking technology involves still a considerable number of problems, we must emphasise that the gear chuck is very far from completion as compared with others. We will thus discuss quickly the gear chuck in the following.

Reportedly elsewhere, a fatal problem in the motorcar industry at present is the jaw chuck for gear manufacturing. Actually, we must use gearing either in the traditional transmission, or in car wheel even in the era of hybrid and electric cars, provided that the gear quality should be enhanced to reduce the frictional energy loss. This new wave necessitates the development a new chuck for mass-production of precision gear; however, up to now we cannot have an effective remedy, which will be able to replace the chuck of spring and diaphragm types.

Figure 5.15 reproduces a spring chuck being grasped the gear in the 1980s. More specifically, the chuck is for facing and boring the gear and can be realised the reproducibility of chucking accuracy better than 0.0005 inches by using the (1) independent travelling jaw, and (2) to grasp the gear at the average pitch circle.

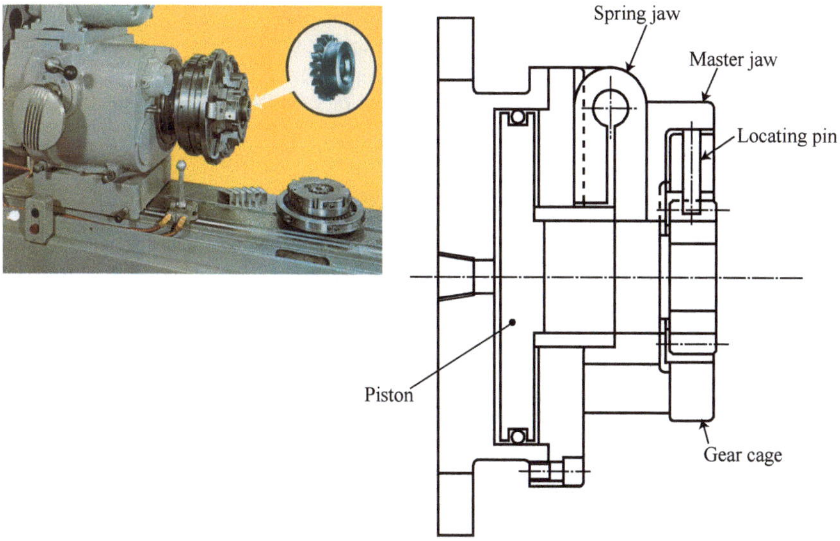

Fig. 5.15 Gear chuck of spring jaw type (by Erickson Tool Co., in 1980s)

By the latter remedy, it is possible to minimise the pitch circle error caused by the heat treatment as well as previous machining.

Importantly, the gear should be grasped correctly at the pitch circle of gear as the reference, more specifically "*reference index being over-pin diameter*", so as to guarantee the best machining accuracy. In contrast, such a grasping method is not rigid. In fact, gripping can be performed in the form of point-to-point contact between the Involute curve and sphere across the whole kinds of gears. Such a reciprocal factor is the leading hindrance to develop a new gear chuck, and even now we must use the old-fashioned gear chuck of diaphragm type as shown in Fig. 5.16. Needles to say, the chuck manufacturer has endeavoured to solve such problems as can be seen from Fig. 5.16. In this gear chuck, the locating pin can position accurately and wedge-like pin realise secure gripping, and in addition to these, the swivel clamp of quick changing type can fix the face end of the work.

Of special note, the Fuji Machine contrived the gear chuck shown in Fig. 5.17 for boring the spiral crown (bevel) gear by cBN cutting tool. The machining portion is about 60 HRC in hardness. It is very interesting that the finish surface does not show any fine spot-silky-like chatter mark when using the chuck body made of cast iron. In contrast, we can observe considerably large chatter mark in the case of steel chuck body (based on the memorandum of Ito at the factory tour, 7th December, 1980).

To this end, for the sake of further understanding, Figs. 5.18 and 5.19 show the eccentric chuck and chuck for oil pipe. The eccentric chuck was already contrived in the 1930s, e.g., Japan Patent No. 154764 by Takeo Inoue, in which the tip portion of the jaw can shift perpendicularly to the travelling direction of the jaw by

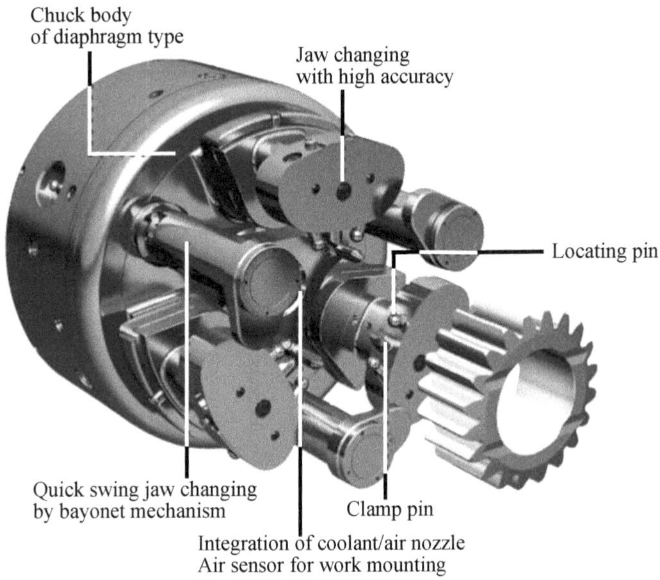

Fig. 5.16 Gear chuck of diaphragm type (by SMW, Type D-KOMBI 2015)

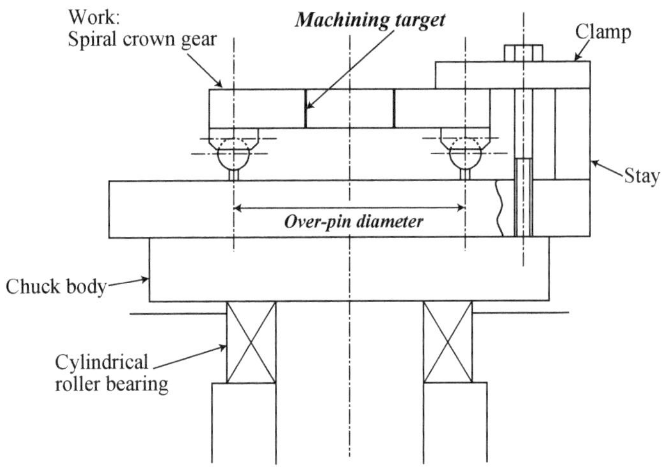

Fig. 5.17 Schematic view of gear chuck placing stress on over-pin diameter"

using the dovetail guide. In case of gripping the oil pipe, we must be aware of the elastic deformation caused by gripping, because the oil pipe is large-sized in diameter and thin-walled.

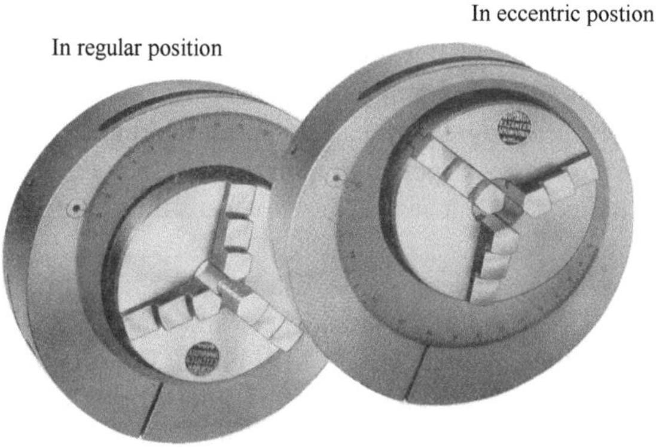

Fig. 5.18 Appearance of eccentric chuck—Horvath-brand, in the 1960s

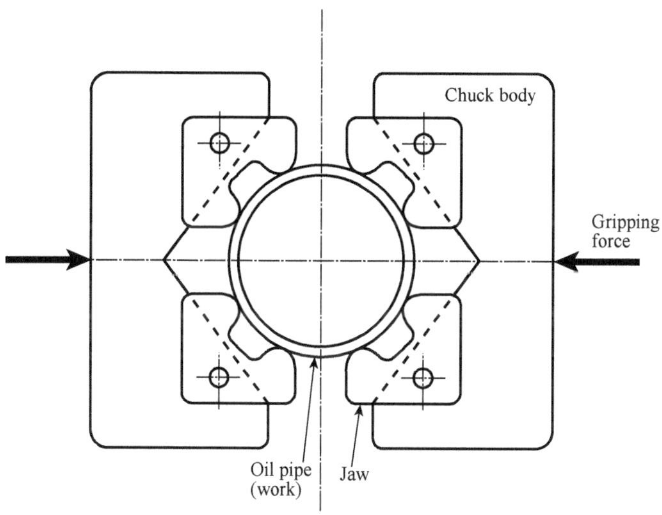

Fig. 5.19 Multi-point self-centring chuck for oil pipe [by Giddings and Lewis, displayed at IMTS (Chicago 1982), Sketched by Ito]

5.1.2 Collet Chuck

The collet chuck was widely used in the traditional automatic to hold the bar work of small size, because of its better chucking accuracy and ease of operation. In the collet chuck, the collet is the kernel component with the preferable spring action to guarantee the better chucking accuracy, and also secure grasping with acceptable rigidity. As a result, the collet has, in general, a considerable number of the small slits or grooves, and this structural configuration induces the manufacturing difficulties to some extent.

In contrast, the collet chuck can hold the work with variety in the cross-sectional shape by only changing the collet itself, which is built within the inside of the chuck body, although the allowable work diameter is very narrow. In addition, we must be aware that the work grasping action can be performed either pushing forward or drawing back the push cone, depending upon the designer's thought, although the holding and positioning accuracies of the work are the same in both the action methods.

On the strength of such advantageous features and with increasing needs to the bar work even in TC, the collet chuck has been more prevailed than ever before, and Fig. 5.20 shows the collet chuck of SMW-brand. From Fig. 5.20, we can see the characteristic features of the collet chuck mentioned above very well.

Accordingly, when replacing the objective from the work to the cutting tool, the collet mechanism is applicable to the milling chuck without any difficulties. With the prevalence of MC, the milling chuck is duly at issue nowadays to hold the end mill, slot drill, drill, tap and reamer. Reportedly, we have, at present, a handful of variants in the milling chuck, some of which are not used the collet with the slit, e.g., those of shrink-fit and press-fit types. Within the milling chuck of press-fit type, Schunk in Germany has been on the market an innovative tool holder (commercial name: TRIBOS) for MC since 2001. In TRIBOS, the tool holder is of thin-walled hollow cylinder and polygon-like cross-section, exactly calling "*die Gleichdicke*". Figure 5.21 illustrates the holding mechanism by using the elastic deformation, and as a result, the following performances can be obtained.

Rotating accuracy of "Out-of-centre": Better than 3 μm.
Reproducibility of positioning accuracy: Better than 3 μm.
Tool changing time: Less than 30 s.

Within the family of TRIBOS, a variant has certain number of cavities staffed with the polymer plastics to increase the damping capacity.

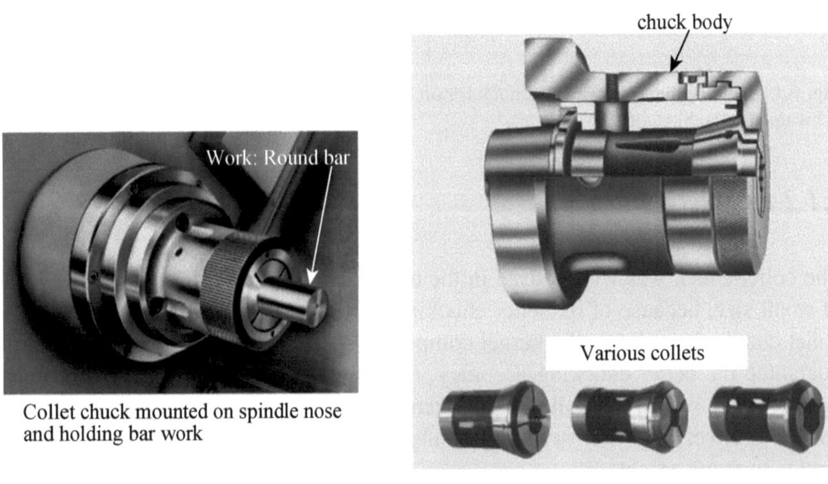

Collet chuck mounted on spindle nose and holding bar work

Fig. 5.20 Typical collet chuck (Type KSZ-MB by SMW 2015)

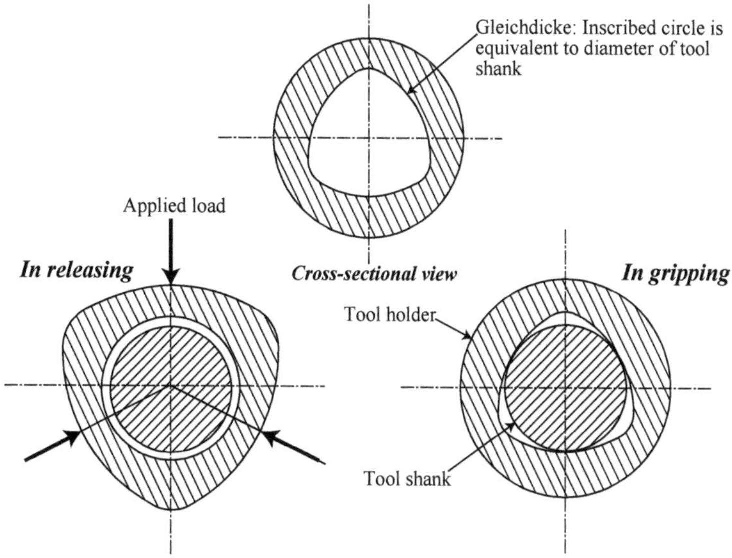

Fig. 5.21 Working mechanism of TRIBOS

Importantly, Fig. 5.22 is a cross-sectional view of the collet chuck of Burnerd-brand for relatively large work in the 1960s. For example, the collet chuck of Burnerd-brand can grasp the bar work with round, square and hexagonal cross-sections. It is expectable that the large-sized collet chuck will be prevailed hereafter with the growing importance of TC of twin-spindle type especially for turning the work with hand-off operation.

5.1.3 Driver

As shown already in Fig. 5.1, we have a handful of the drivers, and of these the driving plate with lathe carrier and face plate were very popular in the era of the engine lathe. With the growing importance of TC, the driver becomes obsolete in general apart from the face driver. In consideration of such the present situation, the face driver and a famous driver in the past, i.e., GF work driver, will be quickly noted in the following.

Face Driver

Figure 5.23 reproduces some examples of the face driver in the past and at present. This work holding devise is for special use, and used mainly in Germany.[3] In the

[3]Face driver with hardened multiple burins is called "der Stirnseitenmitnehmer" in German term. Thus, the face driver appears not adequate translation by nature.

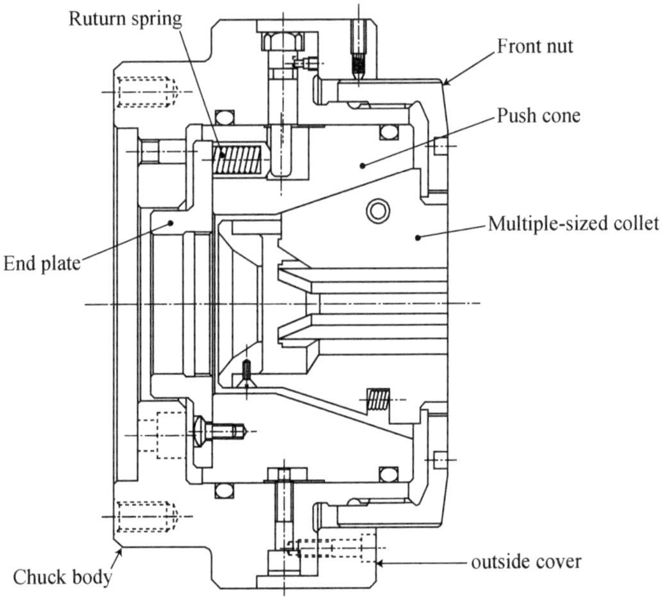

Fig. 5.22 Cross-sectional view of collet chuck for large-sized work—Burnerd-brand, in the 1960s

operation, the work is accurately positioned by centre-guiding, and its face-end is pressed to the hardened multiple pins (burins) by the thrust provided by the barrel of tailstock. The work is thus driven by the hardened multiple pins, and can be machined across the whole outer surface. Obviously, it is desirable to provide the face driver with the cushion mechanism.

For the sake of further understanding, some quick notes will be given for the product of Neidlein Spanzeuge GmbH in the following.

(1) The positioning accuracy, i.e., run-out, of the work is better than 0.005 μm.
(2) The centre is made of sintered carbide insert, and the drive pin is made of high-speed steel or high-speed steel with diamond coating for the hardened work.
(3) The extracting nut can facilitate the quick mounting.

As will be clear from the above, the face driver can also mount on the barrel of the tailstock, and we call it the "*Centre with Burins*". In addition, the face driver is increasing its application area as exemplified by its variants, e.g., flange type shown together in Fig. 5.23 and those for grinding.

GF work driver

GF work driver was an innovative contrivance by Geroge Fischer (GF) in Switzerland to increase the operational efficiency of the copying engine lathe in the

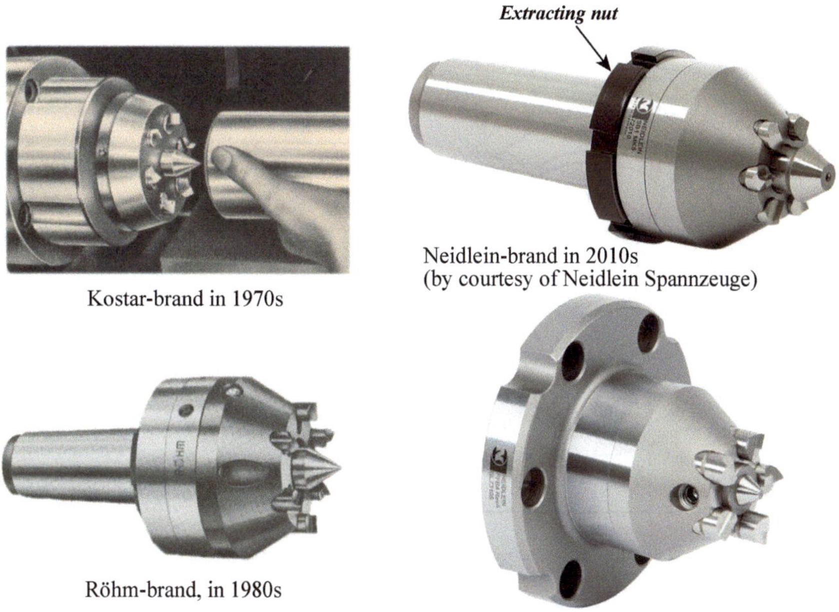

Kostar-brand in 1970s

Neidlein-brand in 2010s
(by courtesy of Neidlein Spannzeuge)

Röhm-brand, in 1980s

Flange type - Neidlein-brand in 2010s
(by courtesy of Neidlein Spannzeuge)

Fig. 5.23 Face drivers in past and at present

1960s.[4] Figure 5.24 shows the mounting appearance and cross-sectional front view of GF work driver (Stau 1963). As can be readily seen, the work can be held by three swing jaws, and supported at the opposite side by axially movable centre, which is spring-preloaded and mounted on the barrel of tailstock. More specifically, the driver has the rid placed within the main spindle for the positive stop of the work, and the swing jaw can open by rotating the outer cap. In the operation, the more increase of rotating moment, the more secure is its grasping. Importantly, the work driver should be used along with the axially movable centre, and by it the work can be positioned in constant even in the medium-volume production. As a result, we can turn the work with the better copying accuracy than that using the conventional centre. In the case of relatively heavy-duty machining, the face driver is accommodated within the centre hole of the main spindle.

[4]This driver was exclusively contrived by GF for its own automatic copying lathe, for example, type KDM-9/80, in the 1960s, in which the allowable work diameter in maximum is 200 mm. The driver is thus called the "*GF Work Driver*".

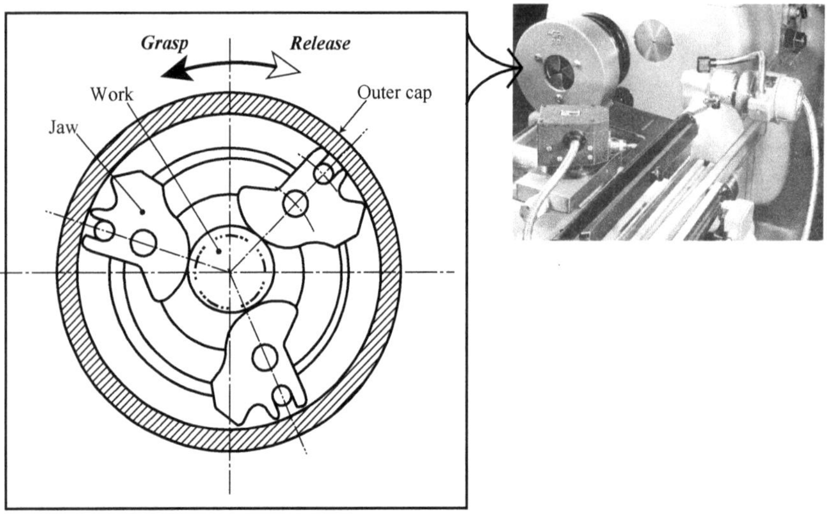

Fig. 5.24 Appearance and cross-sectional view of GF work driver

5.2 Centre and Concerns

In the era of traditional machine tool, the centre work and line boring were very common, when the best machining accuracy was required of. This is because the machining accuracy deteriorates considerably by re-chucking the work and dividing a half around the table. With enhancing the performance of NC machine tool, it enables the better machining accuracy to be obtained without any difficulties even in re-chucking, which is very close to the best machining accuracy performed by the traditional machine tool.

For example, line boring disappears completely from the machining categories in even NC horizontal boring and milling machine. In addition, the machining function-integrated kind accelerates much more the enhancement of the machining accuracy ever than before. For example, TC of opposite allocated twin-spindle type renders the centre work with the best machining accuracy useless in certain cases by using the "*Hand-off*" operation.

Although decreasing the application area, in the very precise machining, the centre maintains its importance even now, and thus a quick note will be given in the following.

In general, the centre can be roughly classified into those of stationary and rolling types as shown in Fig. 5.25, and the work holding accuracy of the stationary type is superior to that of rolling type. The tip of stationary centre is still stand while the work rotates, resulting in a considerable heat generation at the centre. As already shown a variant of live centre in Fig. 4.3, the live centre consists of the dead centre and rolling bearing, and thus there are no relative slip between the tip of centre and

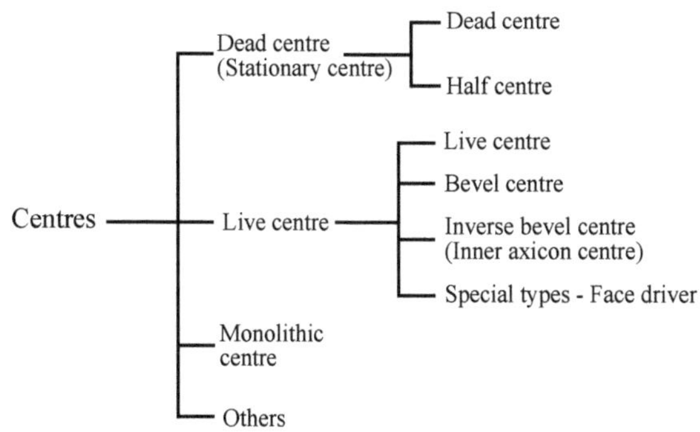

Fig. 5.25 Classification of centres

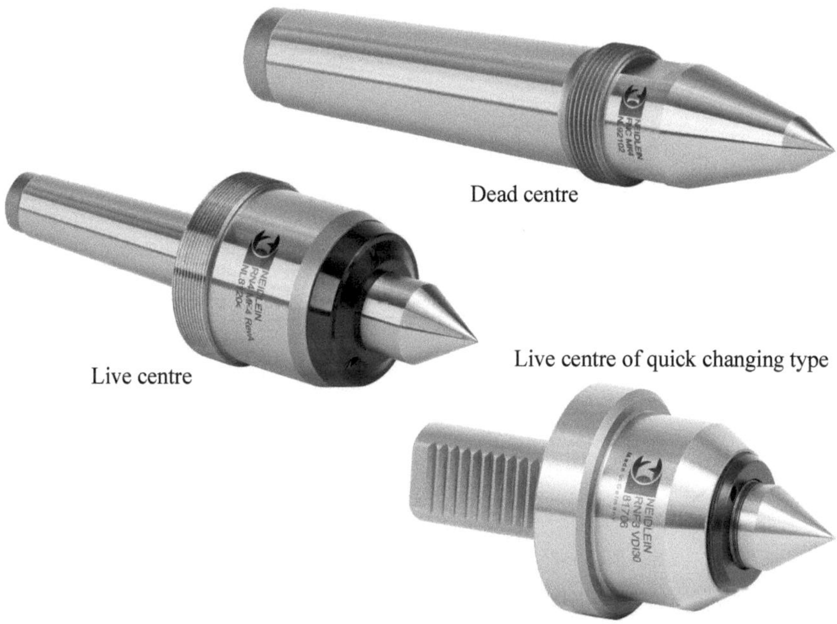

Fig. 5.26 Live and dead centres and a variant for quick mounting (by courtesy of Neidlein Spannzeuge GmbH, 2015)

the work. In addition, it is possible to burden the large thrust to the centre even under the higher rotating condition.

Importantly, it is seldom to use the centre only nowadays, although the centre with high precision is on the market for machining with the utmost best accuracy. Figure 5.26 shows such the centres of rolling type with labyrinth and contact shaft seal, where the run-outs of centres for cutting and grinding are 0.005 and 0.003 mm, respectively.

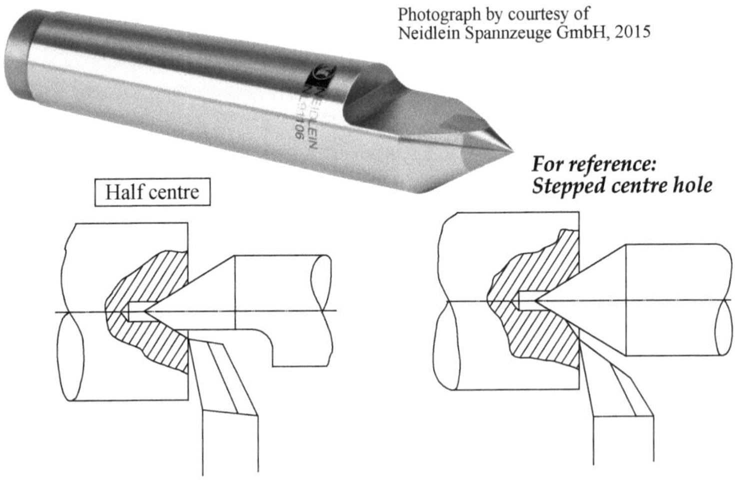

Fig. 5.27 Half dead centre for facing up face-end of work

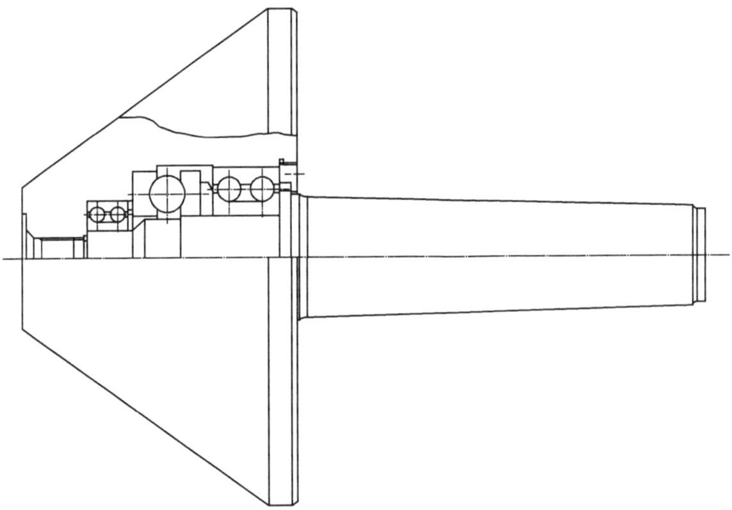

Fig. 5.28 Schematic view for structural configuration of bevel centre (by Röhm, 1970s)

Within the centre context, there are some interesting variants. Figure 5.27 shows the half centre, which is one of the dead centres and for facing up the face-end of the work without any interference of the centre. Reportedly, in the case of large-sized pipe-like work, we must use the bevel centre shown in Fig. 5.28, and furthermore in the case of heavy-duty cutting, the monolithic centre with main spindle is recommended to obtain the satisfactory rigidity in combination with the centre of flange type or centre of flange mounting on the barrel as shown in Fig. 5.29.

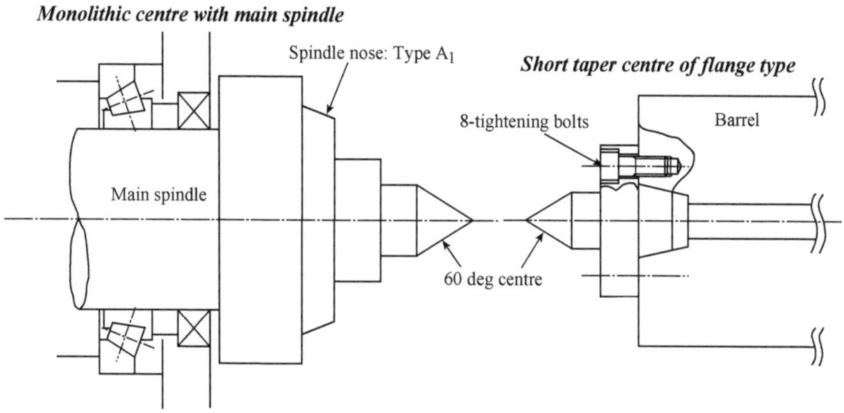

Fig. 5.29 Monolithic centre with main spindle and short taper with flange (sketched by Ito at Dainichi Kinzoku Kougyou 1990s)

5.3 Mandrel and Milling Chuck

The mandrel is, as widely known, for grasping the work, which should be finished with the best machining accuracy, and thus has been contrived at the factory floor to correspond adequately with each machining requirement. As a result, we have now an uncountable number of variants in the mandrel randomly. This is one of the reasons why the family of the mandrel did not show in Fig. 5.1.

In the era of the traditional machine tool, the outstanding technique for the mandrel is the lifeblood for the manufacturing enterprise. With the advance of NC machine tool, however, we can produce, without any difficulties, the component with better accuracy, which may not be achieved without using the mandrel in the traditional machine tool. This is the same story as that in centre, but we need much more mandrel than the centre. In fact, we need to generate the more precise component than ever before with the inevitable quality enhancement in civil and defence supplies. Such machining requirements result in the increasing necessities for the novel and innovative mandrel, and furthermore, the mandrel has been deployed to a new horizon. For example, the precision milling chuck is one of the variants of mandrel as will be discussed later, and its modular system is now prevailing.

Importantly, the mandrel is "*Must*" in machining the component with the best accuracy obtainable on each era, and can be classified into the (1) hole tolerance-basis (reference hole-basis), and (2) shaft tolerance-basis (reference shaft-basis) types. More specifically, the mandrel of hole tolerance-basis type is in the leading position, and it can be furthermore detailed by the expansion method of the thin-walled bush as shown in Fig. 5.30. Obviously, the design principle and fundamental structure of the mandrel are kept in never changed, although the mandrel technology itself has been advanced duly. For example, Fig. 5.31a, b

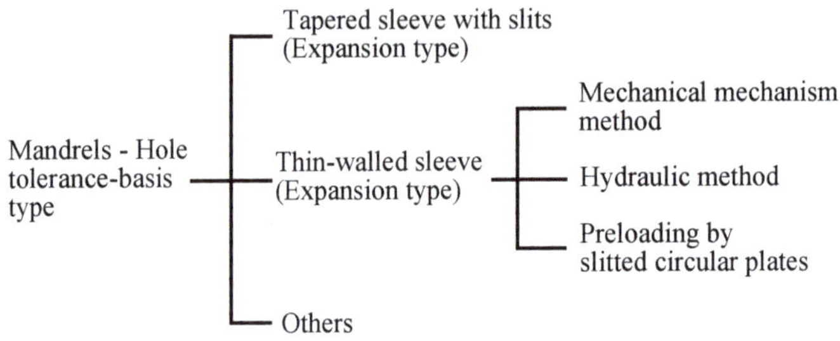

Fig. 5.30 Classification for mandrel of hole tolerance-basis type

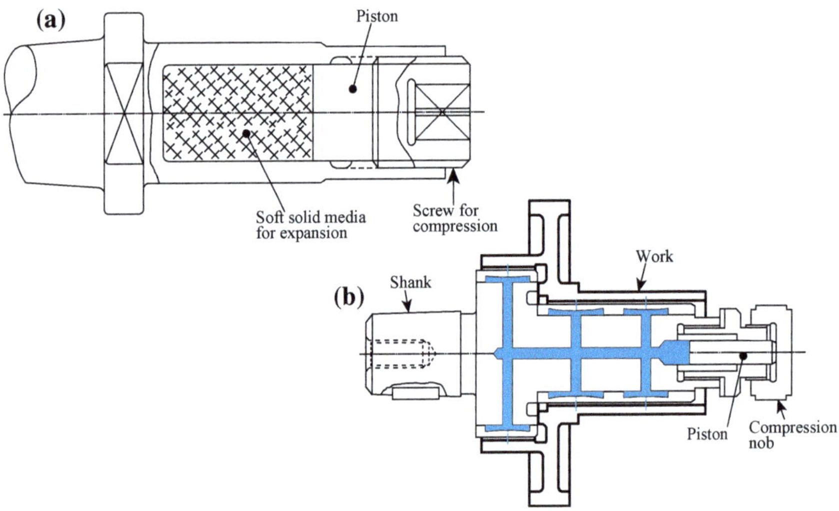

Fig. 5.31 Mandrels of hole tolerance-basis: **a** Mechanical type (by Kienzle 1950s). **b** Hydraulic type (1950s, by Stau)

reproduce the mandrels of mechanical and hydraulic types in the 1950s. In both the mandrels, the nob or screw is first rotated, and then the pressure generating media is duly compressed. Apparently, the thin-walled bush expands to radial direction and the work can be secure-held. Of special note, it is very interesting in the mandrel of hydraulic type that the oil recess is of concave-like, i.e., shallow at centre and deeper toward recess-end, so as to obtain the uniform distribution of the grasping pressure.

In contrast, Fig. 5.32a, b show the mandrels of split bush and hydraulic types at present. Apart from the holding accuracy, there are no marked differences in functional and structural aspects of the mandrels between in the past and at present. As can be readily seen from Fig. 5.32a, the mandrel of shaft tolerance-basis type can be converted without any drawbacks into the milling chuck as similar as that of collet chuck (see Fig. 5.21). In short, the milling chuck of W & F Werkzeugtechnik-brand

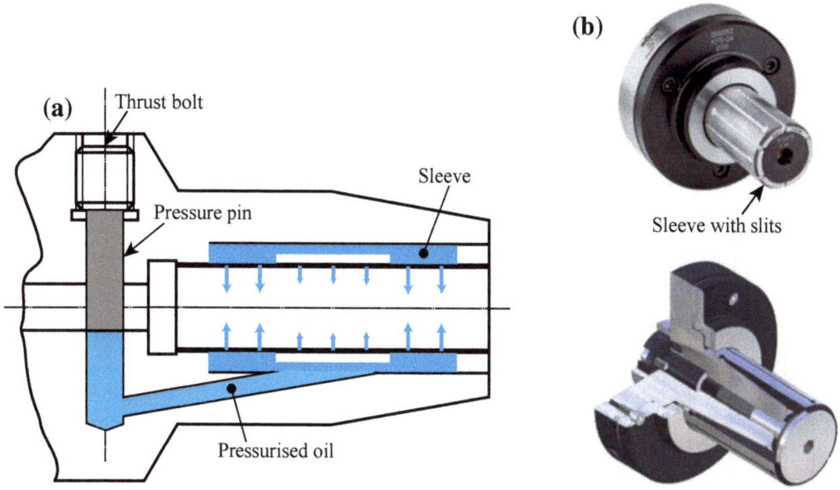

Fig. 5.32 Shaft or hole tolerance-basis mandrel: **a** Hydraulic collet chuck (by W & F Werkzeug Technik 2010s). **b** Thin-walled sleeve with slits—hole tolerance-basis type (by courtesy of Röhm 2015)

can be characterised by two-point grasping the tool shank, when applying the pressurised oil to the thin-walled sleeve. From such a grasping method, we may benefit secure-holding of the cutting tool with higher holding accuracy and rigidity, and also the higher anti-chatter vibration capability (see modular tooling system in Chap. 7). More specifically, W & F Werkzeugtechnik publicises the following data.

(1) The run-out of the tool is better than 3 μm at 50 mm in protruded length.
(2) Allowable maximum rotational speed is 40,000 rev/min.
(3) Allowable maximum driving torque is 210 Nm for 20 mm in diameter of the tool shank with the tolerance of h6.

Importantly, we may obtain now various milling chuck, which can be classified by the shrinkage mechanism of the collet, and these are very handy and high performance as compared with the old-fashioned milling chuck as exemplified in Fig. 5.33.

Admitting the great importance and wide applicability of the mandrel, however, it is very difficult to envision the overall view of the mandrel and its variants in detail. For example, a mandrel of hole tolerance-basis type is for hand-off operation as shown in Fig. 4.21; however, even the production engineer may not anticipate it as the variant of the mandrel in certain cases. In contrast, we face often to seek the innovative mandrel and variants, which are in company's confidentiality, and thus some interesting variants will be discussed in the following as the clue for such activities.

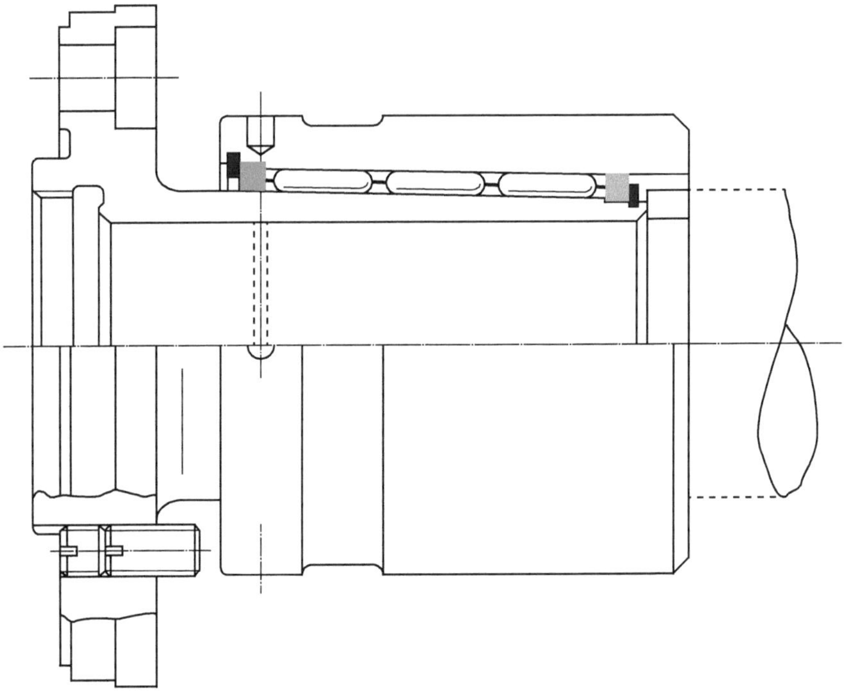

Fig. 5.33 Milling chuck of kneedle roller type in 1950s (by Stau)

Guide Bush

A variant of the mandrel is the "*Guide Bush*" and explicitly exemplified by that for drilling, which is used often in the multiple-drilling head for the *Transfer Centre*. In contrast, such a guide bush has been implicitly employed in the case of deep-hole (gun) drilling. For example, SIG + TBT displayed an interesting CNC gun drilling machine (Type T 30 KTN FTS) on the occasion of EMO Show in 1987, which was capable of drilling up to 40 mm in diameter. The machine can be especially characterised by its main spindle, which consists of two-layered structure of line allocation. More specifically, the front and rear spindles are for tapping and stab-drilling, and for gun drilling, respectively, and furthermore the hole-end of the front spindle plays the role of the guide bush for the gun drill. As will be clear from this example, it is, in general, very difficult to understand correctly the overall view of the mandrel.

Milling Chuck of Shrink Fit Type

It appears that the shrink fitting for the cutting tool is one of the variants of the mandrel as shown in Fig. 5.34. As can be seen from its configuration, this chuck is suitable for grasping, for example, end mill of small size less than 25 mm in diameter, which is applicable to machining the precise piece of the combination

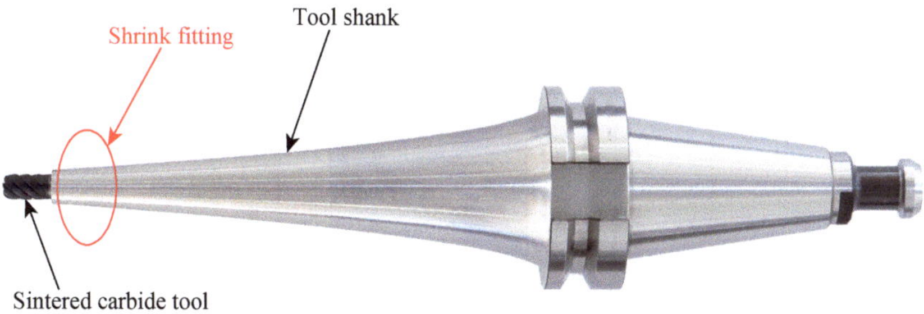

Fig. 5.34 Tool holder of shrink-fitting type (by courtesy of MST corporation, 2014)

mould die. In fact, the shrink fitting is very simple and handy, whereas we must learn both the heating and fitting tolerance techniques. For example, the shrink fitting shown in Fig. 5.34 is to be in reality by the skillful application of differing thermal expansion between the sintered carbide tool and the tool shank made of special steel.

5.4 Perspective for Research into Jaw and Collet Chucks

Within the attachment context, we have contrived and improved the device on the basis of the experience in the factory floor to a large extent. It is however very interesting that the jaw chuck is, at present, a leading objective of the academic research among attachments, because the jaw chuck increases its importance with the prevalence of TC and also the advance of the higher-speed machining. In short, the jaw chuck can be regarded as a rotating plate with protruded piece, and thus to unveil the essential feature, we need to apply the knowledge about aerodynamics, mechanical dynamics and sound engineering. In other words, some facing problems observed in the jaw chuck are very attractive for the researcher, and more importantly, it is very difficult to enhance the functionality and performances of the jaw chuck without having the authentic academic knowledge.

Apparently, there have been, as already mentioned, a considerable number of the research and development activities on the jaw chuck while especially rotating higher-speed, and they are four-fold as follows.

(1) Bell-mouthing phenomenon in chucking. A considerable number of the contrivances in the factory floor have been carried out, and a famous remedy is, for example, to employ the pendulum-like movement of the top jaw by supporting the master jaw with spherical seat.
(2) Recovery remedies for reduced clamping force by the centrifugal force simultaneously developing a chuck suitable for higher-rotational speed

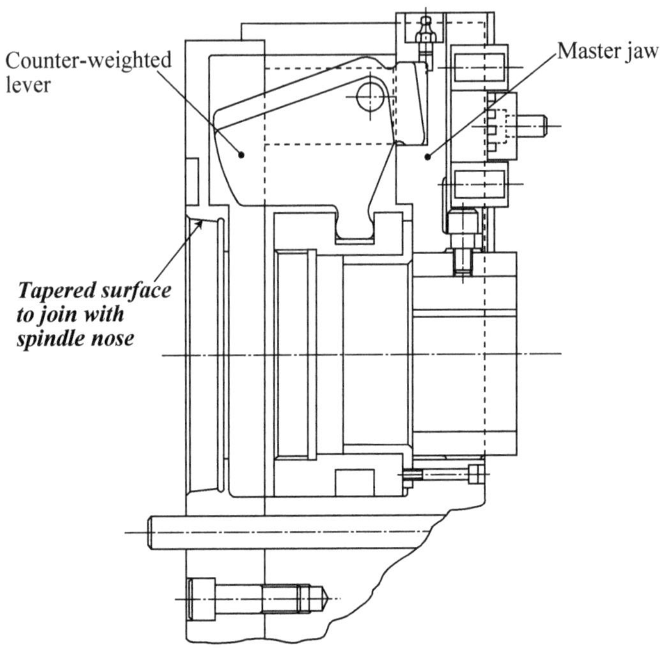

Fig. 5.35 Compensation for gripping force reduction by centrifugal force by using counterweight (by S-P Co., in 1975)

(Thornley and Wilson 1972).[5] In retrospect, the research into the jaw chuck concentrated on the reduction of clamping force, and consequently a considerable number of the research papers were publicised. With the prevalence of higher-speed cutting, however, the centrifugal force looses its influence considerably, because the thermal expansion of the work in radial direction becomes larger and thus apparently the clamping force increases more than that reducing by the centrifugal force.

Importantly, a representative remedy in practice is to compensate the influence of the centrifugal force by the counterweight. For example, Röhm Co. enabled to recover completely the reduction of the grasping force up to 7000 rev/min in the 1980s. Figure 5.35 reproduces that of S-P Co. in 1975 (Tooling & Production 1977). In this case, the master jaw is not let go of a work even at

[5]IWF (Institut für Werkzeugmaschinen und Fertigung; Institute for Machine Tools and Fabrication) of Technical University of Berlin once developed a prototype of higher-speed chuck made of CFRP. This chuck is available for machining of the work of 90 mm in diameter while rotating up to 15,000 rev/min. The chuck is furthermore designed in consideration of the directional-orientation properties of CFRP adequately.

Rudolph und Stelzer (1993). Faserverbundkunststoff erweitert die Systemgrenzen für Werkzeugmaschinenkomponenten. ZwF; 88–10:475–478.

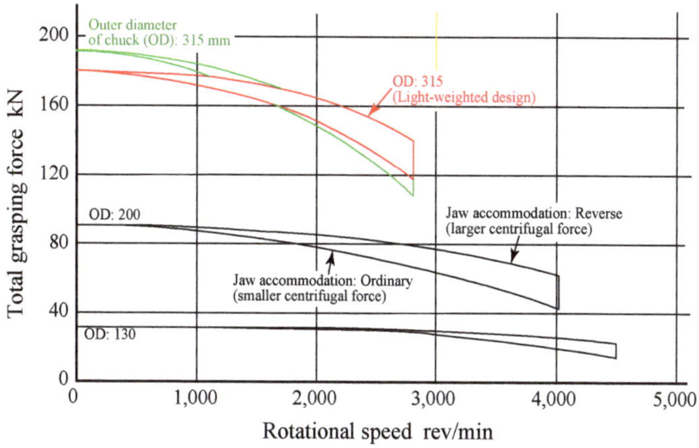

Fig. 5.36 Grasping force—rotational speed diagramme for self-centring three-jaw power chuck—wedge type (by courtesy of Röhm, Type KFD 2015)

higher-speed, because the same centrifugal force, which pulls the jaw away from the work, also pulls the counterweighted lever.

Of special note, Fig. 5.36 reproduces the grasping force-rotational speed diagramme in the case of self-centring three-jaw chuck of wedge type provided by the chuck manufacturer at present. More specifically, the grasping force was measured experimentally, when the top jaw is either ordinary or reverse accommodation at the master jaw to observe the differing influence of the centrifugal force. In fact, we can observe considerable reduction of the grasping force with rotational speed notwithstanding the number of jaws; however, the diagramme is, in general, for non-cutting state. In addition to it, the light-weighted design of the chuck may improve the grasping performance at the higher-rotational speed as shown also in Fig. 5.36.

(3) Appearance of turbulent flow of air and noise by the protruded jaw with the increase of rotational speed.
(4) Secure grasping of work in both the higher-speed and heavy-duty cutting.

In contrast, these facing problems in the jaw chuck are not so serious in the collet chuck. It is however crucial problem in the collet chuck to realise the better machining accuracy than ever before with the advent of TC with twin-spindle. In such a TC, the hand-off becomes popular and thus the skim-cut for truing should render useless. Paraphrasing, we need now an innovative collet chuck for the best machining accuracy together with relatively heavy-duty cutting.

In short, primary concerns in the research at present are (1) the interfacial pressure distribution between the jaw or collet and the work, and also (2) the reduction of the air-born noise especially in the higher-rotational speed as will be discussed in the following. In this context, we must be furthermore aware the growing importance to clarify the interface pressure distribution between in the collet and the chuck body.

Interfacial Pressure Distribution

In chucking by the jaw chuck of self-centring type, a serious problem is how to guarantee the correct and secure grasping the work across the whole contact face of the jaw with the work. This is because the grasping face of the jaw is liable to depart from the squareness with the chuck face, i.e., prancing against the axis of the main spindle, and to hold the work only with its back portion, i.e., bell-mouthing. In addition, each jaw shows the differing grasping force and interface pressure distribution one another.

Figure 5.37 shows some measured results of the interface pressure distribution for the three-jaw scroll chuck reported by Kadowaki (1985). He measured the interface pressure by using the pressure sensitive paper, and thus the measured interface pressure differs from the real one; however, we can observe something definite mentioned above from it. Importantly, we have used correction grinding as the utmost effective remedy even now for such a bell-mouthing problem of the jaw.

With the advance of such a jaw chuck of ball-rock type, the work can grasp with the draw-in and rocking motion of the jaw as can be seen, for example, from Fig. 5.11. As a result, we may not have the bell-mouthing problem; however, nobodies have verified it so far by measuring the interface pressure distribution.

As well as the ball-rock chuck, we have not visualised the interface pressure distribution between the collet and the work so far. In the collet chuck, furthermore, the contact pattern between the collet and the wedge sleeve or chuck body may affect to the work grasping performance as can be imagined from Fig. 5.20. In fact, Hainbuch has merchandised an interesting collet chuck of modular type, in which

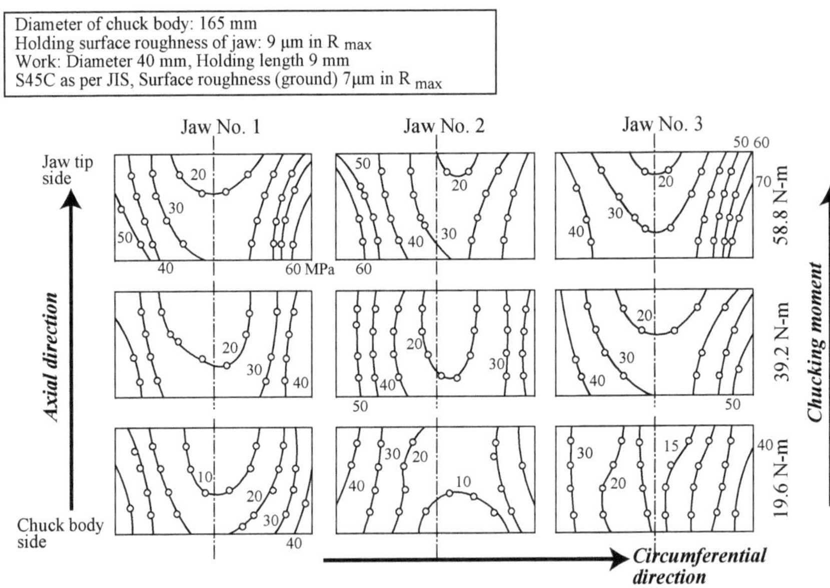

Fig. 5.37 Visualisation of grasping pressure distribution (by courtesy of Kadowaki)

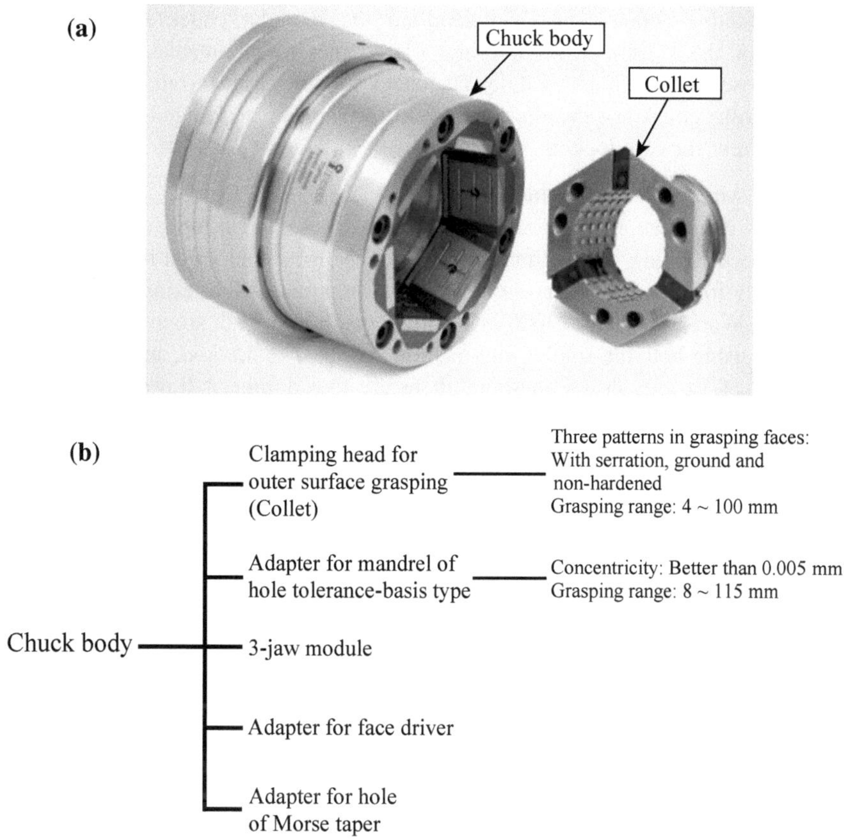

Fig. 5.38 An innovative collet chuck: **a** Appearance. **b** Modular systems in work holding (by Hainbuch 2016)

the work holding module ranges from the collet and mandrel, through the adapters for jaw and face driver, to adapter for taper exchange as shown in Fig. 5.38a, b. In addition, the chuck itself is capable of quickly changing, if necessary. Importantly, the chuck can be, as can be readily seen, characterised by the pyramid arrangement of the wedge face, and it appears as to be the inverse configuration the round collet with hexagonal hole.

As exemplified that the jaw travelling mechanism, i.e., one of the contact patterns of machine elements, has considerable effects to the interface pressure distribution, i.e., contact pattern, between the jaw and work, the collet moving mechanism may vary the interface pressure distribution between the collet and the work. In short, both the contact patterns play the important role in chucking, and thus it is desirable to visualise both the contact patterns. As already publicised elsewhere, the ultrasonic waves method may facilitate such a measurement to a large extent (Ito 2014).

Importantly, Forkardt has merchandised the modular chuck system, which consists of the (1) chuck module and (2) adapter to intermediate between the spindle nose and the chuck module. The chuck module ranges from the jaw chuck, through collet chuck and mandrel, to face driver, and can be changed quickly by the central screw for cam locking.

Flows of Air and Noise Reduction

In the jaw chuck, another serious problem is the noise due to the air sharing while higher-speed rotating, and with the increase of the rotational speed in TC and MC, the remedy for reducing the noise becomes an important engineering problem. For example, Warisawa et al. (2000) visualised the flow of air around rotating jaw chuck by using both the smoke wire method and the *Taft Method*, and unveiled the existence of various flows. In short, there are five dominant flows as shown in Fig. 5.39 and on the basis of such observations, he and his co-workers proposed a remedy to place the small channel at the root of jaw. Supposedly, by such aremdy, the zone of negative pressure flow of air may be eliminated, resulting in the considerable reduction of the noise. That of Warisawa is being in the patent claim.

To this end, it can furthermore suggest that in the jaw chuck, one of the grass root-like knowledge is in the raw material preparation. In fact, the top jaw should be made of the raw material, in which the "*Direction of Grain Fibers*" produced by hot rolling coincides with the radial direction of the jaw, otherwise the top jaw is liable to rapture while rotating.

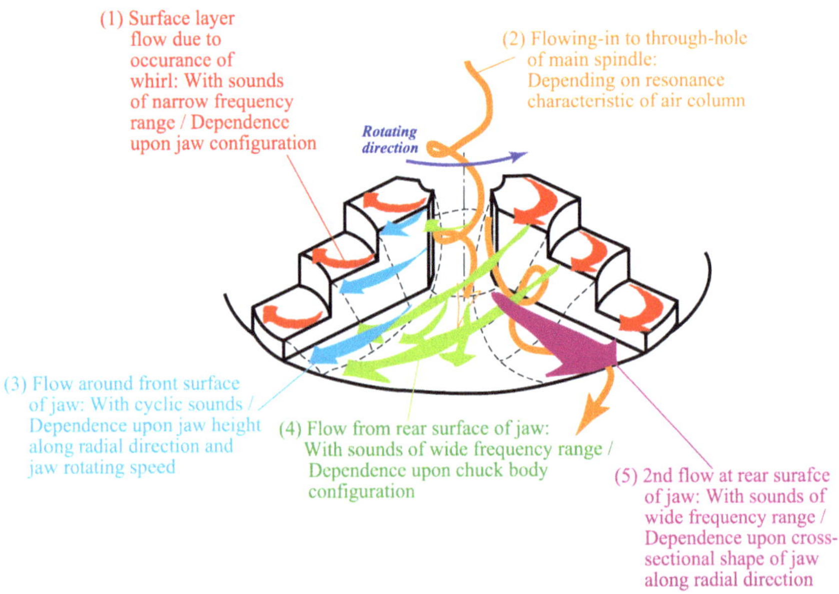

Fig. 5.39 Observation of flows of air in rotating jaw chuck

References

Ito Y et al (1989) A newly developed flexible chuck with quick changing mechanism of top jaw. J JSPE 55–11:2081–2084

Ito Y (2014) Chapter 3 ultrasonic waves method for interface pressure meaurement: what is acceptable definition of interface pressure. In: Ito Y (ed) Thought-evoking approaches in engineering problems. Springer, London, pp 41–70

Kadowaki Y (1985) Development of identification sensor for gripping condition in three-jaw scroll chuck. Trans JSME 51–466:1372

Mueller G (1977) Let's talk about chuck gripping forces and speed: Part 1. Tooling Prod 43:76–78

Owen JV (1995) Strategic chucking. Manuf Eng 114–6:35–38

Stau CH (1963) Die Drehmaschinen – Drehbänke und verwandte Werkzeugmaschinen Grundlagen –Bauteile-Bauarten. Springer-Verlag: 107–119

Thornley RH, Wilson B (1972) A review of some of the principles involved in chuck design. Prod Eng 87–97

Tooling & Production (1977) Counter-centrifugal chucks for high speeds

Warisawa S, Kiyokawa T, Ito Y (2000) Investigation into generation mechanism of aerodynamic noise caused by rotating lathe chuck and its reduction remedies. Trans JSME 66–649: 3174–3180

Chapter 6
In-Process and On-the-Machine Measurement

In discussing the sensor (transducer) and concerns in the machining space, we must first understand the fundamental and mandatory knowledge about the measurement of general concerns as follows.

(1) The measurement can be roughly classified into the "On-the-machine" and "Off-the-machine" ways. By another way, the measurement can be detailed to the "Between-process", "In-process" and "Post-process" types. Of these, the between-process and post-process types are carried out either on-the-machine way or off-the-machine way; however, the in-process type belongs to the on-the-machine way.
(2) Measurement technology consists of the sensor, transmission device for either in-put and out-put signals in the active sensor, or out-put signal in the passive sensor, and the signal recorder and processing devices. If necessary, the out-put signal is supplied to the adaptive control (AC) system.
(3) The differing environment for the measurement in the machining space from that of environment-controlled measurement room.
(4) Generally speaking, the numerical control (NC) machine tool can produce the component with satisfactory quality according to the indications in the part drawing, and there are no necessities of any inspections in the case of civil supplies.

Figure 6.1 shows an adaptive control system for the feed rate of cutting tool to understand the importance of the sensor and out-put signal processing in the measurement. By detecting the out-put signal of the piezoelectric sensor, we can identified the states of the cutting tool, and duly control the feed rate of the cutting tool. In addition, Fig. 6.2 shows the flexible manufacturing cell (FMC) for the component of the armoured personnel carrier (APC). As can be seen from the cell layout, the co-ordinate measuring machine (CMM) is installed for the inspection of all the finished components, although the machining centre (MC) can finish, in general, the component with acceptable quality (Ashley 1983).

Y. Ito and T. Matsumura, *Theory and Practice in Machining Systems*,
DOI 10.1007/978-3-319-53901-0_6

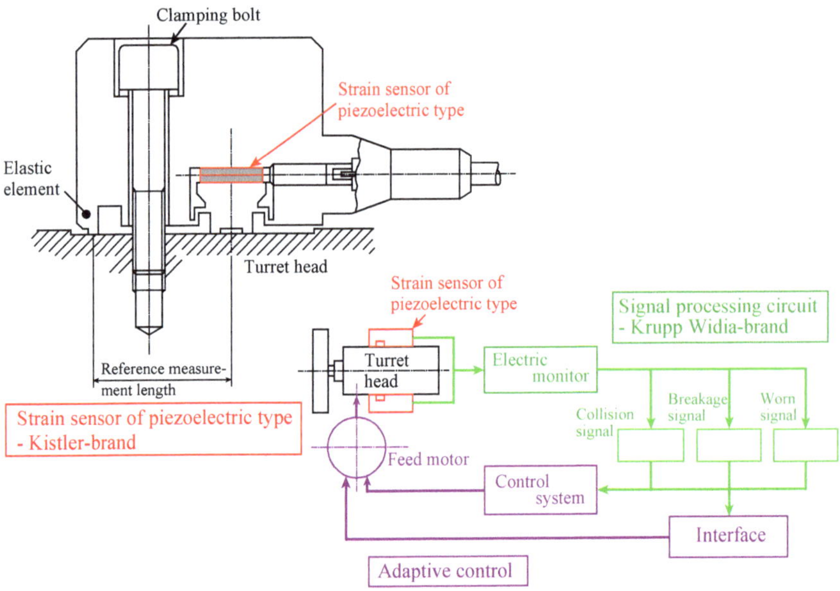

Fig. 6.1 Strain sensor of piezoelectric type and signal processing circuit

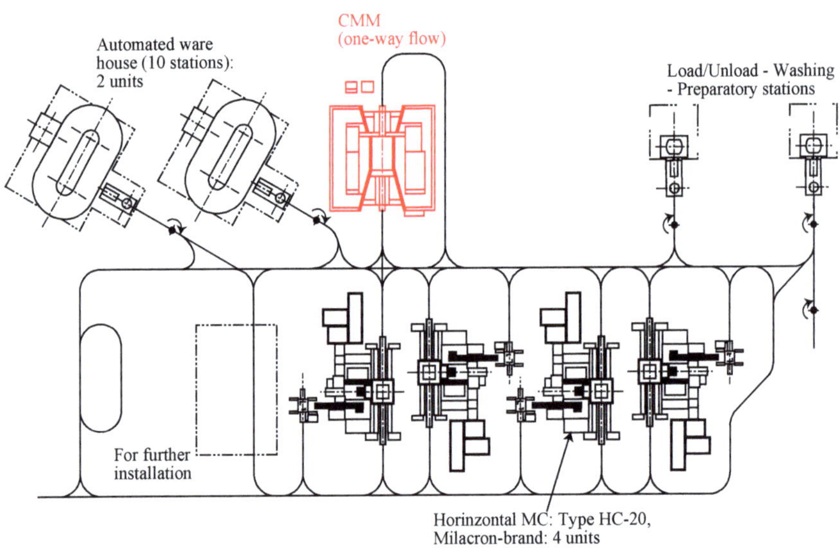

Fig. 6.2 Flexible manufacturing cell for APC components installed at Aiken S.C. Plant in FMC Co. in 1980s

We must then clarify the necessity of to what extent the on-the-machine and in-process measurement in the machining space. When considering the automatisation to large extent like FMC, it has been believed that the in-process measurement is necessary for detecting and evaluating the (1) dimensional and shape specifications of the work being finished, (2) work setting and holding, (3) tool holding, tool wear and damage, and machining force, (4) swarf configuration, and also (5) healthy condition of the machine tool itself. Figure 6.3 summarises all the measuring objectives and corresponding sensors necessary.

As will be clear from Fig. 6.3, a crucial issue is to determine preferably the measuring objective in consideration of the machining requirement and economic benefits. More specifically, we must discuss the following issues.

(1) To obtain the reliable information to control the machining space, we need the much more number of the sensors, whereas we could have unexpected troubles with increasing the number of the sensors. As well known, the machining space is the ill-defined environment for the sensor, because of the oil misty, dusty and swarf scattering.
(2) To clarify the beneficial aspects when placing the sensor in the machining space together with the estimation of the minimum number of the sensors required.
(3) To investigate the effectiveness of the sensor fusion and problems to be solved.

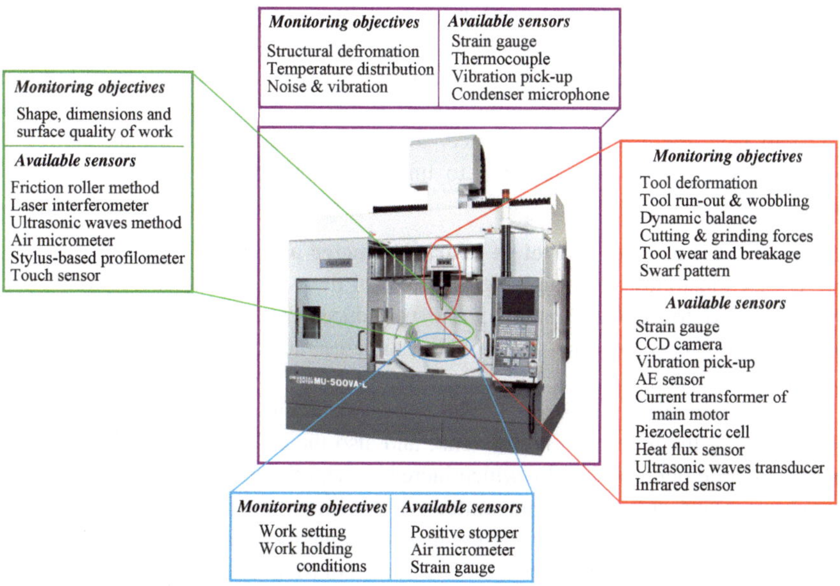

Fig. 6.3 Various objectives and sensors for in-process measurement while machining (machine photograph by courtesy of Okuma)

(4) Leverage between the reliability of NC and the necessity of in-process measurement. Of note, Japanese machine tool manufacturer is not so keen to install the in-process measurement function within the machine tool.

Actually, nearly all machine tool users are interested in whether they can benefit by using the in-process measurement or not, and thus some quick notes will be given in the following for the present and near future perspectives in on-the-machine and in-process measurement.

6.1 In-Process Measurements in General

In retrospect, the in-process measurement was vigorously investigated, and tried to use in practice in the 1980s especially in Germany. In fact, there have been a myriad of research papers and technical reports; however, of these, we may apply a less number of the in-process measurements to the practical work. Paraphrasing, both the manufacturer and the user appear not so keen to use the in-process measurement at present, because of the economisation of machining.

Conceptually, it is desirable that the in-process measurement can facilitate the authentic detection of the state information in the machining space, and that the machining condition can be fully automatic-regulated on the basis of such information to finish the component according to the indication of the part drawing.

Obviously, fully automatic machining is far from the fruition as yet, and a root cause of difficulties lies in the reliability of the out-put signal of the sensor. We may understand such the causality from the fact that nearly all sensors shown in Fig. 6.3 are applicable to the laboratory environment and only in the stage of the prototype.

Apart from the usability in practice, some interesting sensors so far publicised will be discussed in the following to unveil the shortcoming of the in-process measurement.

Ultrasonic Transducer to Detect Work Diameter and Tool Wear

Figure 6.4 illustrates the measuring principle of the work diameter and tool wear, and also two trials. Conceptually, one transducer can detect the two different out-puts, i.e., those in time and voltage domains, and thus this method can be evaluated highly.

As shown in the top of Fig. 6.4, the work diameter and tool wear can be detected by measuring the reflected time of echo, and also the reduction of the echo height on the cathode ray tube (CRT), which increases with the flank wear, because of the increasing magnitude of the transmitted waves. Importantly, the tool wear can be measured without any influence of the thermal elongation in the tool shank, and of course it is very beneficial; however, we must contrive out-put signal processing, because of its weak magnitude.

In contrast, the preferable transmission of the ultrasonic waves is, in general, very difficult as shown in the case of measuring the work diameter, where we need

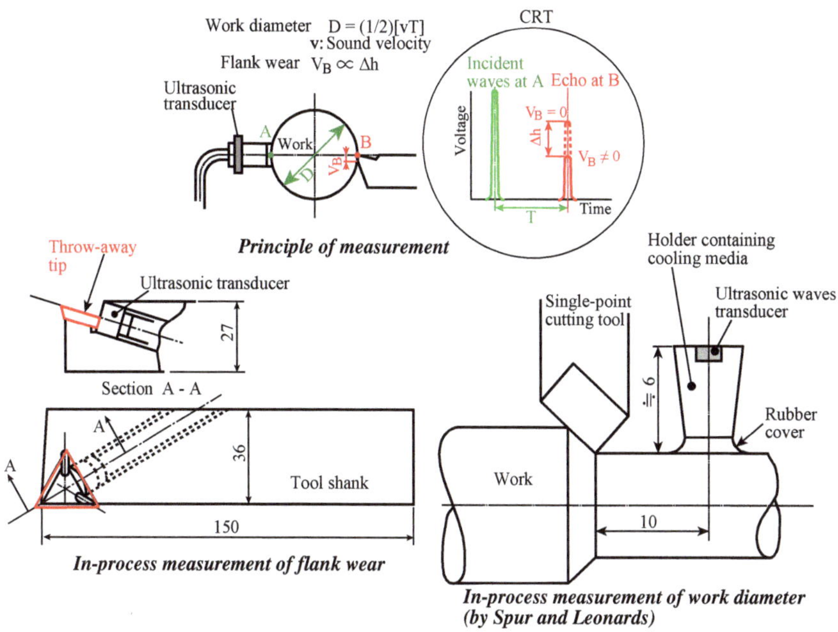

Fig. 6.4 Ultrasonic waves transducer for in-process measurement

to provide the media for the effective sound transmission (Spur and Leonards 1975).

Tip with Multiple-function Integrated Sensor and Built-in-thermocouple in Grinding Wheel

Figure 6.5 shows the tip with sensor and the grinding wheel with built-in-thermocouple. The former can measure both the crater wear in the tool and the cutting temperature. More specifically, the measuring pattern like the printed circuit is formed on Ti-layer within the multiple coated layers of sintered carbide tip (TiC-TiCN-TiN) by using sputtering. As shown in left-hand side of Fig. 6.5, the patterns are placed closely to the cutting edge, and obviously, the pattern can be dexterously designed and sputtered on the tip in accordance with the measuring objective. In addition, the pattern is secure-sealed with the insulation layer, and the crater wear can be detected by the effective area reduction in sensing spot (Klocke and Rehse 1997).

In the latter, the grinding wheel is special purpose-oriented, and not applicable to the practical use, but research only (Rucker 2005). Of note, Noritake produced once a cBN grinding wheel with built-in AE sensor in 1993 to detect grinding-related factors, and also on the strength of such information as follows, to choose the desirable time interval for dressing.

(1) Grinding force and surface roughness of finished work from the root mean square value of out-put signal.

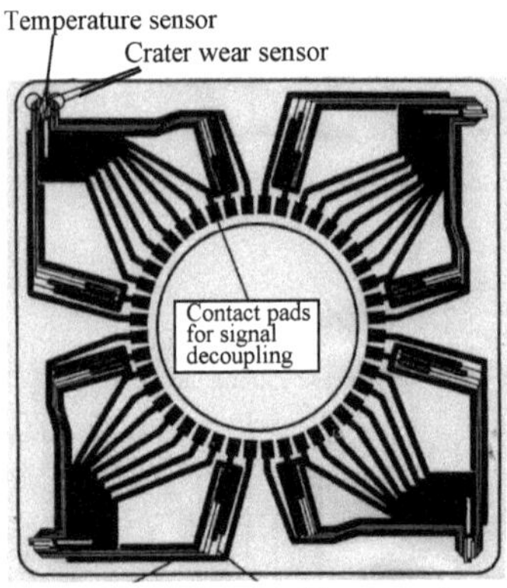

Sensor layout produced by sputtering to detect crater wear of tool and cutting temperature (by Klocke and Rehse)

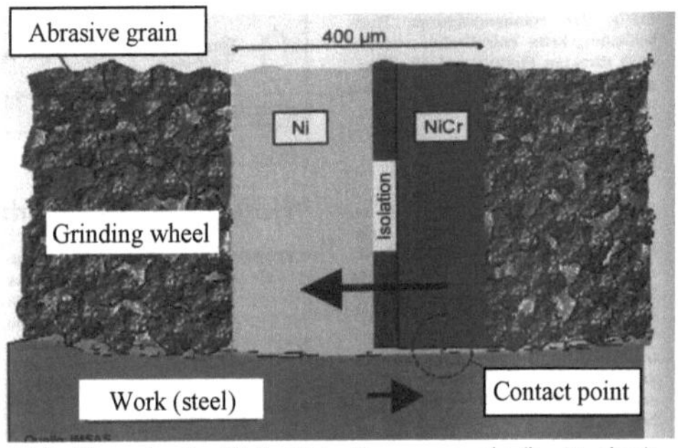

Grinding wheel with built-in-thermocouple (by Rucker)

Fig. 6.5 Multiple-function integrated sensor and grinding wheel with built-in thermocouple

(2) Chatter vibration from the increase of the specific frequency in the power spectrum.
(3) Touch sensing from the increase of the amplitude of out-put signal.

As will be clear from the above, out-put signal processing becomes complicated, although we can obtain a handful of important information. In addition, another issue is to effectively transmit the out-put signal of AE sensor from the rotating wheel. In that of Noritake, the slip ring was employed, and in general, the noise may include in this method.

In-process Measurement for Swarf Configuration with Piezoelectric Transducer

The piezoelectric transducer of Kistler-brand has been widely employed so far to measure the cutting and grinding forces across the whole world, and especially to identify the machining state by the German machine tool manufacturer as will be discussed in Sect. 6.2.

Figure 6.6 shows the signal processing systems to detect the swarf configuration, and an example of signal processing (Chung et al. 1989). As can be seen, once we can determine a suitable cut-off frequency, the valuable information can be obtained as shown together in Fig. 6.6. In fact, the dynamic component indicates clearly the sinusoidal-like variation after band-pass filtering. Importantly, we can identify the cutting condition, by which the swarf with long comma-like configuration, i.e., swarf with better disposability, can be formed.

As exemplified by this research, it is emphasised that the piezoelectric transducer benefits much more than our expectation in the academia. Within the industrial application context, however, the piezoelectric transducer and concerns should be furthermore improved especially from the aspects of the thermal drift and cutting

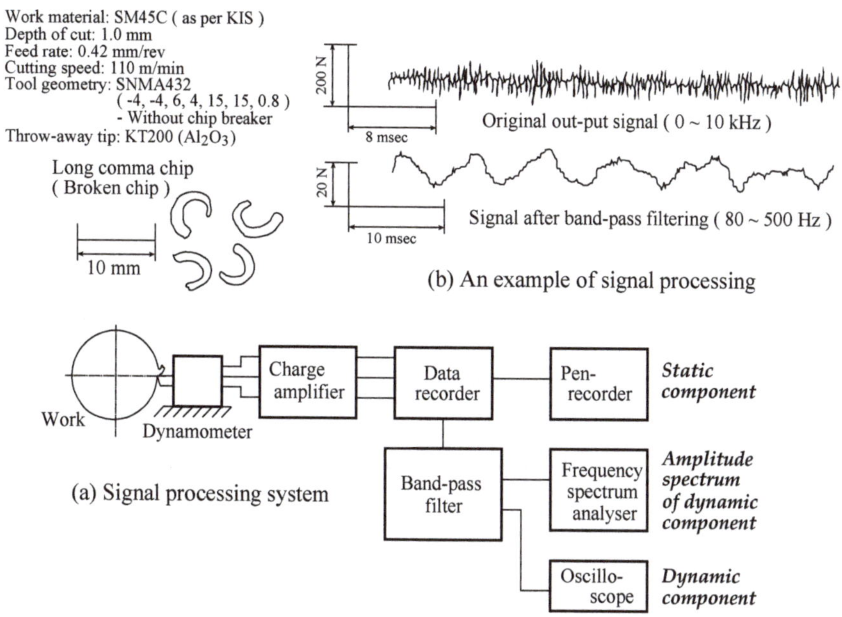

Fig. 6.6 Signal processing system to detect swarf configuration

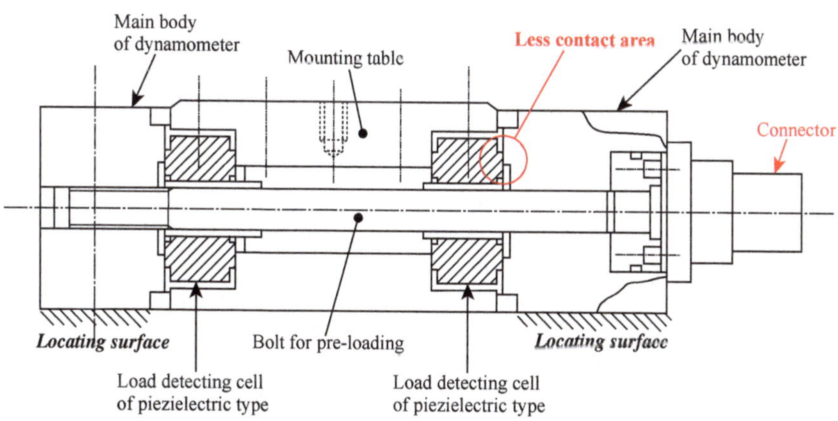

Fig. 6.7 Concept of load detecting cell of vertical mounting type—In case of "Type MiniDyn 9256A1" (by courtesy of Japan Kistler)

fluid tight. In due course, Kistler conducted the necessary remedies in the middle of 1990s as shown in Fig. 6.7, in which the noteworthy remedies are as follows.

(1) To suppress the thermal drift, the load cell is mounted vertically and with a less contact at the inner circle of locating face than ever before by recessing the dynamometer body. In addition, the number of tightening bolts for the load cell

reduces from 4 to 2 by simultaneously clamping a couple of load cells by a single bolt.

(2) To secure the cutting fluid tight, the cover sleeve of bay-type flange is employed instead of the bayonet connector.

Admitting that the piezoelectric transducer and concerns are, at present, sophisticated to large extent, we must be aware that the transducer should be furthermore improved as can be readily seen from its structural configuration shown in Fig. 6.7. In short, the transducer in the form of dynamometer is too large as a sensor for in-process measurement at the machining space. In addition, the oil and water tightness appears not to be perfect, resulting in certain dangers of the transducer down. To be compatible with the machining space, thus, a remedy is to integrate the small-sized piezoelectric transducer within the machine component as will be discussed in Sect. 6.2.

In-process Measurement Without Transducer

It is common sense that the core of the in-process measurement is the sensor itself; however, the sensor is one of the fatal causalities leading the machine tool to down. In addition, we must use, as will be discussed in Sect. 6.2, the sensor-fusion to obtain the reliable and authentic out-put signal from the machining space. Admitting such shortcomings of the sensor, thus, another idea is to detect the necessary out-put signal by using some facilities within the machine tool.

As reported elsewhere, the increasing magnitude and its irregular changes in the motor current are detected by using the current transformer, and from these we may understand the collision between the work and the cutting tool, over-load of main spindle, tool wear and tool fracture. Importantly, this method is very handy; however, the sensing accuracy depends upon the inertia of the driving motor.

In this context, Shinno et al. (2003) proposed a sensor-less monitoring for the cutting force in ultra-precision machining. In fact, the cutting force gives us with the valuable information for the state of the machining space. Figure 6.8 shows the measuring principle, in which they use the out-put signal from the disturbance observer within the position controller. Actually, the proposed method is applicable to the accurate measurement of the very small magnitude of cutting force in ultra-precision machining. In short, the determinants of the disturbance observer are, the system matrix, input and output vectors.

Importantly, the output of the disturbance observer includes unfavourable disturbance such as a force ripple of the linear motor. As can be readily seen, such a reproducible system-related disturbance can be subtracted from the out-put. As a result, we can measure, for example, around 2–6 N in cutting force, while cutting Al–Mg-alloy with diamond tool and 10 μm in depth of cut.

To this end, it is worth suggesting that the sensor may be regarded as the mimics of human's five senses, i.e., hearing, sight, touch, smell and taste. Of these, the smell is not sublimated to the sensor for machining as yet, although the sensor for fish freshness has been in practical use. In short, we can recognise the following states in machining by the smell.

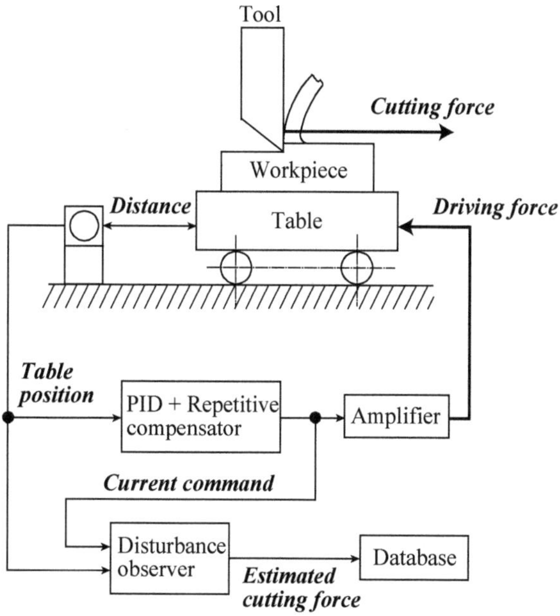

Fig. 6.8 Concept of sensor-less status monitoring (by courtesy of Shinno)

(1) Rotating condition of driving system, overload of main spindle and supply condition of cutting fluid by the smell of oil burning.
(2) Overload condition of main spindle and over-currency of power electric cable by the smell of smoke.
(3) Bacterial degradation of water-soluble cutting fluid and humidity of machining space by the smell of ambient air (see Chap. 3).

It is thus expectable to be some innovative sensor in reality before not too distant future.

6.2 On-the-Machine and In-Process Measurements in Practice

As can be imagined from Fig. 6.3, it emphasises that we have, without no doubt, investigated all the sensors applicable to the detection of the necessary information in the machining space. This interpretation is, in part, correct, and in part wrong. Obviously, we have so far tried, developed and contrived an uncountable number of sensors in both the academia and the industrial sectors, and now we have no difficulties to choose the preferable sensors under no restrictions. Against this context, we have faced to the serious shortages for the sensor, which is reliable, robust and with high operability within the machining space, i.e., ill-defined environment for the sensor.

In both the on-the-machine and in-process measurements, the measuring objectives are in situ, and thus we can benefit considerably from many aspects. Obviously, it is expectable to gain much more benefits when employing the in-process measurement as compared with the on-the-machine measurement; however, in addition to the ill-defined environment for the sensor, we must solve the following problems to enhance the usability in practice.

(1) Deterioration of measuring accuracy by temperature changes and thermal deformation: Even in CMM installed within the temperature- and humidity-controlled room, the measuring objective should be placed still-stand, at least, over 24 h to guarantee the measuring accuracy. In the machining space, the temperature distribution is not uniform and the reference for the measurement thermally deforms often. In short, the thermal deformation affects considerably the measuring accuracy, especially in the in-process measurement, and it is, in general, difficult to eliminate the error derived from the thermal deformation (see Fig. 6.4).
(2) Improvement of measuring accuracy by using sensor fusion: In general, one sensor detects on signal; however, to obtain the reliable and authentic out-put signal, the sensor-fusion has been often employed. As widely known, the sensor fusion can be, in wider scope, classified into the three types as shown in Fig. 6.9, and the hardware-oriented type can, in general, facilitate the accurate measurement for single physical state property by comparing necessary information obtained from various sensors. As reported so far elsewhere, the sensor and concerns are one of leading causalities of the down time of NC machine tools, and thus the sensor fusion with one transducer has been and being prevailed, although we must contrive suitable signal processing to various and

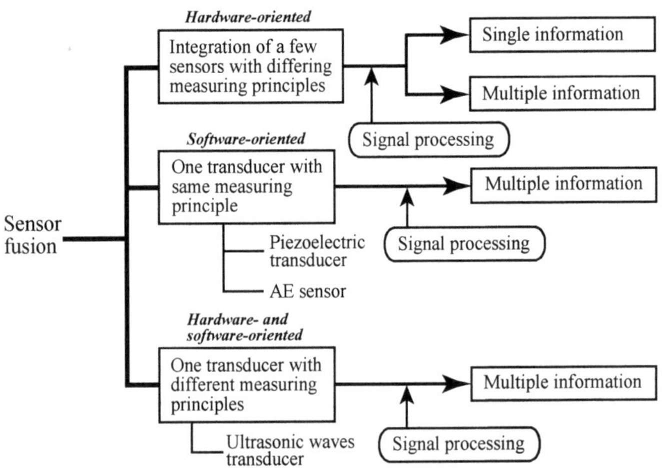

Fig. 6.9 Three representative types in sensor fusion

large extents. Of special note, it is desirable to employ one transducer of hardware- and software-oriented type.

(3) Transmission of out-put signal from rotating element: In machining space, the sensor should be often placed on the rotating structural component such as the main spindle and grinding wheel. It is thus necessary to use, for example, slip ring and frequency modulation (FM) telemeter, for the signal transmission, which induces the noise.

(4) Choice of sensor mounting place: The sensor should be placed at the most suitable position within the machining space, otherwise the sensor cannot detect any reliable signals. For example, at the two bodies-in-contact, the signal decays its strength and mingles unwelcome noise.

As a result, we may use a very few measuring systems in the machining space with the satisfactory performance, i.e., (1) current transformer, (2) piezoelectric transducer, (3) acoustic emission (AE) sensor and (4) bearing load sensor of strain gauge type, as will be described and discussed in the following.

On-the-machine Measurement With Touch Sensor

In the on-the-machine measurement, the touch sensor is being prevailed, although being either the between-process or post-process type. In on-the-machine measurement, we can make the (1) automatic detection for the marking out of the work and tool position, (2) phase adjustment of tool to work, and (3) evaluation of the dimension and shape of the finished work by using together with the laser displacement sensor.

Figure 6.10 shows the principle of the touch sensor on the market. The touch sensor is the most popular equipment in MC and TC, because of its simplicity, high operability and economisation. Actually, the detection coil can identify the induced current, which flows in the closed loop when the touch probe contacts with the work, provided that the oscillation coil is active. For example, the work dimension and centring can be measured and carried out by reading the corresponding NC co-ordinates when the touch probe indicates the contact signal. In certain cases, the tool breakage can be detected if the touch probe does not give any signals after travelling body moves the expected length.

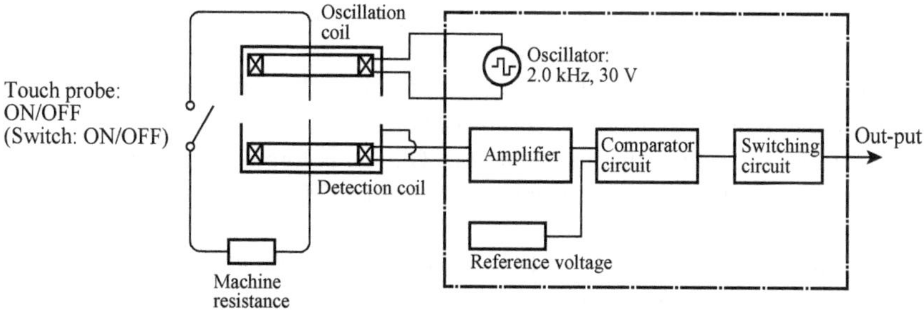

Fig. 6.10 Principle of touch sensor

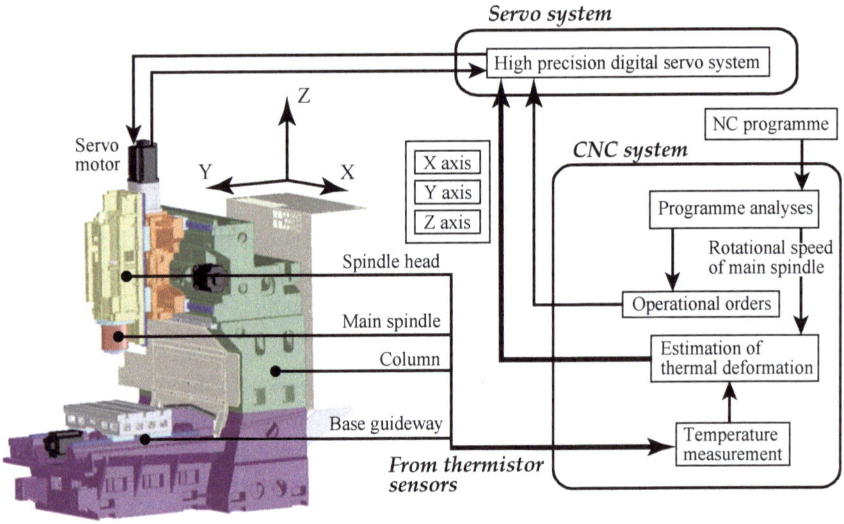

Fig. 6.11 Compensation system for thermal deformation and allocation points of sensors (based on material provided by Okuma 2007)

In-process Measurement for Temperature and Compensation of Thermal Deformation

Okuma has merchandised a highly evaluated compensation system for the thermal deformation as shown in Fig. 6.11, and its utmost characteristic feature is to allocate the thermistor sensors directly within the machining space. As can be expected, the system runs very well, and for example, the machining accuracy is better than 10 μm, when the ambient temperature changes 8 deg.

Importantly, the noteworthy reduction of the thermal deformation can be, to certain extent, facilitated with the synergy of hardware and software remedies, i.e., (1) symmetrical structural configuration with large thermal inertia (thick wall), (2) equalisation of the surrounding temperature of the machining space with the front cover, (3) compensation of the heat sources in machining space by the counter heat source generated by the control cabinet of rear allocation, (4) unification of temperature distribution by the double-oil cooling jacket, and (5) highly accurate control technology. In addition, the main motor of in-house manufacturing contributes greatly the reduction of the heat generation as compared with that of conventional motor. In short, these remedies result in the simple elongation of the main spindle, which can ease the compensation.

It is worth suggesting that the system of Okuma is forerunning and superior to that has been conducted by Technical University (TU; Technische Universität) Hannover in corporation with MAG Powertrain and Siemens (Denkena and Scharschmidt 2009) under one of EU Projects. In that of TU Hannover, the temperature sensors are placed at base, column, and linear scales for three axes, and on the strength of the measured temperature, the thermal deformation is estimated by using the finite element method (FEM) model. Although the thermal deformation can be reduced, there are no remedies from the structural design. In retrospect, KTM of England merchandised an MC (Flex-Matic V Series) with similar compensation system in 1988; however, the sensors were allocated around the machine body, but not in the machining space, and the main spindle only was cooled.

In-process Measurement With AE Sensor

AE sensor is very handy and of high operability together with oil and water tight, and thus has been and being prevailed. In contrast, AE sensor is of passive type, but not the Anti-Submarine Detection Investigation Committee (ASDIC)-like, and thus we can obtain surely the out-put signal; however, there is no guarantee for the one-to-one relationship between the out-put signal and something changed in the measuring objective. In general, AE signal includes uncertainties to various and large extent, and thus out-put signal processing is primary concern to obtain the reliable information as will be shown in Fig. 6.12.

Although aiming at the detection of the wear and damage in the cutting tool, one is to use the threshold in signal discriminator, and the other is to use the maximum magnitude in vibrational amplitude, and depending upon such the detection index, the signal processing circuits differ duly from each other.

In not only AE sensor, but also other sensors for the in-process measurement, another issue is to sublimate the sensor itself to the built-in type. In general, AE sensor on the market is much larger than that capable of using as the built-in type. With the growing importance of the tool with built-in-sensor and also advance of

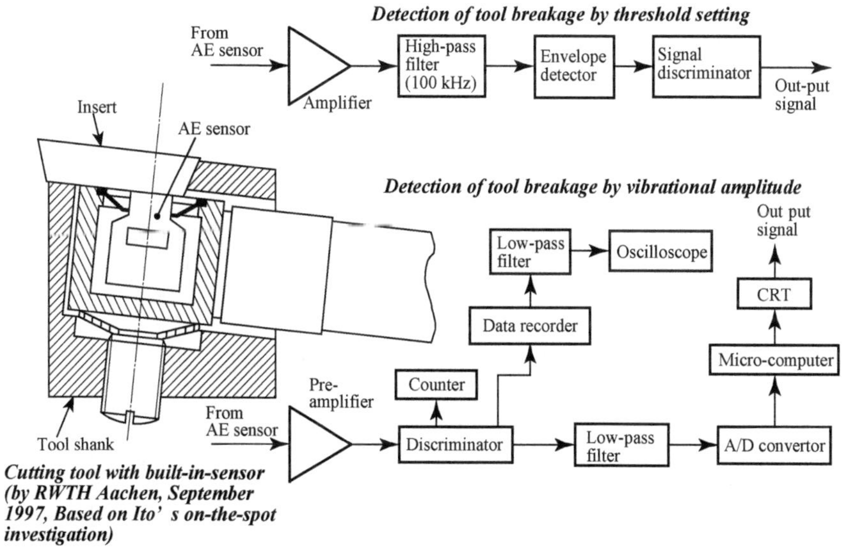

Fig. 6.12 Detection of wear and damage in cutting tool with AE sensor and its signal processing

the related technologies, the research into the compact sensor became active from the 1990s.

Figure 6.12 reproduces also the single-point cutting tool with built-in-sensor for detecting the AE signal. Technical University of Aachen (RWTH Aachen) conducted an investigation into this built-in-sensor in the late 1990s. Lindberg tried furthermore to use the thin film with piezoelectric effect for the measurement of the segmentation frequency in the swarf generation. In this trial, the thin film was placed under the throw-away tip, resulting in the very compact sensor; however, he did not report its practical availability (Lindberg et al. 1983).

Small-sized Piezoelectric Transducer for Shear Force Detection

As having been verified, the piezoelectric transducer of Kistler-brand appears as to be defacto standard in the measurement of the cutting and grinding forces across the whole world. In due course, the piezoelectric transducer can facilitate the in-process measurement for the machining states to various and large extents.

Importantly, the piezoelectric transducer can be characterised by its outstanding property, i.e., measuring the force with negligible small elastic deflection. In addition, that of Kistler can measure the shear force by cutting the crystal along its electrical axis as shown in Fig. 6.13. As a result, Kistler can merchandised a load cell for detecting the three components of the cutting force, and eases to built-in it within the structural body component.

Against this context, the out-put signal of the piezoelectric transducer is, in principle, static electricity, and thus liable to leak. The piezoelectric transducer should be, thus, very cutting fluid tight, and one of the remedies is integrated in the structural body component, e.g., main spindle of MC as shown in Fig. 6.14. In this case, the piezoelectric transducer for measuring the shear force is used. As will be clear from the illustration and by comparing that of Fig. 6.7, the cutting fluid-tight may be guaranteed to some extent together with realising the compactness of the transducer.

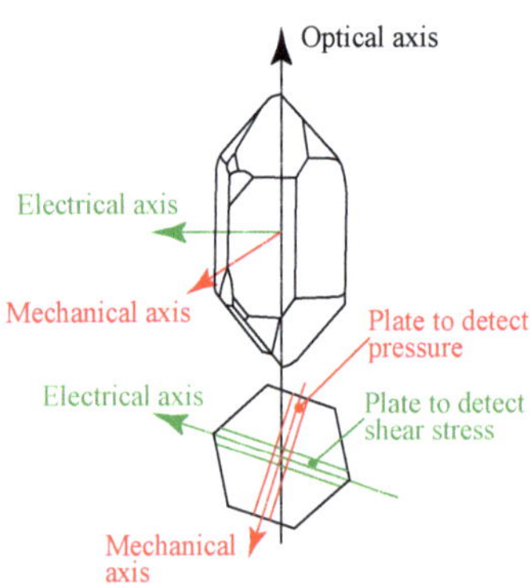

Crystal and its piezoelectric effects

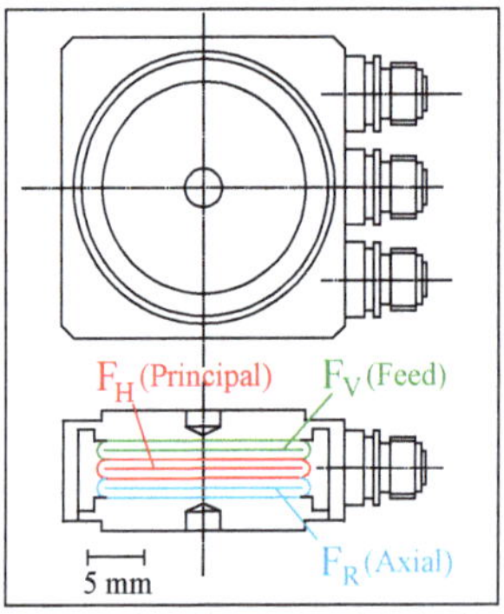

Structural configuration of load cell

Fig. 6.13 Load cell of piezoelectric type (by Kistler AG in 1970s)

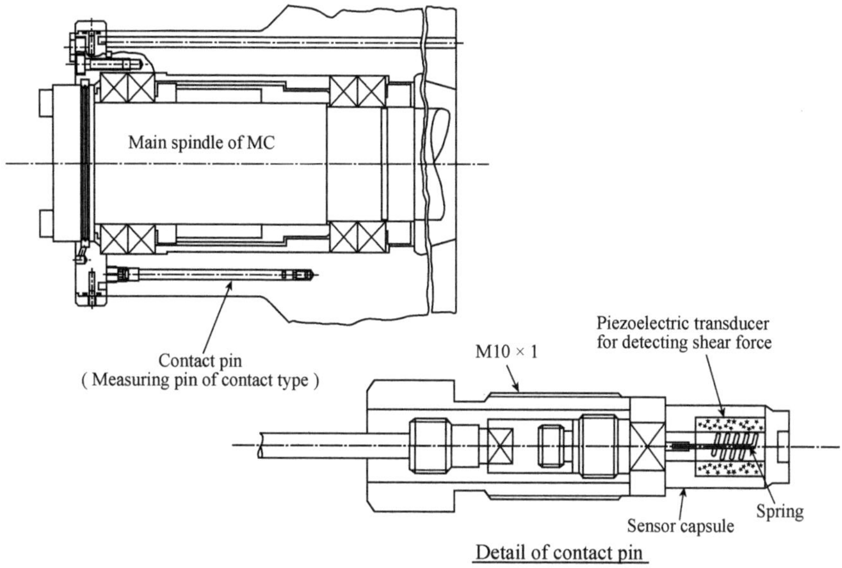

Fig. 6.14 Contact pin of built-in type to measure axial component of cutting force (by report of Kirchheim on occasion of IMS Meeting at October 1994, by courtesy of Japan Kistler)

Table 6.1 New requirements for measurement of cutting force

Accumulation of design data for complex loading in machining method-integrated kind like "Mill turn"
Visualisation of machining space by using information for cutting force while higher-speed machining within oil mist and dust
Establishment of in-process measurement for finish cutting with noisy states
Recognition of cutting condition in new method like long slotting with Trochoidal locus
Influence of tool run-out centrifugal force on cutting force while high-speed drilling and end milling
Superiority analysis for drill with functionality
Clarification of machinability of new materials and cutting performances of new tool

In short, the piezoelectric transducer is the most reliable at present when carrying out the in-process measurement, and also has certain potentialities. Table 6.1 summarises thus some engineering subjects newly arisen and closely related to the measurement of the cutting force.

References

Ashley S (1983) FMC's machining system seen as springboard to automation. Am Metal Market/ Metalworking News 9–10

Chung E-S, Namgung S, Ito Y (1989) In-process recognition of chip forms using dynamic component of cutting force. Trans JSME (C) 55(518):2632–2636

Denkena B, Scharschmidt K-H (2009) Modellbasierte Temperaturkompensation für Werkzeugmaschinen. ZwF 104-9:698–702

Klocke F, Rehse M (1997) Intelligent tools through integrated micro systems. Prod Eng 4-2:65

Lindberg B et al (1983) Measurement of segmentation frequency in the chip formation process. Ann. CIRP 32-1:17

Rucker J (2005) Optimalem Schleifprozess ein Stück näher gekommen. ZwF 100-11:661–663

Shinno H, Hashizume H, Yoshioka H (2003) Sensor-less monitoring of cutting force during ultraprecision machining. Ann. CIRP 52(1):303–306

Spur G, Leonards F (1975) Sensoren zur Erfassung von Prozesskenngrößen bei der Drehbearbeitung. Ann CIRP 24(1):349–354

Chapter 7
Concept of "Platform", Its Application and Modular Tooling System

As will be clear from a quick note for the *platform* in Chap. 2, the *platform* is an outstanding concept from the viewpoint of the user-oriented modular design. As widely known, the modular design is one of the leading methodologies and skills for the engineering design in machine tools since 1930s, although changing its term to be compatible with the engineering environments on each occasion. In fact, we used to call it by the terms, "*Unit Construction*" between in the 1930s and 1950s, "*Building Block System* (BBS)" between in the 1950s and late 1960s, and "*Modular Design*" since 1970s.

On the strength of our long-standing academic research and engineering experience, we can establish a rational system for the modular design at present, i.e., "*Different Kind-Generating Modular Design of Hierarchical Type*", and benefit considerably from it. Importantly, in accordance with the user's myriad requirements, the machine tool manufacturer must produce the product with various functionalities, and also performance and dimensional specifications' by using the modular design. It is however worth suggesting that the modular design at present is on the basis of engineering knowledge and concerns for the traditional machine tool, i.e., those obtained from the machine tool with limited machining function, to a large extent (Ito 2008).

With the advent of (multiple) machining function-integrated kinds, e.g., "*Mill-Turn*", which is a synergy of the turning centre (TC) and the machining centre (MC), the machining sceneries have been changed drastically beyond our expectation. In fact, we can observe a handful of contrivances and remedies to be compatible with such new sceneries, and in addition to it, the kinds of machine tools reduce considerably (see Figs. 1.10 and 1.11).

In retrospect, TC and MC is the first stage of the integration of the machining function. Figure 7.1a, b display obviously that a TC or an MC can replace efficiently and effectively various machining methods carried out by a group of traditional machine tools. It is thus natural that TC and MC play now the leading role in machining, and following these kinds, the grinding centre (GC) and gear production centre are growing up.

Y. Ito and T. Matsumura, *Theory and Practice in Machining Systems*,
DOI 10.1007/978-3-319-53901-0_7

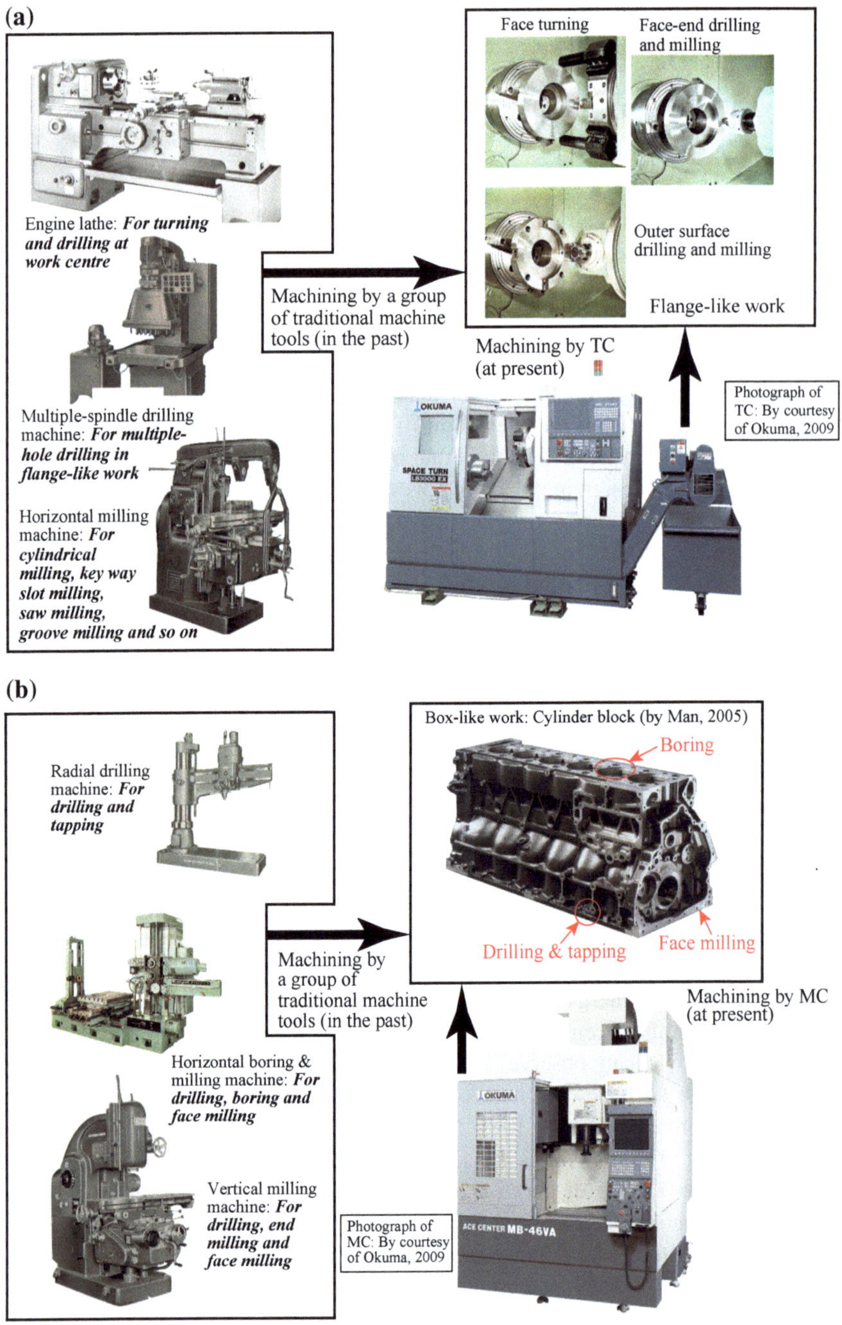

Fig. 7.1 Integration of machining functions of traditional machine tools to TC and MC: **a** In case of TC. **b** In case of MC

In consideration of the necessity of the modular design, there is however a serious discrepancy as follows.

> TC and MC are of typical machining function-integrated kinds, and thus have satisfactory versatility and flexibility in the machining methods. In contrast, the modular design provides the machine tool with the versatility and flexibility in the machining method; however, even TC and MC are modular-designed, if necessary.

In addition, the modular design at present has been, as already suggested, established on the basis of the structural design for the traditional machine tool; however, we have applied it to NC machine tools and developed some new technologies so far.

Importantly, we must understand the necessities of the modular design even in the *mill-turn*, and also the demands for the *platform* newly arisen. The *platform concept* is not new, but already proposed in the early 1990s; however, with the growing importance of the machining function-integrated kinds, the *platform concept* is going to play a leading role in the design of their structural configurations. Paraphrasing, we may consider it a new "Raison d'être" for the modular design. Intuitively, the *platform concept* is for the machining space-oriented modular design, and thus has certain potentialities to enhance the beneficial aspects of the user.

In this Chapter, some quick notes will first given for the modular design at present, and then a "*Platform-based Unit Construction*" is conceptualised by investigating some forerunning trials for the *platform concept*. To advance the *user-oriented platform concept*, we will discuss the applicability of such the concept to the machining function-integrated kinds, and also suggest some core research and engineering development subjects. Within such subjects, furthermore, the modular tooling system will be discussed to deepen the understanding for the *user-oriented platform concept*.

Conceptually, the *user-oriented platform* will be able to apply the design of another kinds for remanufacturing and localised globalisation era, and thus a quick note will be stated later (Ito 2011).

7.1 First-Hand View of Whole Concept for Modular Design at Present

Figure 7.2 shows a whole concept of the "*Different Kind-Generating Modular Design of Hierarchical Type*". This concept has been established on the basis of two outstanding achievements (Brankamp and Herrmann 1969; Koenigsberger 1974) together with considering some forerunning trials in the industry, e.g., those of Ikegai Iron Works in 1962 and VEB of Karl-Marx-Stadt in the 1970s. As can be readily seen, we can deploy various modular design, depending upon to which layer the basic module belongs, e.g., either unit or unit complex layer, and also to what

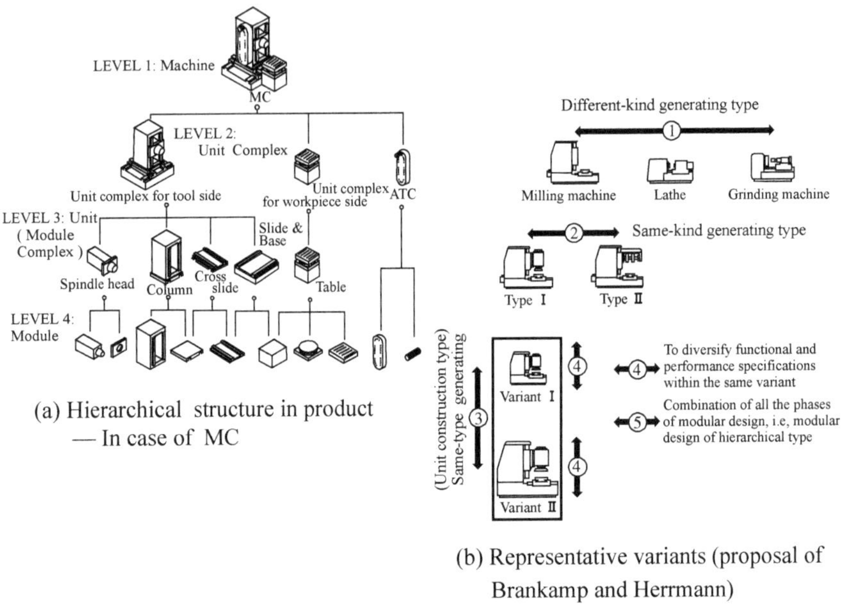

Fig. 7.2 A whole concept of modular design—different-kind generating type with hierarchical structure

extent the modular design should be ranged, e.g., either within the same kind or within the same type.

In consideration of the technological difficulties, economisation, practical applicability and so on, reportedly, the modular design of unit type has been and is being prevailed, and accordingly, such the unit construction can be defined as follows.

> Once a group of the basic units can be pre-determined, we can produce a considerable number of the variants, which have multifarious functional, performance and dimensional specifications', by combining these units in accordance with user's requirements.

Eventually, we need to have certain design guides to create the variant on the strength of this definition. In this context, Doi of Toyoda Iron Works is credited to the first proposer of "*Four Principles for Modular Design*" in the 1960s, i.e., *Principles of Separation, Unification, Connection and Adaptation* (Doi 1963). Having in mind the importance of the proposal of Doi, Ito modernised and detailed the contents of each principle in his proposal together with discussing the state-of-arts after then (Ito 2008).

In short, the "Principles of Connection and Adaptation "have been sublimated to the corresponding engineering spheres, i.e., "*Machine Tool Joint Problems*" and "*Machine Tool Description*" (see Chap. 2), respectively, to some extent. In contrast, "*Principles of Separation and Unification*" have not been established in the form of some design methodologies and engineering guides yet. Importantly, at burning issue is to seek a clue for remedies with special respect to "*Principle of Unification*". More specifically, the "*Principle of Unification*" can facilitate to determine and form a group of the basic modules, which is an essential prerequisite in conducting the modular design in practice.

In the meanwhile, "*Four Principles for Modular Design*" can be regarded as the implicit design flow of the modular design, which is even in 2013 at issue. For example, Schuh et al. of Technical University (TH; Technische Hochschule) Aachen has investigated such a design flow within a project of German Federal Ministry of Education and Research (BMBF), and it consists of the following four phases (2013).

1st phase: Identification of attributes, which can clarify the application range of the modular design to differentiate the product from those made by the competitor and give rise to the product simplicity.
2nd phase: Possibility of module standardisation and clarification of characteristic features of the product to be created.
3rd phase: Creation of much more suitable structural configurations with a less number of the modules.
4th phase: Embodiment design for structural configuration together with formation of organisational structure for the management.

Although Schuh et al. consider the incorporation of the supply chain management (SCM), the original equipment manufacturer (OEM) and manufacturing in subsidiaries, in principle, their proposal appears not to be new, but warmed-up as compared with that of Doi.

7.2 Some Practical Applications of "*Platform Concept*"

Admitting that the "*Different Kind-Generating Modular Design of Hierarchical Type*" has been established to some extent, we must be aware that the *platform concept* becomes to prevail with growing importance of the machining function-integrated kind. In retrospect, Gleason Pfauter Hurth conducted a fore-running application of the *platform concept* to the production of a series of the hobbing machine, gear shaping machine and gear grinding machine in the early 1990s (Metternich and Würsching 2000). After then, the *platform concept* was applied to the "*Mill-Turn*" and MC as shown in Figs. 7.3 and 7.4. In the former case, the *platform* is of monolithic structure by integrating the base, bed, and so on, and some variants can be produced by assembling the headstock, turret head and milling head of tilting spindle type, and if necessary the feed unit, with the platform in accordance with the user's requirements.

To understand in detail, Fig. 7.4 shows an application of *platform concept* to MC of Grob-brand (Dreer 2008). Importantly, there are three basic types in accordance with different dimensional specifications', and a platform in each basic

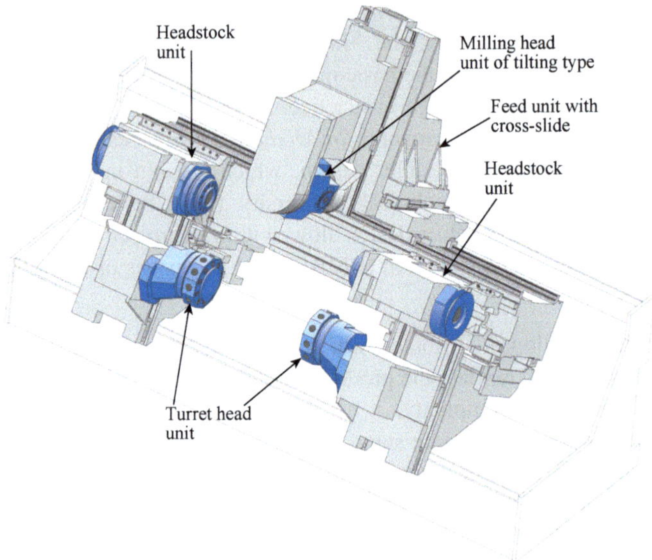

Fig. 7.3 "Platform Concept" in "Mill-Turn"—Type TNX 65/42 (by courtesy of Traub 2009)

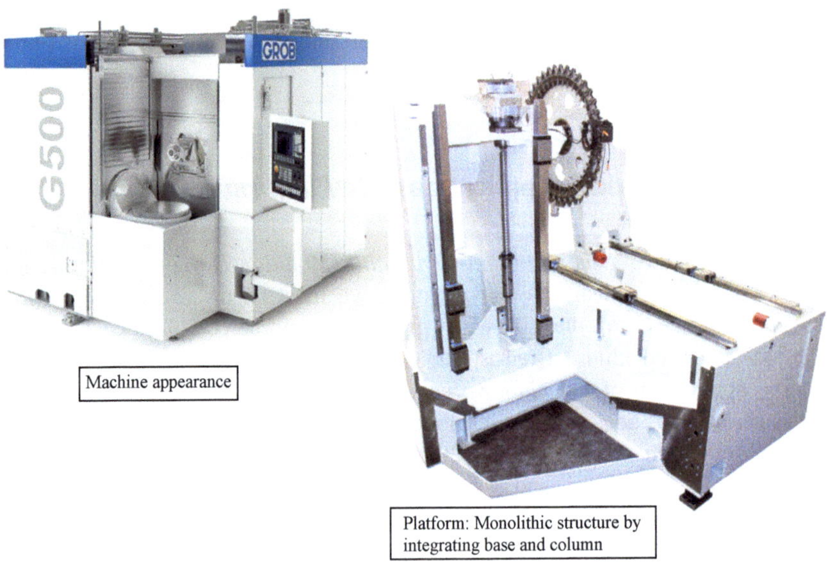

Fig. 7.4 "Platform Concept" in MC—Grob 2008

type consists of the monolithic base-column unit complex of steel welded structure. We can thus produce a considerable number of the variants by integrating the spindle head (single- or twin-spindle with two different tapered holes; either

HSK-A63 or HSK-A100) and table (single- or twin-type) with the *platform*. In addition, the spindle is of modular design, where the speed range is between 6000 and 18,000 rev/min, and maximum allowable torque is between 34 and 1270 Nm.

All MCs are quinaxial NC controlled and for 5-face machining, especially aiming at the motorcar component machining. More importantly, the machine can be characterised by its form-generating movement, i.e., movement of spindle head being X- and Z-axes, whereas table of bridge type being Y-axis, respectively, where the table travels within the column of two-pillar type.

It is very interesting that Kellenberger has applied the *platform concept* to the cylindrical grinding machine in 2013 as shown in Fig. 7.5. As can be readily seen, we can produce both the universal and production types by assembling the grinding wheel headstock, work spindle headstock, tailstock and table with the platform. It is worth suggesting that the grinding machine manufacturer has not employed so far the modular design apart from that of Schaudt in 1965. This is because the grinding machine is for work finishing, and thus there are, in general, no necessities to employ the modular design. In addition, Mitsui Seiki and Wera have recently applied the *platform concept* to produce the vertical thread grinding machine and gear cutting machine, respectively.

Reportedly, there is another machining function-integrated kind, i.e., compact flexible transfer line (FTL), which is applicable to the batch-like production in motorcar industry. The compact FTL is one of the variants of flexible

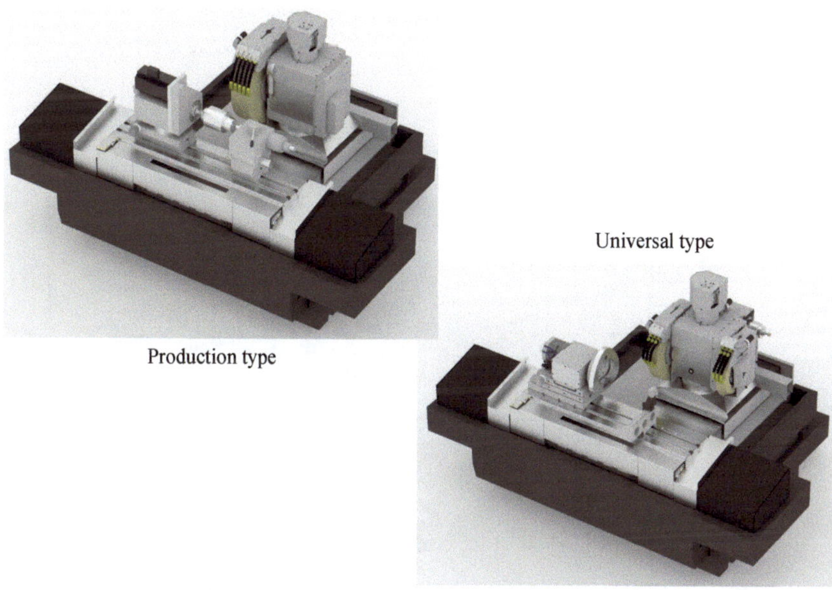

Fig. 7.5 "Platform Concept" in cylindrical grinding machine (by courtesy of Kellenberger 2014)

manufacturing cell (FMC)-integrated flexible manufacturing system (FMS), and as literally shown, of machine-like structural configuration such as the rotary indexing machine. In fact, the compact FTL can be classified into three types, i.e., MC-integrated type, advanced head changer (*Commercial name: Transfer Center*, see Fig. I.4), and rotary indexing machine. Of these, we can suggest that in MC-integrated type of Icon-make in 2011, the main structural body can be regarded as the *platform*, and the machining space consists of a group of the machining units with quick changing function as shown in Fig. 7.6. More specifically, the machining space can allow to place up to 8 three-axis controlled machining units. Thus, the machine appears as to be a way machine.

From these forerunning trials mentioned above, it may be concluded that the *platform concept* has been employed extremely aiming at the design rationalisation for machining function-integrated kind and also emphasising the "*Modular Design Specified to Machining Space*". Paraphrasing, we can expect that the machining function-integrated kind may render the modular design useless as already mentioned above, because of its much wider flexibility in the machining method than ever before. It is however very interesting that we need the *platform concept* even in such a kind to meet neatly the user's requirements. More specifically, the user does not like to purchase the expensive machine with over-specifications, which results in the redundancy in functionality and performance.

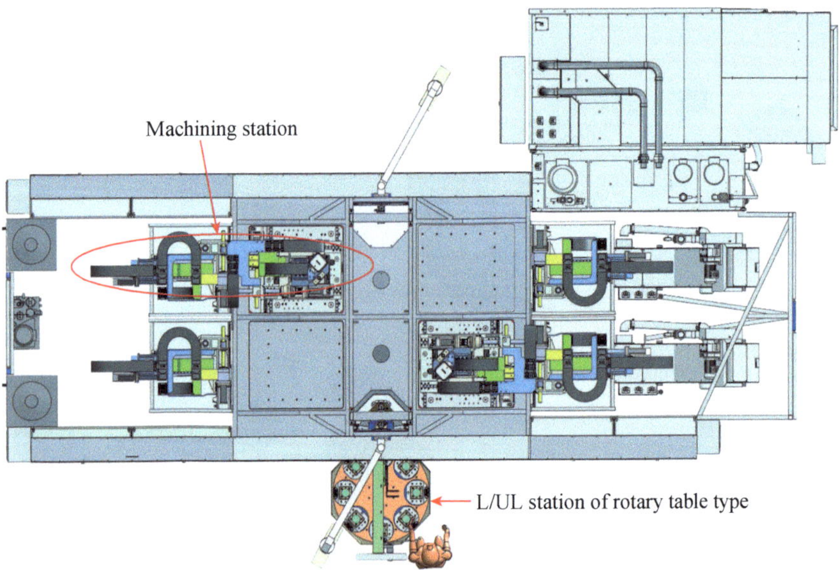

Fig. 7.6 MC of multiple-station type—Type Icon 6-250 (by courtesy of Icon 2012)

7.3 Concept of "Platform-Based Unit Construction"

On the basis of the forerunning trials mentioned above, a whole concept of "*Platform-based Unit Construction*" is proposed herein as shown in Fig. 7.7, which is especially available for machining function-integrating kinds ranging from TC and MC including that with grinding function, through GC and gear processing centre, to "*Mill-turn*" (see Sect. 2.2.2). Importantly, this concept can be characterised by the obvious separation between the platform and a group of the structural configuration entities, i.e., structural body components for form-generating movement (see Chap. 4). In fact, the feed unit can facilitate the intermediation between the platform and the structural configuration entity by regulating the number of the feed axes as like as the adapter in the transfer line (TL). In due course, the feed unit might be driven by the double-pinion mechanism to ease the adaptation of the structural configuration entity rather than that with the ball screw.

In the "*Platform-based Unit Construction*", it is furthermore necessary to clarify the advantageous aspects in detail. For example, TC of opposite allocated twin-spindle type is capable of driving both the spindle simultaneously without any speed difference as well as supplying the same driving force like the car wheel lathe (see Fig. 1.3b). Obviously, such a driving method does not apply any torsional load to the work, and results in the better machining accuracy especially in the case of the slender work, which deforms easily by torsional load. Importantly, NC technology eases the simultaneous driving of both the spindles without any differences in rotational speed, rotational angle and driving force.

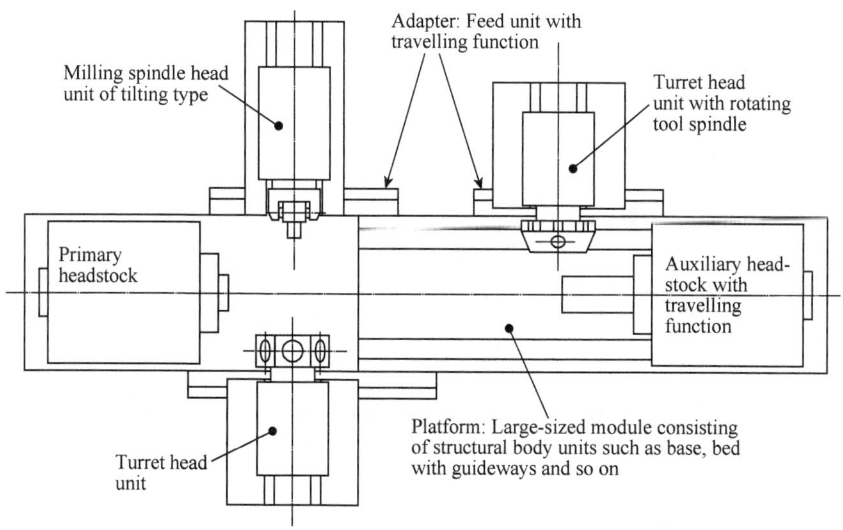

Fig. 7.7 Concept of "Platform-based Unit Construction"—a group of structural configuration entities forming basic machining space

As can be readily seen, the "*Platform-based Unit Construction*" may be regarded as "*Different-kind Generating Type of Unit and Unit Complex Layers*" in Fig. 7.2, and in retrospect, such a modular design was once investigated, provided that the structural configuration entity can be varied, while other components maintain their fixation states (Shinno and Ito 1987). As a matter of course, the definition of the traditional unit construction already stated elsewhere is available, provided that we allow the settlement of the large-sized module, i.e., *platform* within the concept. Of special note, an advanced "*Platform-based Unit Construction*" was conceptualised as shown in Fig. 7.8 (Abele and Wörn 2004). In their proposal, the module for form-generating function is first defined by the combination of the work mounting phase and movement function phase, after classifying broadly the work into the rotational and box-like shapes. Then, they consider the technology phase module and tool layout phase module as same as those in the traditional modular design. In short, Abele and Wörn determine first the form-generating movement module and then detail such a functional module to the corresponding structural modules.

Obviously, that of Abele and Wörn is noteworthy in some extent; however, in paving a way to its fruition, a root cause of difficulties lies in the absolute conversion procedure from the functional module to the structural module, i.e., "*One-to-one Conversion Problem*" (see Chap. 2). Substantially, one functional module can be converted, at least, a handful of structural modules as already suggested

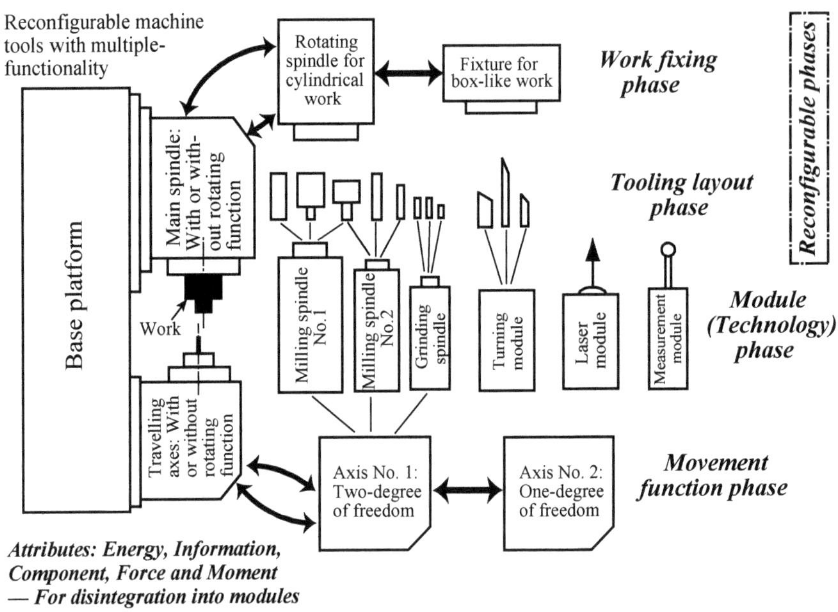

Fig. 7.8 Reconfigurable phases in machine tool structure and details of each phase (by Abele and Wörn)

elsewhere; however, they did not state anything about this problem. Reportedly, we have the same problem in the computer aided process planning (CAPP), where the geometrical information on the drawing should be converted into the manufacturing-related ones. Conceptually, the former and latter correspond with the functional and structural attributes, respectively.

Now let us discuss potential variants in the "*Mill-Turn*", which are capable of producing on the strength of the "*Platform-based Unit Construction*" together with suggesting the corresponding research and engineering development subjects necessary to carry out for their establishment.

7.3.1 Variants for Practical Use—in Case of "Mill-Turn"

There is two-pronged way to discuss the potential variants possible to produce from the concept shown in Fig. 7.7. One is to discuss the structural configuration obtainable from the combination of the conventional structural configuration entities as follows.

(1) Type I: Combination of the principal headstock, milling spindle head and turret head.
(2) Type II: Combination of the principal and auxiliary headstocks and milling spindle head.
(3) Type III: Combination of the principal and auxiliary headstocks and turret head with tool rotating mechanism.

The other is to discuss the variation of the structural unit by facilitating it with various functionalities as follows.

Auxiliary Headstock (Secondary Headstock)
The auxiliary headstock may be, in certain cases, converted into the tailstock. It is thus preferable to provide it with mechanised barrel driven by electric, pressurised air or oil hydraulic power in addition to the travelling function of the headstock itself. This function may assist the long shaft machining with better accuracy. It is furthermore desirable to employ the chuck of modular type, although such a chuck is rare at the moment as already discussed in Chap. 5 (see Fig. 5.38). In nearly all cases, we can find the chuck with modular type of preliminary stage as exemplified by the following.

(1) Quick changing systems of top jaw as shown in Fig. 5.13.
(2) Quick converting systems of top jaw from external to internal gripping as shown in Fig. 5.11.
(3) Milling chuck with extension adapter (Micro-chuck Extensions) for the small end mill of Albrecht-brand (2012).

Milling Spindle Head

The milling spindle head at present remains that of universal head in MC, and thus its functionality should be preferably enhanced by enabling the small-sized turret and cassette for work holding to be accommodated. In this context, we may recall the special work fixture shown already in Fig. 5.3, and also recommend certain technology transfers from the traditional TL of rotary indexing type, i.e., hanging square turret for work holding as shown in Fig. 7.9. In addition, we must eye the attachment manufacturer, e.g., Rückle Group, who provides us with the milling head of quick changing type, e.g., 2-axis simultaneous control milling head of tilting type. Obviously, such milling heads can be mounted to MC and ram of 5-face processing machine.

Turret Head

The turret head can be classified into "Polygon", "Drum", "Disk" and "Conical" types. Of these, the polygon type has been prevailed in the "*Mill-Turn* "(for further detail, see Chap. 4). The turret head itself is, in general, one of the basic modules; however, the tool seat is not standardised as yet. In fact, there are various tool seats, e.g., tapered hole for HSK and BT, dove tail type, flange type with locating shoulder or tenon slot and flange type with recess and T-nut. Generally speaking, the machine tool designer chooses the preferable tool seat in accordance with the product deployment of the attachment manufacturer. From the viewpoint of the

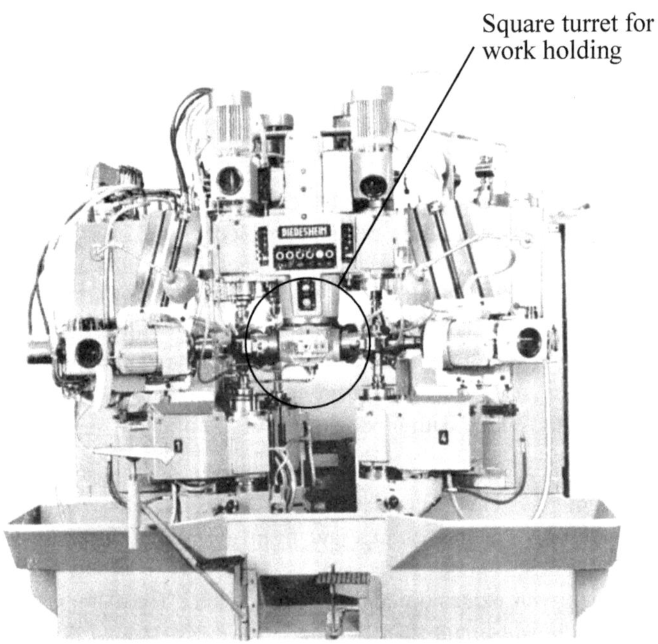

Fig. 7.9 Rotary indexing machine of trunnion type—Type AM, Diedesheim-brand 1960s

modular design, it is recommendable to regulate the structural configuration and dimensions of the tool seat as like as the spindle nose in TC.

To this end, it is worth suggesting a percussion effect of the "*Platform-based Unit Construction*" on remanufacturing. More specifically, we can produce the "*Mill-Turn*" with functionality for the self-innovation in specifications, for example, by replacing some structural units with the up-to-date and re-use ones. Of note, such a proposal was already publicised in 1989, i.e., an idea for remanufacturing of machine tool (McMaster 1989).

7.3.2 First-Hand View for Facing Research and Engineering Subjects

We can first suggest that the "*Platform-based Unit Construction*" should have the better adaptability with the attachment and tooling of modular type. Although having been no proposals, such a total modular system in the machining space is capable of enriching and enlarging the obtainable benefits. In consideration of such a precondition, we will discuss the leading research and engineering development subjects in the following.

(1) Establishment of configuration generating methodology
Although "*Principles of Separation and Unification*" become simple, a facing problem is, on the strength of the machine tool description, to establish the configuration generating methodology for the variant, which is capable of changing the corresponding structural configuration entity at user's factory. In certain cases, the methodology must determine the priority order among design attributes with special respect to the machining method to avoid unnecessary complexity.

(2) Choosing methodology for a preferable configuration of machining space
In the modular design, a root cause of difficulties lies in the final choice of the preferable configuration from a handful of outputs, although the "*Platform-based Unit Construction*" reduces such a burden to some extent.

(3) Preferable structural design for platform
It is vital that the platform is designed and manufactured in consideration of multifarious cutting force vectors (magnitudes and directions of loading) and various heat sources and their magnitudes. In principle, we must use positively the "*Directional-orientation Effects in Rigidity*", although having considerable difficulty. More specifically, the platform should be designed by leveraging the reciprocal attributes and by considering differing resultants of the cutting force and their acting directions. We can see such a structural design in an MC of Hermle-brand as shown in Fig. 7.10. As can be readily seen, the resultant of

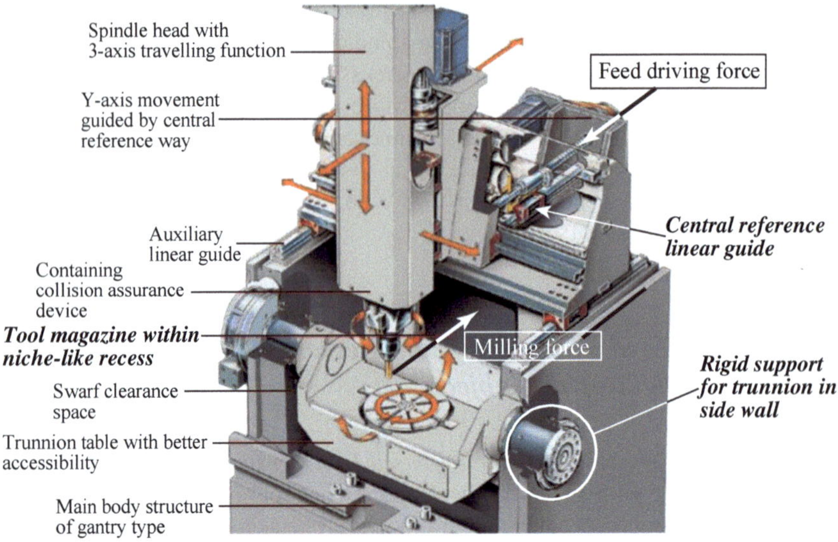

Fig. 7.10 Characteristic features of structural design in quinaxial-controlled MC—Type B 300 (by courtesy of Hermle)

cutting force in milling directs to the niche-like recess in the column, and also is in the balance with the feed force in the same vertical plane. In addition, the better guiding accuracy of the cross-beam can be facilitated by the three-way linear guide, in which the central way placed very closely to the feed screw regulates the guiding reference.

Figure 7.11 implies furthermore the difficulty of the structural design of the platform, i.e., linkage of structural bodies surrounding the machining space, because its processing capability ranges from drilling, through jig grinding, to co-ordinate measuring, which corresponds with the machining and measuring accuracies ranging from 0.1 mm to 0.1 μm. Obviously, we cannot establish the desirable structural design without appointing the well-experienced and talented chief designer.

(4) Development of multiple function-integrated unit

To enhance the beneficial aspects of the "*Platform-based Unit Construction*", it is desirable to develop a function-integrated structural unit. Figure 7.12 reproduces a feed unit for the internal grinding machine in the past, and as can be seen, the guideway can be facilitated by the outer surface of the hydraulic cylinder, whereas the piston placed inside of the cylinder provides the feed driving force. Of special interest, the cylinder should be for the reference of the better guiding accuracy, and thus a crucial engineering problem is to suppress the thermal deformation of the cylinder within the allowable magnitude.

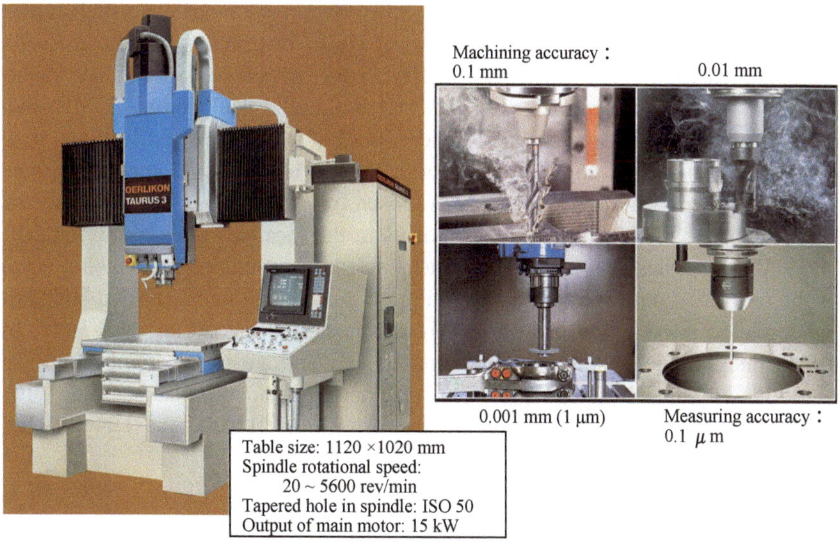

Fig. 7.11 Multitechnology Centre—"TAURUS 3" (by Oerlikon 1984)

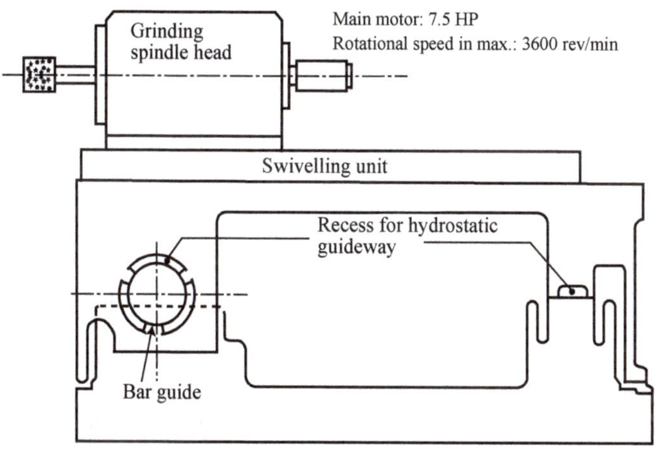

Fig. 7.12 An example of function integrated unit—bar guide by using outer surface of feed driving cylinder in internal griding machine (Type ICF, Heald-brand in 1960s)

To this end, it is worth suggesting that the "*Platform-based Unit Construction*" has higher potentiality to be compatible with the advanced kinds and also with the local region-oriented "*Mill-Turn*" in the year 2010 and beyond. As exemplified by an Asian region-oriented "*Mill-Turn*", we must extremely investigate such further possibilities of applying the "*Platform-based Unit Construction*" to those for the small- and medium-sized enterprise (SME) across the whole world. SME cannot afford, in general, to purchase TC and MC with full specifications. SME must however install them with acceptable prices, but not with over-specifications, so as to overcome the fierce competition in both the world and the domestic markets. Accidentally, an idea has been already proposed for the modular design, which will be applicable to the machining environments in the localised globalisation era (Ito 2011). Within Asian region-oriented kinds, we must remember the growing needs for the retrofitted machine tool with keen price in Mainland of China and Vietnam.

Of special interest, we must be aware of the new horizons in "*Mill-Turn*" by further integration of both the functions of GC and gear processing centre as exemplified by the advanced "*Mill-Turn*" of R series of Index-brand (see Fig. 1.10). In fact, the "*Mill-Turn*" of Index-brand shown in Fig. 7.13 is quinaxial-controlled and can be characterised as follows.

(1) Application of the "*Platform-based Unit Construction*".
(2) A variant of twin-spindle type by off-set allocation of 1st and 2nd spindle heads and also by placing two milling heads of independent driving system.
(3) To guarantee the necessary function and performance, the milling head can travel by the bar-guide of hydrostatic type.
(4) Grinding is a primary form-generating function as well as turning and milling, and the machine is capable of cylindrical, internal and surface grinding.
(5) The milling cutter of annual circular type can facilitate spiral bevel gear machining (ranging from 0.6 to 4 in Module) from bar stock, which is compatible with the tooth profile of Klingelnberg Zyklo-Palloid, so that the interchangeability can be guaranteed to some extent.

In addition, Monforts has merchandised the quinaxial-controlled "*Mill-Turn*" with laser processing (Commercial name: Hybridmaschine RNC 400 Lase Turn) in 2014 (Monforts 2014). The machine can be especially characterised by the hydrostatic guideway, which guarantees 10-year long durability.

Paraphrasing, a concept of "One-machine Factory" is to be in reality to certain extent on the basis of such an advanced "*Mill-Turn*", although the structural design of the platform will become much more complex than those for MCs shown in Figs. 7.10 and 7.11. Obviously, it is very important to apply positively the concept of the linkage diagramme for the machining space to the design for such an advanced "*Mill-Turn*" (see Figs. 2.24 and 2.28).

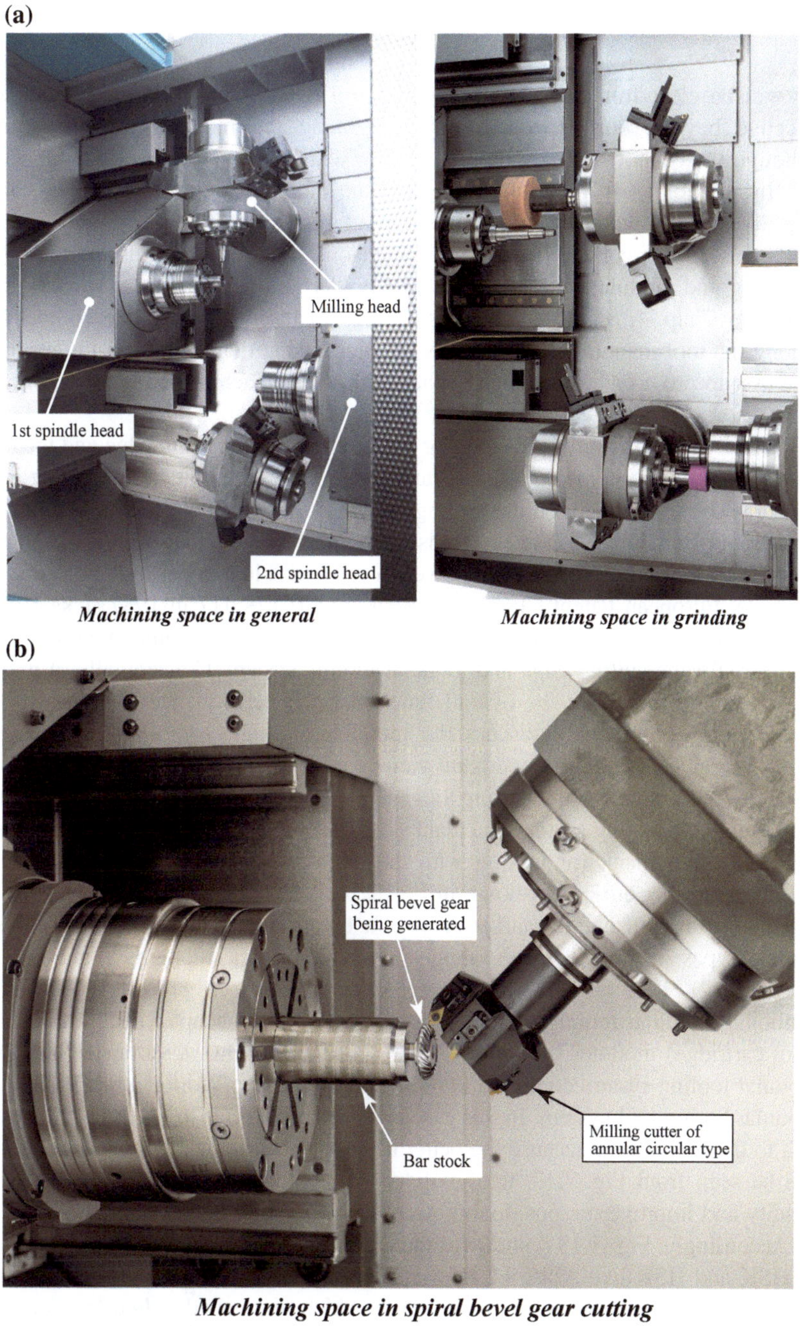

Fig. 7.13 Machining spaces in "Mill-Turn" capable of grinding and gear cutting: **a** Machining spaces in general and in grinding. **b** Machining space in spiral bevel gear cutting (R Series, by courtesy of Index 2016)

7.4 Modular Tooling System

As will be clear from the above, the *platform concept* will become much more effective beyond our expectation by integrating the modular tooling system, although we cannot find any reports dealing with such an engineering subject as yet. In addition, it is very interesting that the modular tooling system has, accidentally or incidentally, been developed to a large extent.

Against to this context, we are now in chaotic situation in discussing the quick changing and modular systems in the tool holder and cutting tool as already suggested in Chap. 4. Supposedly, this situation is caused by the fierce competition among the tool manufacturers, and also the increasing demands from the machine tool user to provide the machining space with wider machining flexibility, but not costly.

Actually, both the quick changer for throw-away tip or cutting edge and the total modular tooling system are at issue, and dare to say, the matters of cutting edge concern, i.e., simplified modular tooling system, is dominant at present. In addition, the kernel of modular tooling system is, without any doubt, that of quick changing mechanism, and in due course, there were a considerable number of the patents, for example, European Patent Claims 0 169 543 and 0 178 417 in 1987 (Icks 1989).

For the ease of understanding, Fig. 7.14 demonstrates some popular quick changers for the cutting edge and modular tooling system. Conceptually, it is thus preferable to grasp once the overall functional and performance attributes of the cutting tool, and then to discuss the modular tooling system. In this context, Figs. 7.15 and 7.16 are the first-hand views for all the functional and performance attributes in the single-point cutting tool and drill. For example, in the establishment of the modular drilling system, the main stress should be placed on the cutting edge, drill body and shank for joining with the main spindle together with implicitly considering other attributes shown in Fig. 7.16.

On the strength of the market survey, it emphasises that there are now three leading types in the modular tooling system with quick changing function, i.e., (1) total, (2) HSK, and (3) cutting edge-oriented types. In general, a total (conventional) modular tooling system consists of the shank, adapter and cutting edge (tool cartridge) modules (see Fig. 4.11). Figure 7.17 shows a variant of the total modular tooling system contrived by Nikken Kousakusho Works in the 2000s, i.e., modular boring tool system. In this case, a facing problem is to fix the cutting edge unit to the tool shank, because the better machining accuracy is required of, and as can be seen from Fig. 7.17, the auxiliary support block may increase the fixing rigidity and improve the positioning accuracy of the cutting edge.

Accordingly, Fig. 7.18 reproduces three examples of the modular tooling system of HSK and HSK-like types, i.e., those of application of DIN, and also of Krupp Widia-brand in the beginning of the 1980s, which was called "*Multiflex Coupling*

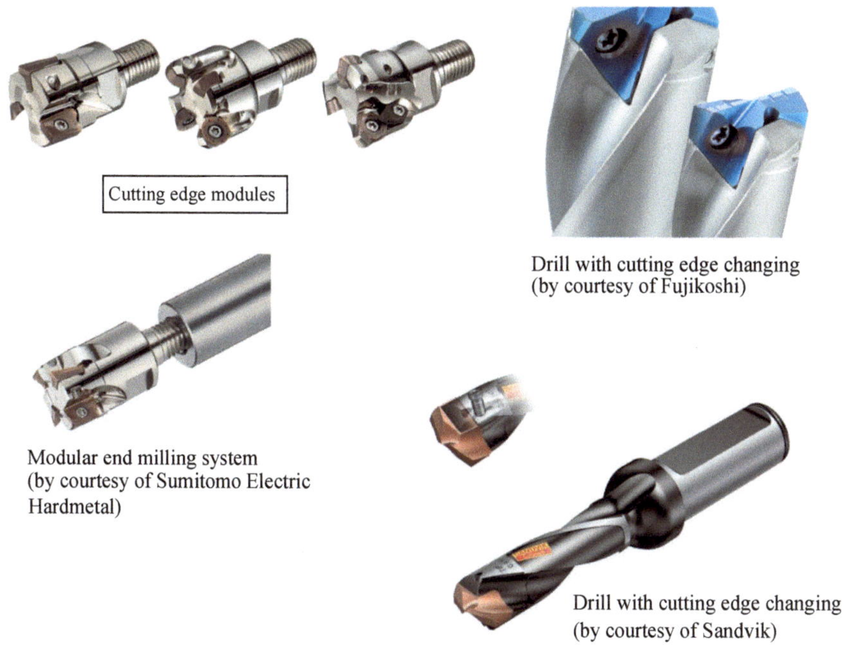

Fig. 7.14 Modular tooling system and quick changing cutting edge around 2015

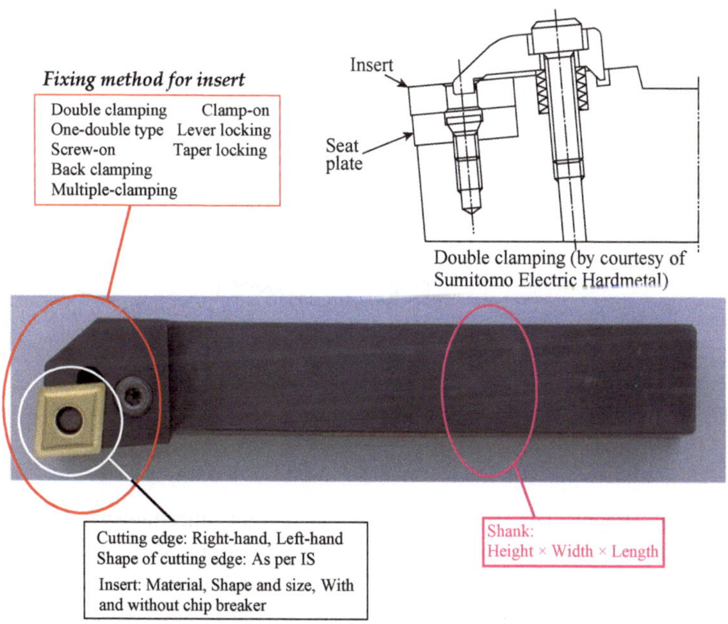

Fig. 7.15 Functional and performance attributes in single-point cutting tool

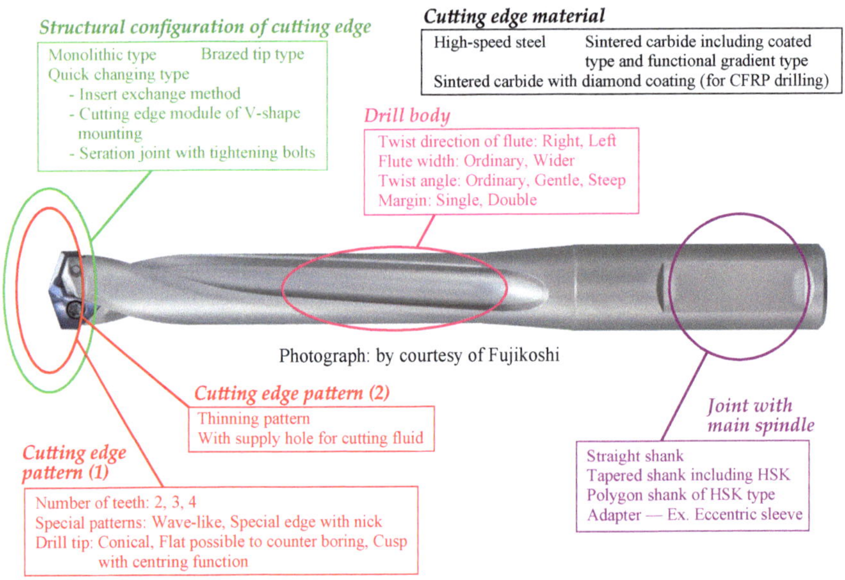

Fig. 7.16 Functional and performance attributes in twist drill

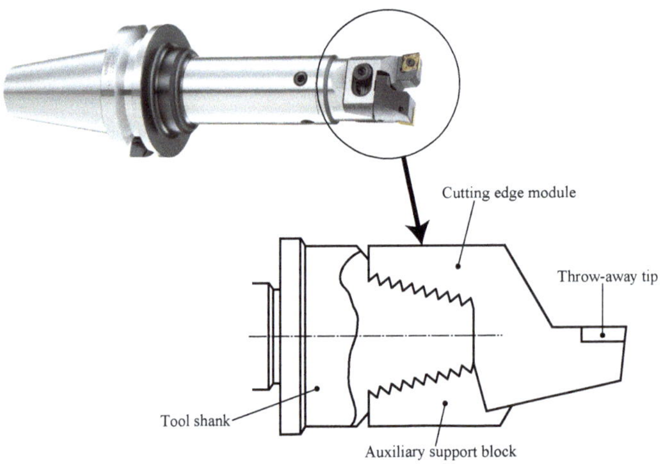

Fig. 7.17 An innovative contrivance for joining cutting module to tool shank (by courtesy of Nikken Kousakusho Works)

and Locking" (Pegels 1987). Although the exchanging mechanism of the cutting edge becomes complicated as compared to that of cutting edge-oriented type, the two surfaces-in-contact can obviously facilitate position setting of the cutting edge module within the allowable accuracy. Kennametal merchandises thus the

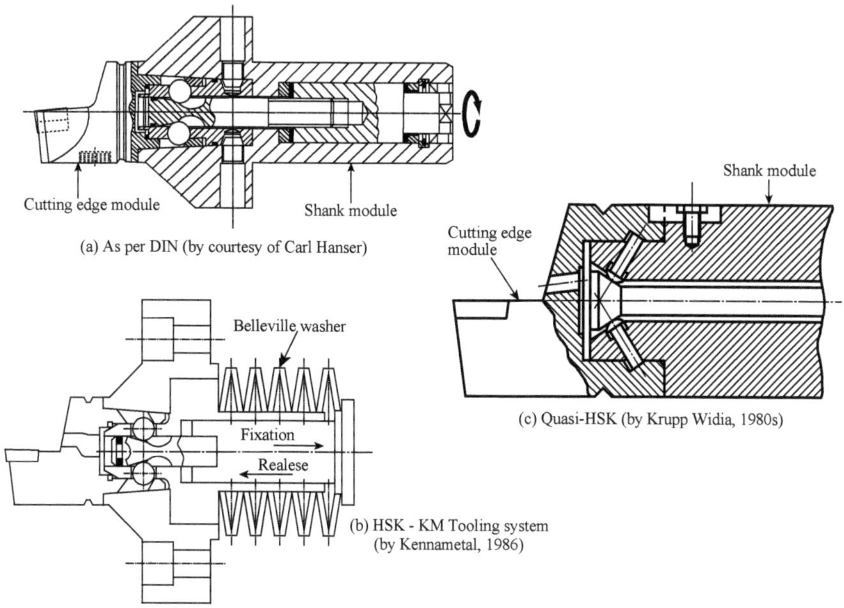

Fig. 7.18 Modular tooling system of HSK type

monolithic drill body module with HSK in the 2010s, in which the insert is quickly changeable.

As will be clear from these modular tooling systems, it is very handy and convenient to place the stress on the cutting edge module only in the establishment of the modular tooling system. In retrospect, Sandvik merchandised such a modular tooling system called the "*Block Tool System*" in the 1980s to apply it to FMC of robot type. Paraphrasing, the *Block Tool System* can be regarded as ATC in MC, and is duly very convenient to automatise the tool changing operation by using the robot. Figure 7.19 shows furthermore such a modular tooling system in the past to understand the jointing method between the cutting edge and the shank modules. In general, a core of quick changer is to use the wedge effect at the joint.

On the basis these earlier contrivances, the cutting edge-oriented type becomes dominant in the advanced modular tooling system such as shown in Fig. 7.20, and it is being developed with the advance of the cutting tool technology, especially those related to the handy management for the throw-away tip and insert in choice of tool material, tool life control, re-grinding, disposal and so on.

As can be readily seen from Fig. 7.20, there are various jointing methods for the cutting edges for end mill and drill. For example, that of Kennametal can be characterised by (1) the two surfaces-in-contact and (2) double cone surfaces combined with the third contact area in back, resulting in the larger rigidity of the cutting tool. In this context, those of Tungaloy and Kyocera do not indicate the thread profile; however, it is worth suggesting that the threaded joint appears as to

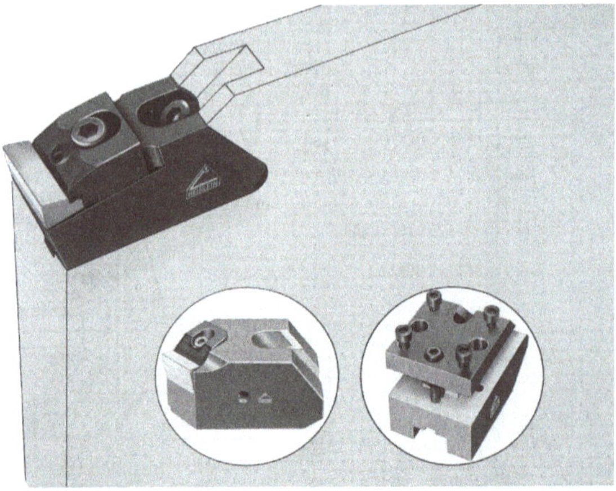

Fig. 7.19 Cutting edge module with quick changing function (Displayed at 3rd EMO Show by Sitzmann and Heinlein 1979)

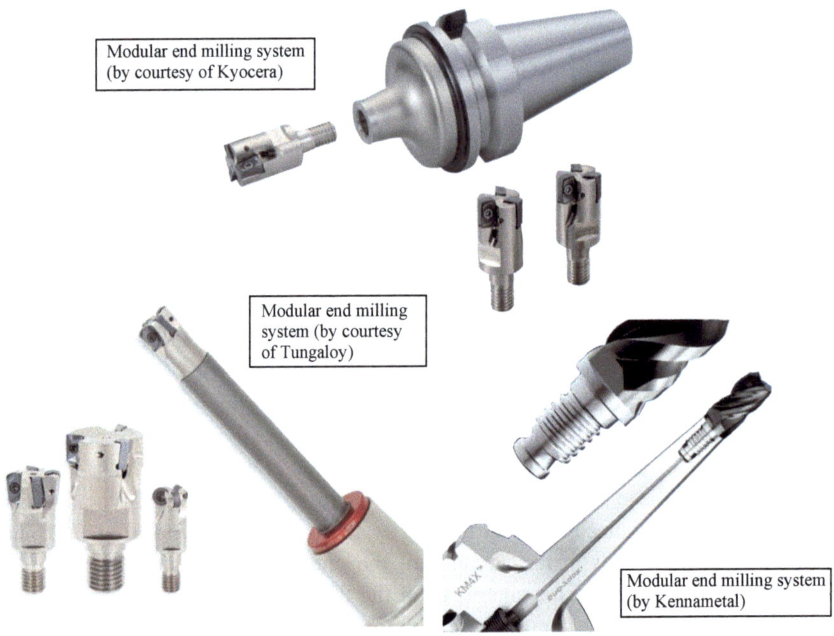

Fig. 7.20 Modular tooling system of cutting edge-oriented type around 2015

be of "*Buttress Type*". This thread is very famous by especially using in the defence supplies from the past and being used, for example, at the screw breech of howitzer (see Fig. 1.3c). The profile of "*Buttress Thread*" is saw-tooth-like and thus is ease of tightening, but hard to loosening.

For further understanding, there are another joining methods as follows.

(1) OSG-brand: Cutting edge is fixed in the form of two surfaces-in-contact way by using the threaded joint of "*Buttress Type*".
(2) Sumitomo Electric-brand: Cutting edge is fixed to the face-end of drill body by the serration joint with two-tightening bolt.
(3) Iscar: Cutting edge is fixed by the dove-tail with self-locking function.

Although the research into the modular tooling system is not active so far and at present, at least, we must investigate the effect of jointing method of the cutting edge module to the adapter or shank module on the static and dynamic behaviour. Importantly, the chatter vibration with higher frequency may be suppressed by placing the damping source closer to the cutting point as exemplified by Fig. 5.17. We can observe the same behaviour when placing the tool slide made of concrete closer to the cutting point in the turning machine. As widely known, the internal damping capacity of concrete is larger than that of cast iron. It is thus thought that we may benefit the increase of the anti-chatter capability from the modular tooling system of cutting edge-oriented type.

References

Abele E, Wörn A (2004) Chamäleon im Werkzeugmaschinenbau. ZwF 99–4:152–156
Brankamp K, Herrmann J (1969) Baukastensystemtik—Grundlagen und Anwendung in Technik und Organisation. Industrie-Anzeiger; 91–31: 693–697 und 91–50: 133–138
Doi Y (1963) On application of BBS. Toyoda Tech Rep 4–3: 22–32
Dreer R (2008) Modularity is the new standard. Eur Prod Eng: 55–57
Icks G (1989) Modular tools. J IProdE: 17–18
Ito Y (2008) Modular design for machine tools. McGraw-Hill
Ito Y (2011) A proposal of modular design for localized globalization era. J Mach Eng 11–3:21–35
Koenigsberger F (1974) Trends in the design of metal cutting machine tools. Proceeding of 1st International Conference on Production Engineering (ICPE), JSPE, Tokyo
McMaster P (1989) Renaissance in remanufacturing. J Inst Prod Eng: 23–24
Metternich J, Würsching B (2000) Plattformkonzepte im Werkzeugmaschinenbau. Werkstatt und Betrieb 133–6:22–29
Monforts (2014) Hybridmaschine, RNC 400 Laser Turn. ZwF. http://www.zwf-online.de/ta001/na201203201254240/pi
Pegels H (1987) Werkzeugtechnik für eine flexible automatisierte Fertigung. Werkstatt und Betrieb 120–10:875–878
Schuh G et al (2013) GiBWert—Gestaltung innovativer Baukasten- und Wertschöpfungssysteme. ZwF 108–11:813–817
Shinno H, Ito Y (1987) Computer aided concept design for structural configuration of machine tools: variant design using directed graph. Trans ASME, J Mech Trans Autom Des 109: 372–376

Part III
Practice and Theory at Machining Point

Chapter 8
Various Machining Methods and Tools in General

In Chap. 1, some representative metal removal processing methods are shown, where the emphasis is placed on the growing importance to understand the removal processing within the machine-attachment-tool-work system. Of special note, Fig. 1.2 is very helpful to grasp at glance the total view of metal removal processing; however, each processing is not detailed systematically. Thus, in this chapter, the fundamental knowledge of each processing will be given in consideration the tiers 3 and 4 in Fig. 1.2. In short, we used to classify the removal processing in accordance with the terms belonging to these tires.

Cutting is a leading processing method to form the part with removing the unnecessary allowance (undue proportion) in the raw material or semi-finished work. The material removal rate in cutting is relatively higher than those in other mechanical machining processes such as grinding and polishing. Importantly, the accuracy of the part is controlled easily in cutting; and cutting, therefore, is applied to manufacturing a wide variety of shapes as well as realising an economical machining manner. In due course, cutting has a handful of leading variants such as turning, milling, drilling and so on, and usually can be classified using these variants (see Fig. 1.5). Thus, we must first discuss their outlines to deepen our understanding.

8.1 Turning

Turning is, in principle, performed by the engine lathe, turning machine and turning centre, and for machining of cylindrical shapes as shown in Fig. 8.1. The examples of the cutting tools used in turning are also shown in Fig. 8.2. In turning operations, the insert tools have mostly been used to save the tool material costs and increase the machining efficiency. The cutting tool parameters, in general, are defined as shown in Fig. 8.3. Of special note, we must be aware that turning is the utmost

Y. Ito and T. Matsumura, *Theory and Practice in Machining Systems*,
DOI 10.1007/978-3-319-53901-0_8

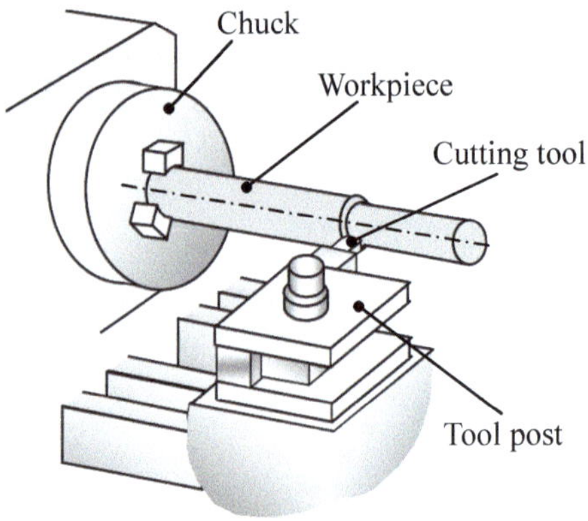

Fig. 8.1 Turning operation

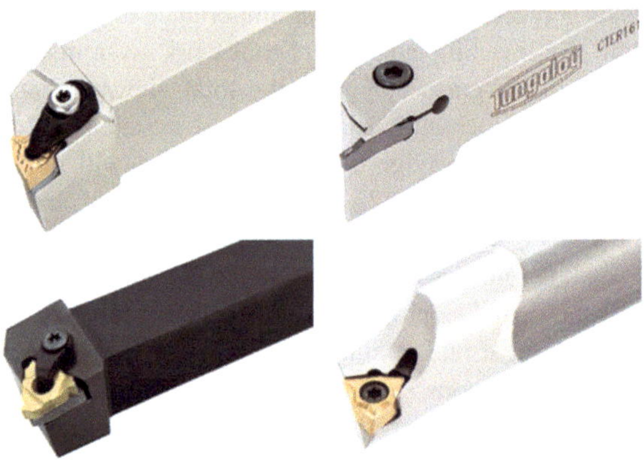

Fig. 8.2 Turning tools (by courtesy of Tungaloy)

fundamental within all the metal removal processing methods, and thus it is recommendable to learn it in detail.

The material is removed by travelling the cutting tool along the specified contour while the workpiece is rotating. The radius of the part is controlled by setting the relative position between the rotating central axis of the part and the tip of single-point cutting tool, which determines the removal shape at a cutting point. When the forming tool is used, the shape of the cutting tool is copied onto the workpiece. More specifically, turning involves facing, boring, drilling, grooving, parting-off, threading and knurling, and these are characterised quickly as follows (see Fig. 1.6).

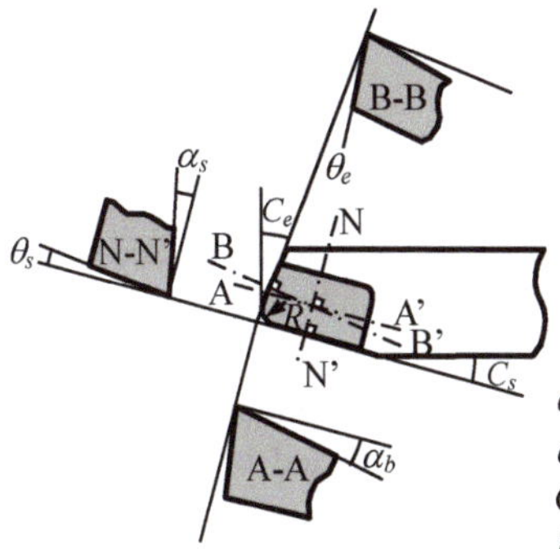

α_b: Back rake angle
α_s: Side rake angle
C_e: End cutting angle
R: Nose radius
θ_e: End clearance angle
θ_s: Side clearance angle
C_s: Side cutting angle

Fig. 8.3 Tool geometry

(1) Facing facilitates to produce a surface at the end of the part.
(2) Boring enlarges a hole machined by a previous process or produces internal grooves.
(3) Drilling produces a hole, which may be followed by boring.
(4) Parting-off, also called cutting-off, cuts a piece from a part as a whole.
(5) Threading produces external and internal thread in workpiece.
(6) Knurling produces a regular shaped roughness on cylindrical surfaces, so that the part has a gripping area for sure-handed use.

Figure 8.4 shows the typical turning process, in which a tool is fed in the workpiecc axis at a feed rate of f in a depth of cut of d (refer to Form-generating Movement in Chap. 2). When the tool removes the material in a workpiece of D in diameter, the cutting speed, V, is associated with the spindle speed of N as:

$$V = N \times \pi \times D \tag{8.1}$$

The removal volume rate, R_{vol}, is then estimated as:

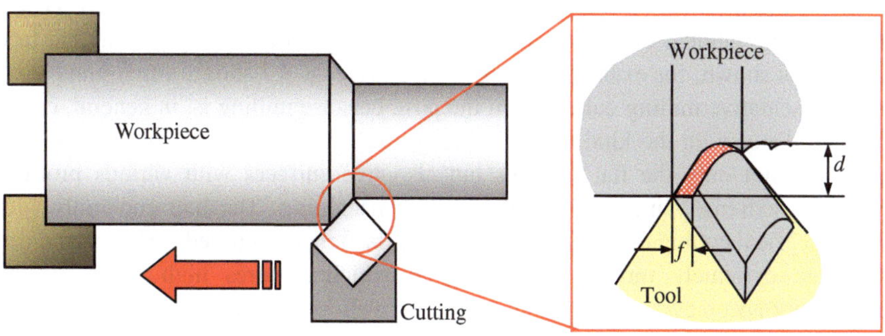

Fig. 8.4 Turning process

$$R_{vol} = V \times d \times f \quad (8.2)$$

Therefore, the production rate could be promoted with the increase of V, d and f. However, the tool life, which mainly depends upon the cutting speed, becomes short with increasing the cutting speed. On the other hand, the force loaded on the machining point mainly depends on the depth of cut and the feed rate. When a machining process is performed in a large depth of cut and/or at a high feed rate, the tool damage or the chatter vibration occurs; and it leads to deteriorate the machining qualities e.g., the machining accuracy, surface finish and affected layer in subsurface. Therefore, the cutting parameters should be determined considering the tool and the surface damages in the workpiece.

The cutting is classified into the steady and non-steady processes in terms of the change in the removal parameters, V, d and f, during cutting. When the turning is conducted in constant V, d and f in a cutting tool path as shown in Fig. 8.4, the material is removed in the steady process. For example, when the turning is conducted to produce a taper shape with changing the depth of cut, the material is removed in the non-steady process.

The turning operations are also conducted with repeating cutting and non-cutting, which is called as "interrupted cutting". Because the tool is not in constant contact with the workpiece, the tool damage promotes due to fluctuation in thermal and mechanical load.

8.2 Milling

Milling is to machine a variety of configurations with rotating a multiple-tooth tool, and can be, in general, classified into peripheral milling and face milling. In peripheral milling, the cutter is mounted on a main spindle of a machine as shown in Fig. 8.5a, and thus the axis of cutter rotation is parallel to the workpiece surface to be machined. Meanwhile, in face milling shown in Fig. 8.5b, the spindle axis is perpendicular to the workpiece surface.

With the increase of versatility in the machining requirement, we have now a considerable number of variants of the basic milling cutters for peripheral and face milling as shown, for example, in Fig. 8.6. Thus, Fig. 8.7 shows a first-hand view of representative milling cutters from the past, because milling is, in general, called in accordance with the kind of cutter being used. As will be clear from it, milling produces not only the flat surfaces but also free surfaces with various profiles, which is extremely important in die mould machining. The ball end mills with hemispherical cutting edges are also available for finishing curved surfaces, and can facilitate extremely mould manufacturing, which requires high levels in the machining accuracy and the surface finish as well as the high machining rate.

The material removal volume in milling is controlled by the cutting speed associated with the spindle speed and the axial and the radial depths of cut as shown

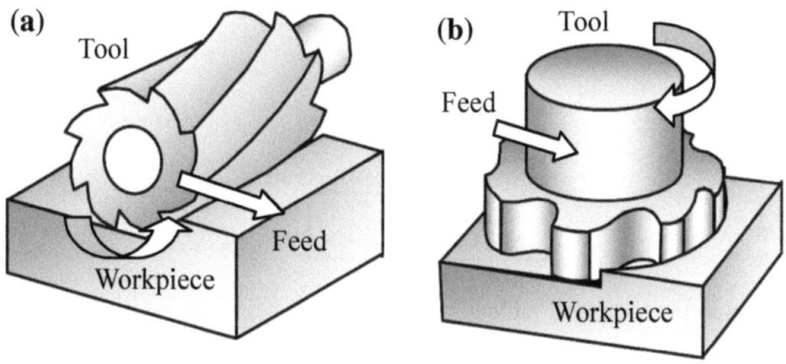

Fig. 8.5 Milling processes: **a** peripheral milling and **b** face milling

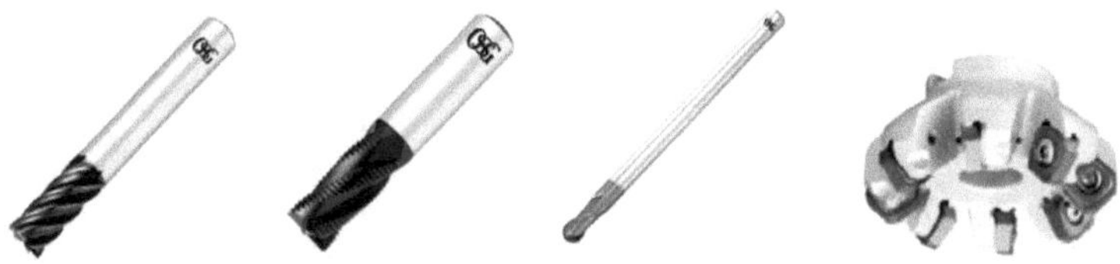

Fig. 8.6 Milling cutters (by courtesy of OSG, Tungaloy)

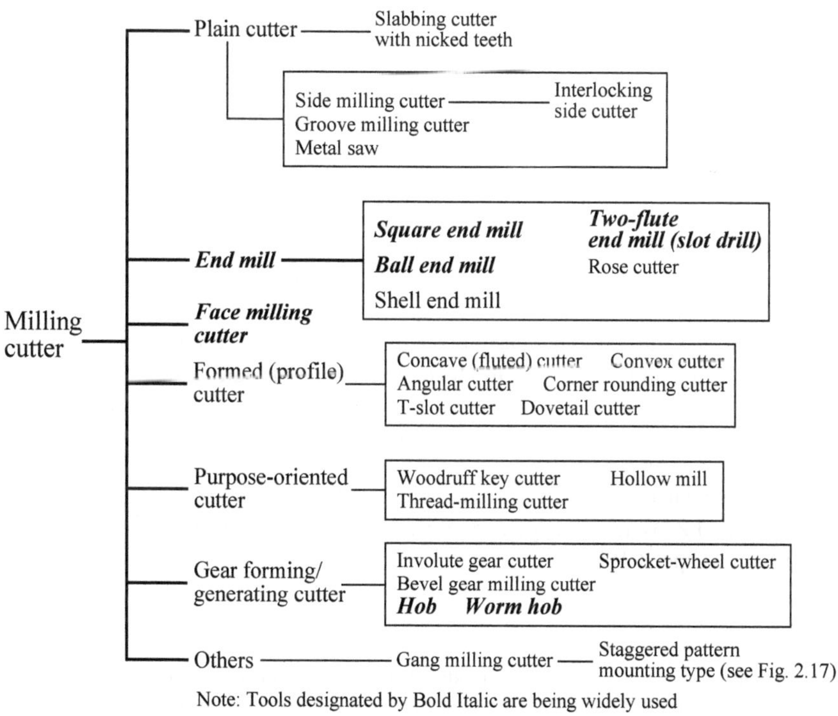

Fig. 8.7 First-hand view for representative milling cutters

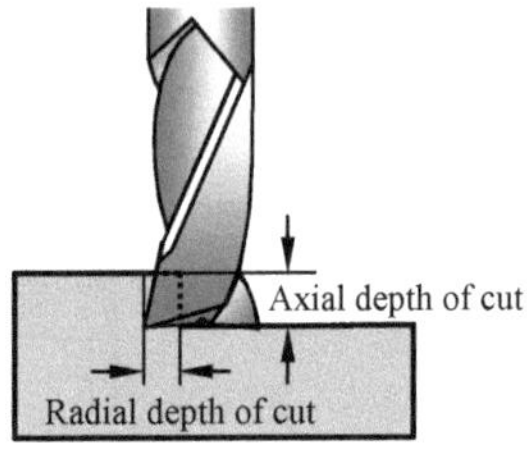

Fig. 8.8 Operation parameters in milling

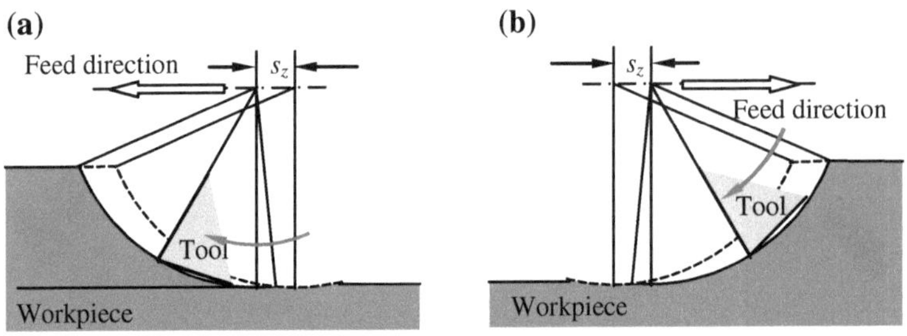

Fig. 8.9 Cutting manners in milling: **a** up cutting and **b** down cutting

in Fig. 8.8.[1] Figure 8.9 shows the edge trajectory viewed from upward in milling. Figure 8.9a is up cutting, in which the cutting thickness increases with the cutter rotation and Fig. 8.9b is down cutting, in which the cutting thickness decreases. The material is removed in the rotation of the edges and the feed motion of the tool axis. As a result, the edge locus is followed by a Trochoidal function, and because the edge repeats cutting and non-cutting, i.e., intermittent cutting, the swarf can be segmented in every cutting (refer to Appendix A). Furthermore, the non-steady process is performed during cutting, because the cutting thickness changes with the cutter rotation angle.

When an end mill having Z in number of cutting edges is fed at a feed rate per tooth of s_z at a spindle speed N, the feed speed per a minute, s, is:

$$s = N s_z Z \tag{8.3}$$

Being associated with the feed rate increases, the material removal rate depends upon N, s_z and Z. When an edge is rotating in a diameter of D at a spindle speed of N, the cutting speed V is:

---

[1]As exemplified in Fig. 8.8, the axial and radial depths of cut coincide with the width of cut and depth of cut in the past. Paraphrasing, the depth of cut should be in direct relation to the generation of Trochoidal locus in cutting tooth movement.

$$V = N\pi D \tag{8.4}$$

The diameter D does not change in milling with square end mill. Meanwhile, when the edge rotation diameter D changes in the height of the cutter such as a ball end mill, the cutting speed V varies on the cutting edge. The actual cutting travel, with which the tool wear is associated, also depends upon the rotation diameter at the edge point. Because multi-axis machine tools control the cutter axis inclinations, the cutting areas on the edges should be determined with the actual cutting speed, which influences on both the surface finish and the tool life.

To this end, it emphasises that the milling cutter with irregular tooth pitch and/or variable tooth lead has been prevailed, because such a cutter is very effective to suppress the chatter vibration. Of special interest, these remedies for the allocation of the cutting tooth have been transferred to the drill, reamer and tap.

8.3 Drilling, Reaming and Tapping

Drilling is performed to machine holes, and its form-generating movement is a combination of rotating tool and its axial movement. The most common tool in drilling is the twist drill as shown in Fig. 8.10. The drilling tools involve, furthermore, the step drill used for machining the step holes, the centre drill for helping the penetration of the chisel into the work and the gun drill for machining the deep holes (see Figs. 1.4a, 2.25 and 2.27).

The twist drill consists of the chisel and the lips as shown in Fig. 8.11, and the lip works as the major cutting edge for removing the material. More specifically, the chisel determines the cutting direction with deforming or removing the material, and thus the thrust, i.e., axial component of cutting force in drilling, largely depends upon the shape of the chisel. As can be readily seen, the larger the thrust in drilling,

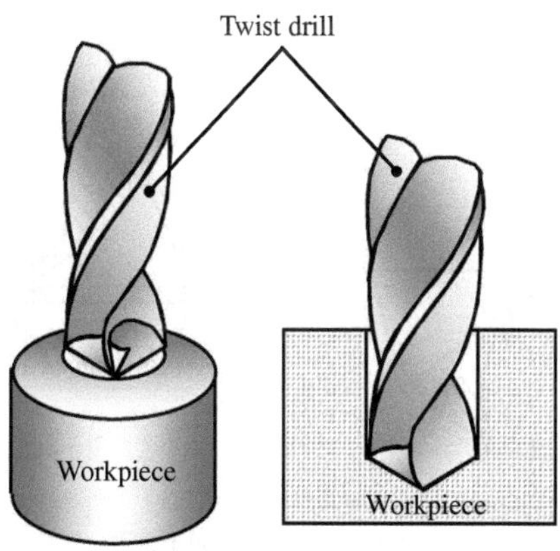

Fig. 8.10 Drilling operation

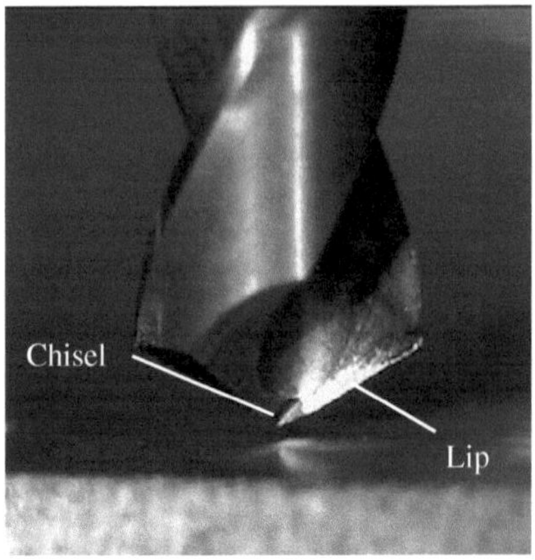

Fig. 8.11 Drill

the more difficult is to finish the good hole. In consequence, thinning is carried out at the chisel to improve the cutting performance by reducing the thrust.

As a matter of course, we have contrived various thinning methods, and Fig. 8.12 shows three representative thinning methods. Of these, point thinning is for general

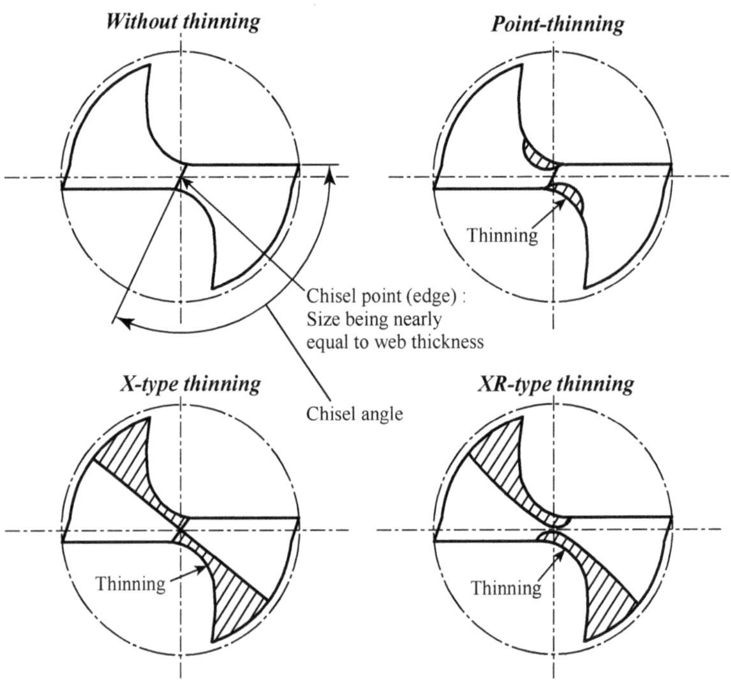

Fig. 8.12 Three representative thinning methods in twist drill

use, ease of thinning and has been prevailed. With the advance of the drilling technology and increase of difficult-to-machine material, another two types, i.e., X and XR types, become popular now, and these can be characterised as follows.

(1) X type is applicable to gun drilling and drilling of difficult-to-machine material, effective when the chisel is thick, and enhances the ability for engagement of the lip.
(2) XR type guarantees the long durability and ease of chisel grinding as compared with X type. In contrast, the ability for engagement is worse than X type, although the lip maintains the higher strength.

The lips remove the material with controlling the swarf flow. When the drill with small diameter is used, the swarf control would be important in terms of the swarf clogging. In principle, the end of the lips can determine finally the quality of the drilled hole. Burr formation in machining of the metal and delamination of carbon fiber reinforced plastic depends upon the cutting performance at the end of the lips.

Figure 8.13 shows the swarf formations at the chisel and lips. Because the tool geometries are different on the chisel and the lips, e.g., rake angle ranging from negative to positive from the centre to the margin, the independent swarfs are formed on an edge. The drilling process includes the steady and the non-steady cutting processes. In the drilling process of a hole, the edges penetrate into the workpiece, the material is removed in a constant removal volume, and the edges exit from the workpiece as shown in Fig. 8.14. In the penetration process, the cutting area increases with the axial feed motion on the tool, which is regarded as the non-steady cutting. Then, the steady cutting is performed in the constant material removal process. After the chisel exits from the workpiece, the cutting area decreases in the non-steady cutting process.

The feed rate and the cutting speed are also given by Eqs. (8.3) and (8.4). The cutting speed changes on an edge with the rotation diameter. Therefore, the swarf velocity increases from the centre to the end of the lips; and, then, the swarf curls with inclination toward the radial direction.

It is worth suggesting that the innovation and contrivance in drilling are amazing with growing importance of hole drilling with higher accuracy and for new materials, e.g., CFRP. Figure 8.15 shows some new drills and we can give their quick notes as follows.

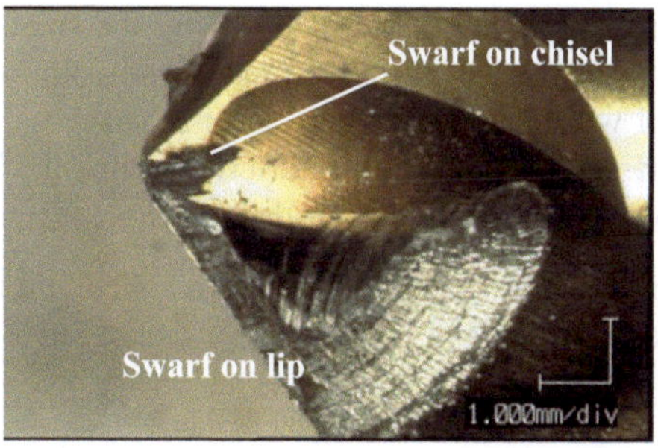

Fig. 8.13 Swarf formations

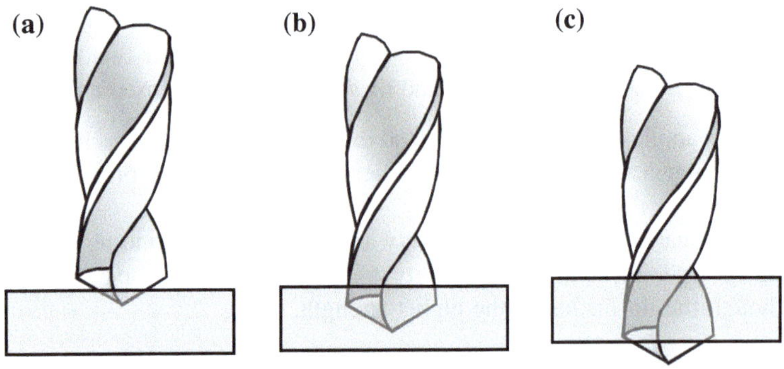

Fig. 8.14 Drilling process: **a** penetration, **b** material removal in hole and **c** exit

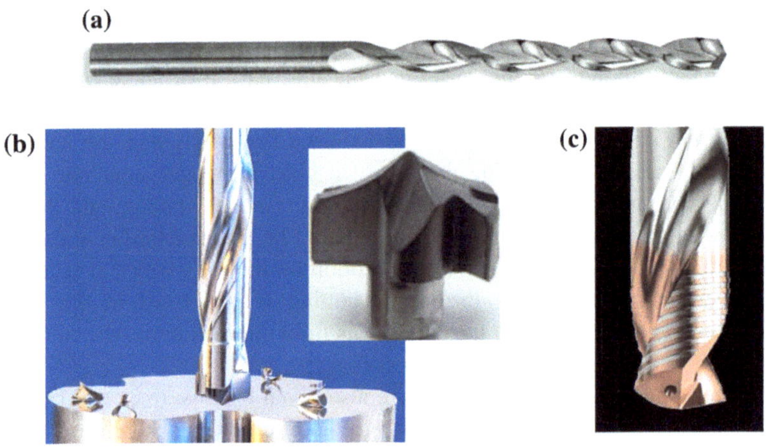

Fig. 8.15 Advanced drill tools: **a** gun drill (Gühring), **b** drill with self-centring function (Iscal) and **c** drill with shallow grooves (Walter)

(1) Gun drill made of sintered carbide for motorcar component (Gühring-brand). In general, we used to machine the small- and medium-sized deep hole by using the gun drill of single-edged drill bit (one-flute type). Against this context, Gühring has contrived a new type by improving the twist drill, which can be characterised by wide flute with mirror-finish to smooth swarf disposal.
(2) Self-centring drill with concave-like lip (Iscal-brand).
(3) Drill with ease of supplying cutting fluid (Walter-brand). This can be characterised by shallow groove at margin of drill end.

Obviously, drill of modular system has been prevailed and there are two types: one is of insert quick changing, and the other is of cutting-edge (head) module (see Fig. 7.14).

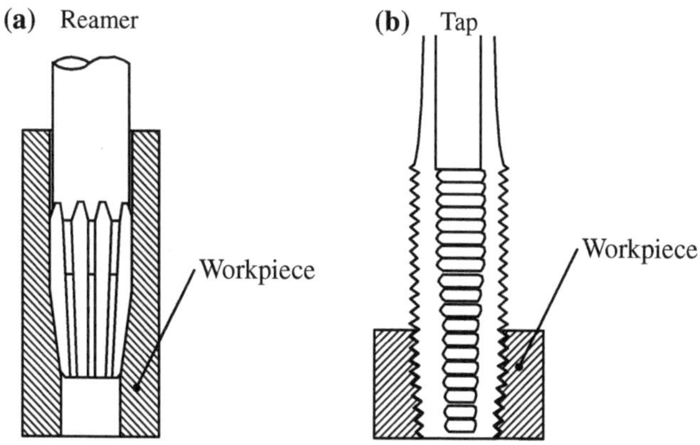

Fig. 8.16 Reaming and tapping: **a** reaming and **b** tapping

Reaming is an operation to make an existing hole dimensionally much more accurate than is obtained by drilling and also to improve the surface finish of the drilled hole as shown in Fig. 8.16a. For even better accuracy and surface finish, the drilled hole may be likely internal ground and honed by reaming. A reamer is a multiple-cutting edge tool with straight or helically flute. Because the removal thickness is small, reaming works not only as material removal but also burnishing.

In contrast, tapping shown in Fig. 8.16b is performed to machine the internal threads. As literally shown, the tap is a threading tool with multiple cutting teeth, and the cutting edges placed at the tapered portion of a tap mainly remove the unnecessary material, whereas the edges placed at the straight rear part works as guide for cutting direction as well as finally finishing thread flanks.

8.4 Planing and Shaping

Planing is a relatively simple cutting manner to machine flat surfaces by travelling the cutting edge along the length of the workpiece as shown in Fig. 8.17a. As a result, planing is usually done for the large-sized workpiece accommodated on the travelling table in the planing machine (planer). In contrast, shaping shown in Fig. 8.17b is used to machine parts like planing does; however, the smaller parts are machined as compared with that in planing. In short, we can distinguish both the machining manners by their form-generating movement. In fact, the cutting tool mounted on the ram repeats the reciprocating motion in shaping, whereas planing finishes the surfaces with moving the workpiece.

The planning and the shaping processes are typically conducted in the steady cutting process at a specified depth of cut in the linear motion of the tool or the workpiece. Although the material removal volume is given by Eq. (8.2), the cutting speed in the linear motion of the table or the ram is relative low as compared with

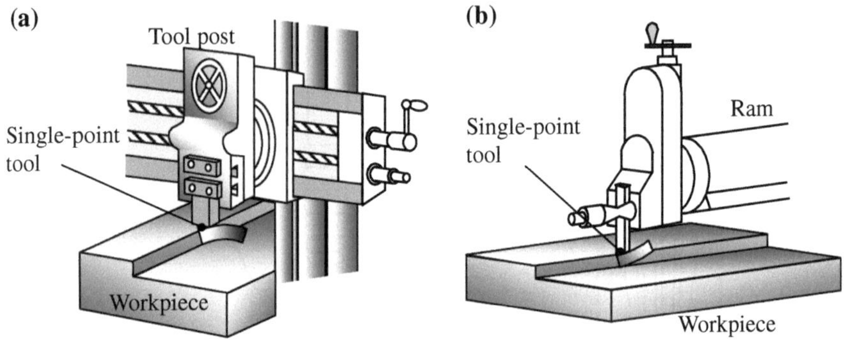

Fig. 8.17 Planing and shaping: **a** planing and **b** shaping

the rotational removal processes such as turning, milling and drilling. Therefore, the material removal rate largely depends upon the cutting area, on which the swarf forms.

In general, it is said that planing and shaping are obsolete, and in fact, these are not employed at present, apart from the special machining requirement. For example, planing is applied to finish the die mould to be very smooth surface in reality, whereas slotting, a variant of shaping, can facilitate the breech machining in the ordnance (see Fig. 1.3c).

8.5 Broaching

Broaching shown in Fig. 8.18 is, in general, performed to machine internal shapes within the workpiece such as serration, spline, polygon and fir tree-like shape by travelling the cutter through holes. A broach is a long multiple-toothed cutting tool for successively deeper cutting. Figure 8.18b shows examples of Christmas broaches. Thus, the total depth of material removed in one cutting stroke is the sum of the depths of cut of each tooth. Figure 8.18c shows an example of the vertical internal broaching machine.

The cutting process is similar to planing and shaping because we can remove the unnecessary allowance by the feed motion of the tool. However, the removal area increases in a tool feed. The removal volume of each tooth can be controlled by the increasing area, which is given by comparing to that of immediately previous tooth, and which does not change in the tool feed, resulting in steady process cutting.

Importantly, broaching can facilitate the flat surface finish of the large-sized workpiece like cylinder block and head of the motorcar, which is far beyond from that finished by large-sized face milling. For the sake of further understanding, Fig. 8.19 shows a surface broaching machine of Fujikoshi-brand.

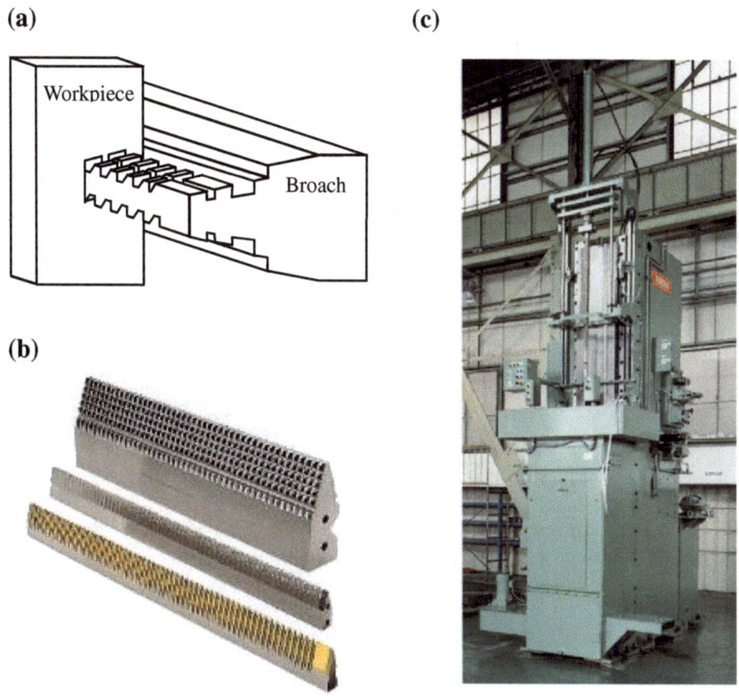

Fig. 8.18 Broaching: **a** operation, **b** Christmas broaches (by courtesy of Mitsubishi Heavy Industry) and **c** vertical internal broaching machine (by courtesy of Fujikoshi)

8.6 Gear Machining

In gear machining, we have two representative methods, i.e., "Forming process" and "Generating process", and nowadays the latter is dominant. Importantly, the tooth profile consists, in general, of Cycloid, Involute, Octoid, and Novikov, i.e., a variant of Circular-arc, curves, and of these Involute curve is dominant. Reportedly, the gear is shaped by the relative movement of the cutting tool to the gear blank, which is capable of generating the Involute curve. Although the form-generating movement is, in principle, the same, there are three methods to produce the gear depending upon the kind of cutting tool as follows (see Fig. 1.3f):

(1) Gear machining with a rack cutter is shown in Fig. 8.20a, in which the cutter reciprocates parallel to the face width of the gear and the blank rotates slowly to be formed the Involute curve. Of note, this machining method is not popular and in the past carried out by the gear planer called "Sunderland type". Figure 8.20b shows the traditional gear planer and as can be imagined, this special term is to remind its first developer.

Fig. 8.19 Horizontal surface broaching machine—Type NSL-35-D61M (by courtesy of Fujikoshi, around 2000)

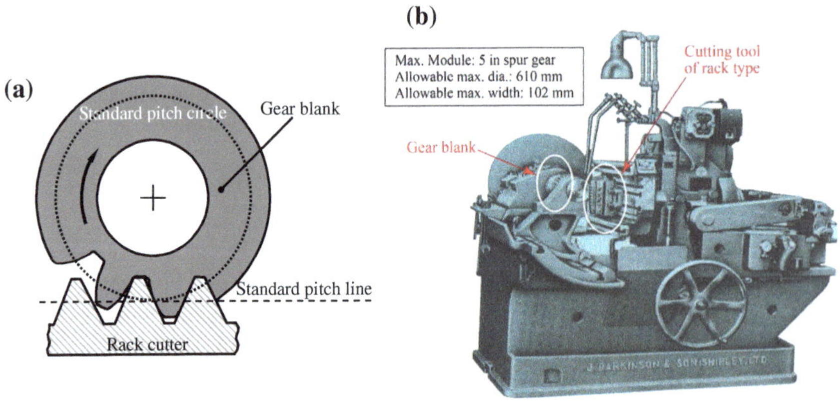

Fig. 8.20 Gear machining with rack cutter: **a** machining manner and **b** gear planer of "Sunderland Type" (No. 5B, Parkinson & Son (Shipley)-Brand, 1968)

(2) Figure 8.21 is machining of the gear shape with a pinion cutter. The cutter has a rotation axis parallel to that of the blank, and then rotates slowly with the blank while doing the reciprocating cutting motion, and also maintaining the same pitch-circle velocity with reciprocating motion. This method has been widely used in gang gear cutting and is carried out by the gear shaper.

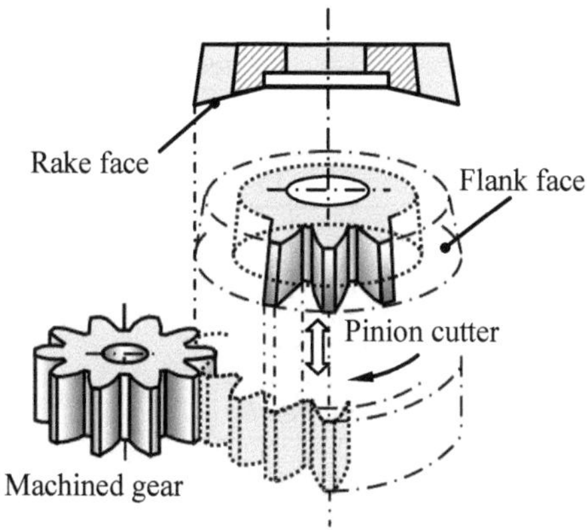

Fig. 8.21 Gear machining with pinion cutter

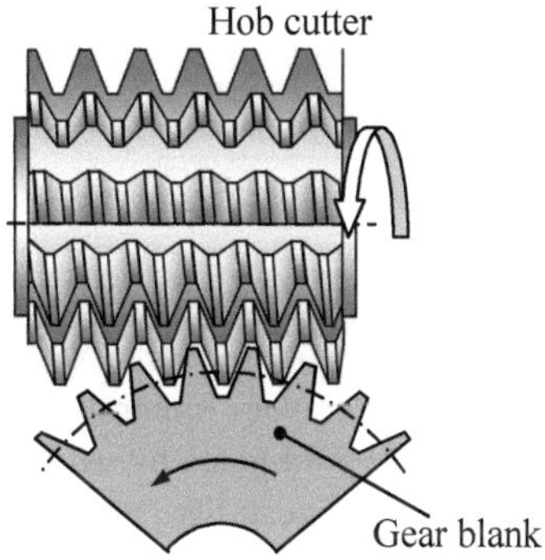

Fig. 8.22 Gear machining with hob cutter

(3) Hob shown in Fig. 8.22 is widely used for machining the gear because of its economisation. The hob and the gear blank rotate together as in two gears meshing, and the hob moves, in general, downwards. In hobbing, furthermore, the worm can be produced.

In gear machinings with the rack and the pinion cutters, the material is removed in a constant depth of cut during a reciprocating motion. Therefore, the steady cutting process is performed in the machining. Meanwhile, in the gear machining with the hob cutter, the edge rotates with the feed of the tool axis. The removal volume, therefore, change with the cutter rotation angle in the non-steady cutting process, as milling process. Obviously, gear shaping and gear hobbing can be characterised by their form-generating movement as can be readily seen from Fig. 1.3f.

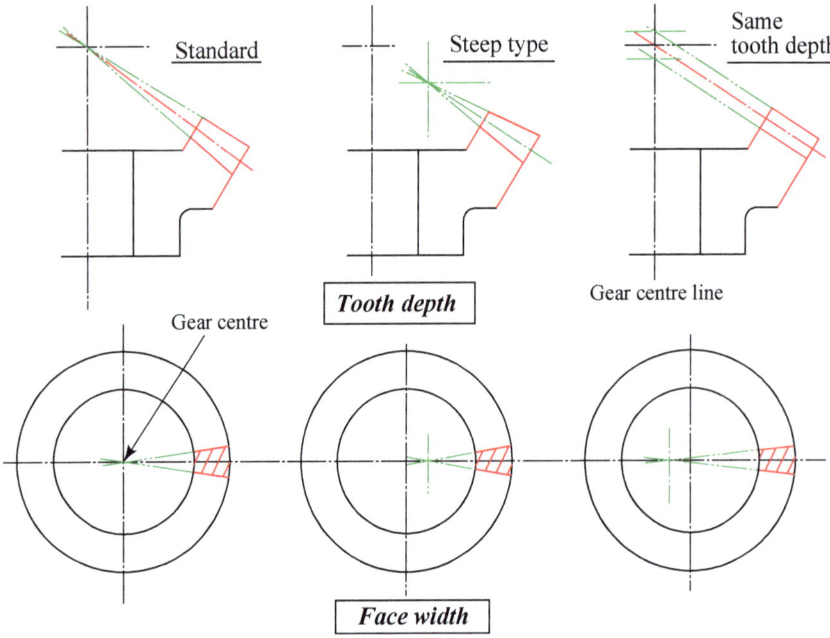

Fig. 8.23 Three configurations in tooth depth and face width of straight bevel gear

In gear machining, it is furthermore very important that we must produce a pair of bevel gears, and also a pair of hypoid gears by using the same machine. This is because of guaranteeing the interchangeability, which is derived from the generating mechanism for tooth profile in the bevel gear cutting machine and hypoid gear generator.

More specifically, there are three types in the tooth depth and face width in straight bevel gear as shown in Fig. 8.23 depending upon the machining method, i.e., differing tooth generation mechanism developed by the specified manufacturer. Importantly, there are three types in the tooth flank (trace) in the case of the spiral bevel gear as shown in Fig. 8.24.

In many respects, we use the hardened and ground gear in accordance with the requirement. In understanding gear grinding, it is very convenient to replace the pinion cutter and hob with the grinding wheel. As exemplified in Fig. 8.25, the former corresponds with the thread-like grinding wheel, and the grinding method with it is called "Reishauer Type". For the latter, the grinding wheel is of "Saucer-like", and the grinding method is called "Maag Type".

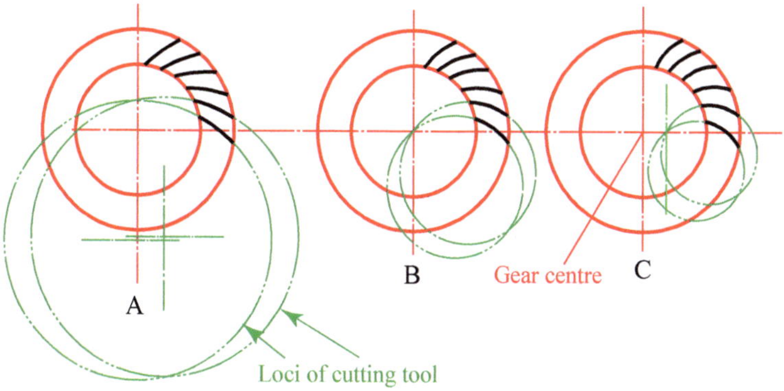

Fig. 8.24 Three types in tooth trace of spiral bevel gear

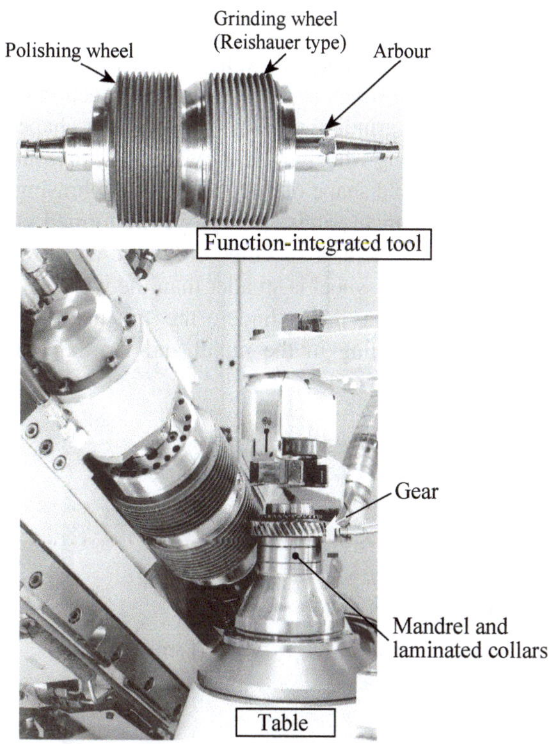

Fig. 8.25 Gear grinding and polishing by function-integrated tool (by Liebherr 1990)

8.7 Grinding

Grinding is usually performed to finish the shape and the surface of the products after cutting and, in certain cases, after hardening and annealing. Figure 8.26 compares the surface finished by turning and the grinding with changing the machining parameters. The surface finish in grinding is much better than that of turning. The material is removed by a rotating grinding wheel in depths of cut of generally less than 0.1 mm. The grinding stone consists of abrasive grains, bonds and pores, whose roles are described in Sect. 9.2.1. Although the material removal is actually performed by the abrasive grains, the grinding process depends on not only the abrasive grains but also the bond material, which controls the stiffness and the damping of the abrasive grains. The grain density associated with the successive cutting edge spacing is given by the ratio of the abrasive grains to the volume of bond.

The grinding operations are conducted in cylindrical, internal, surface and centreless manners (see Fig. 1.4). In the plunge grinding shown in Fig. 8.27a, one of the cylindrical grinding methods, the rotating grinding wheel at a high grinding spindle speed removes the unnecessary allowance of the workpiece surface at a low work spindle speed, where the wheel is fed in the radial direction of the workpiece. Although the general plunge grinding employs the cylinder shape of the grinding wheel, we can form the grinding wheel in arbitrary shape by dressing or truing so as to corresponding to the required shape of the product. Paraphrasing, form grinding, i.e., one of the variants of plunge grinding, employs the formed grinding wheel to finish such the rounded shape with a specified curvature as shown in Fig. 8.27b. When the width of the grinding wheel is smaller than the workpiece length and the uniform surface is required on the cylinder, the traverse grinding shown in Fig. 8.27c is applied with feeding in the radial and the axial directions of the workpiece.

Fig. 8.26 Surface finishes in grinding and turning

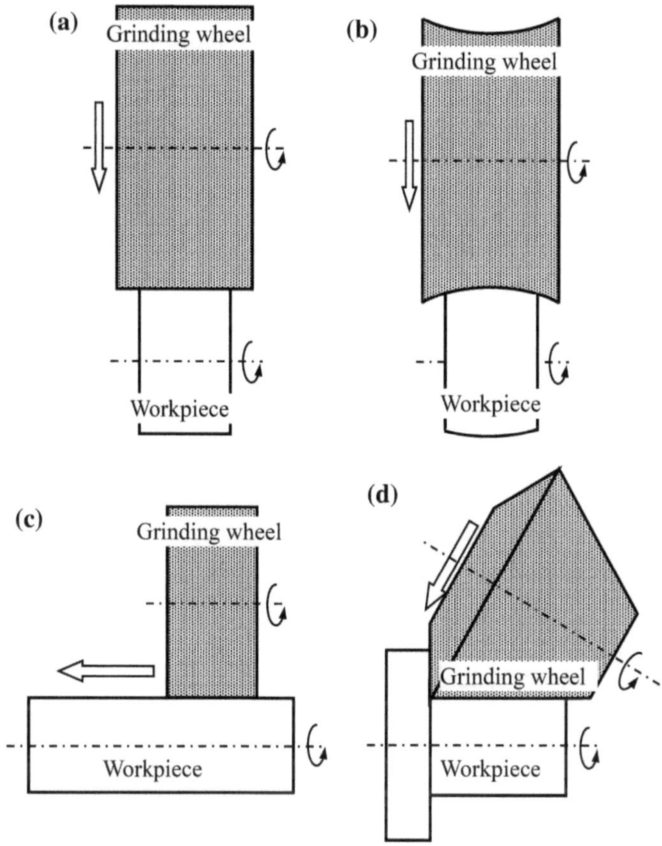

Fig. 8.27 Cylindrical grindings: **a** plunge grinding, **b** form grinding, **c** traverse grinding and **d** angular grinding

In general, the spindle axes of the grinding wheel and the workpiece are aligned parallel with each other in plunge, form and traverse grinding. When the taper shape or the corner of the product should be finished, the spindle axes may be inclined with each other at a specified angle as in angular grinding as shown in Fig. 8.27d.

In internal grinding, the rotating grinding wheel finishes internal surface of the product with rotation of the workpiece as shown in Fig. 8.28a. When the workpiece cannot be rotated due to some reasons such as the large workpiece size, the centre of the rotation of the wheel rotates along the internal surface of the workpiece like planetary movement as shown in Fig. 8.28b.

The surface grinding is applied to finish flat surfaces with reciprocating the grinding wheel on the workpiece as shown in Fig. 8.29, where the rotating wheel is fed to finish the workpiece surface at a specified pitch in the spindle axis direction. As can be readily seen, there are a considerable number of the variants in

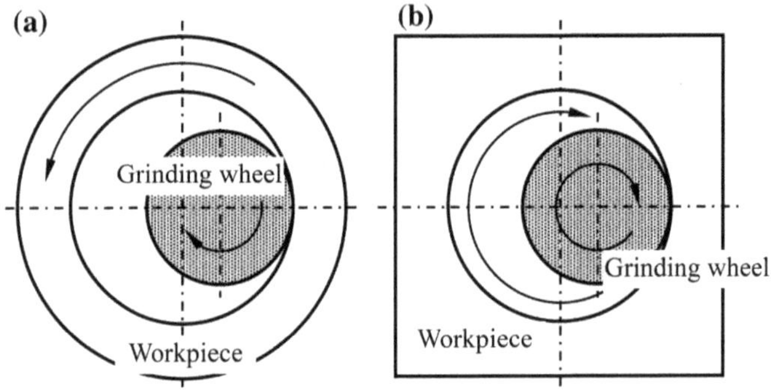

Fig. 8.28 Internal grindings: **a** with workpiece rotation and **b** fixed workpiece

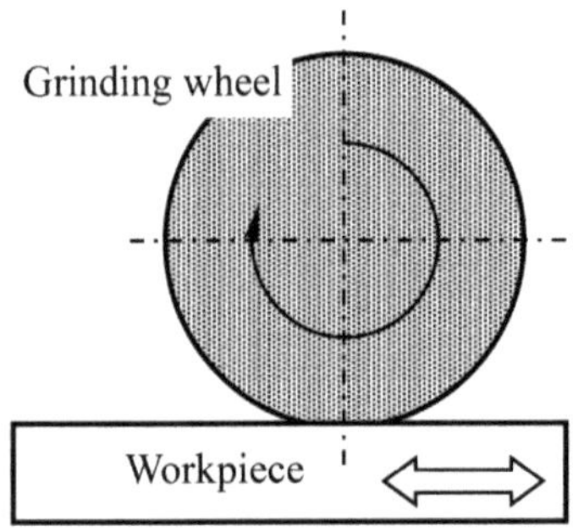

Fig. 8.29 Surface grinding

accordance with the combination of the form-generating movement in grinding wheel and workpiece. For example, the representative variants are as follows.

(1) Grinding wheel rotating around horizontal axis + rotating worktable around vertical axis.
(2) Grinding wheel rotating around vertical axis + worktable with reciprocating movement: This is for heavy grinding with grinding wheel of disc type (grinding wheel of segment stone type), at the face end of which a considerable number of the block-like grinding stones accommodate. Of special note, the segment grinding wheel of pipe-like tip is for ultra-precision grinding of Al_2O_3, Si and sintered metals.
(3) Grinding wheel rotating around vertical axis + rotating worktable: When using the grinding wheel of tapered cup type, this is for grinding Si wafer and sapphire wafer.

The centreless grinding, as shown in Fig. 8.30, does not have the fixed axis of the rotating workpiece. The workpiece surface is removed in a small depth of cut according to the surface profile. Although the material removal rate is relatively low to the other grinding manners, the surface finish becomes better than those manners and the affected layer in subsurface is small.

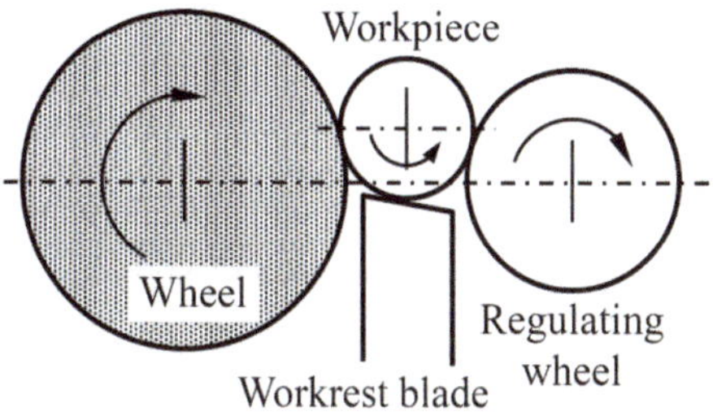

Fig. 8.30 Centreless grinding

Having in mind that there are a myriad of grinding method including special grinding corresponding to the special purposes such as the thread, gear, camshaft cutting tool finishing, a grinding machine as a whole can conduct various grinding like TC as shown in Fig. 8.31. In fact, the CNC grinding machine shown in Fig. 8.31 is for various grinding such as the form, internal and thread finishing. Obviously, it is easy to understand nearly all grinding procedures by replacing the grinding wheel with the cutting tool while turning, when discussing the form-generating movement of grinding.

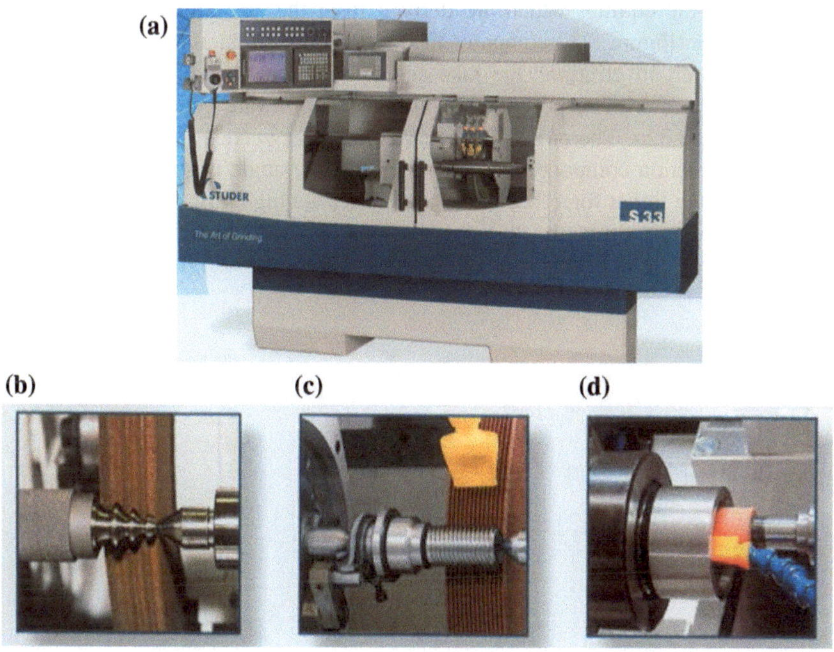

Fig. 8.31 CNC grinding: **a** CNC machine, **b** form grinding, **c** screw grinding and **d** internal grinding

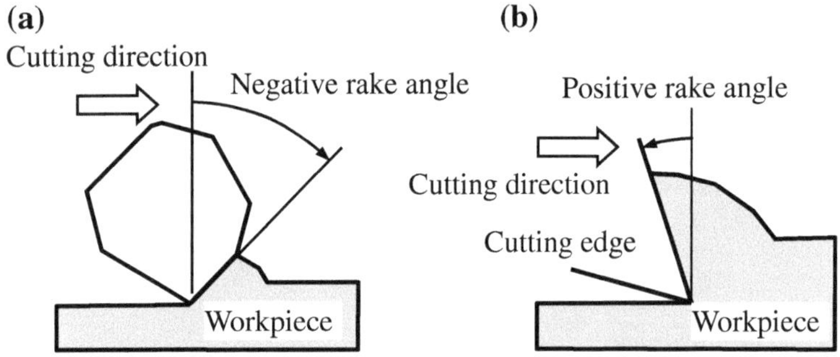

Fig. 8.32 Comparison between grinding and milling: **a** grinding and **b** cutting (milling)

Importantly, in the grinding process, an abrasive grain rotates with the workpiece motion. Therefore, the trajectory follows Trochoidal locus, which is the same as the cutting edge trajectory in milling. As discussed later in Sect. 9.2.1, the shapes and the orientations of the cutting edges in the milling are controlled; meanwhile, those are not controlled in grinding. Therefore, the removal process of each abrasive grain cannot be controlled uniquely. Consequently, the surface finished by each grain changes in the micro-scale order, although the surface roughness is small.

For further understanding, we must discuss the differences of the grinding process from the cutting, which are derived from the size and the shape of the abrasive grains. Figure 8.32 compares the cutting edge of an abrasive grain to that of milling cutter. In grinding, the rake angles of nearly all the grains, which are inclination angles with respect to the cutting direction, are generally negative as shown in Fig. 8.32a. Therefore, the thrust component becomes relatively large to the cutting direction component. Meanwhile, the rake angle in the milling cutter is positive, and the thrust force becomes small. Because higher machining accuracies are required in grinding, the thrust force should be controlled considering the displacement of the grinding wheel axis.

Chapter 9
Mechanisms for Metal Cutting and Grinding

As the major removal processes, cutting and grinding are most commonly applied to manufacturing products at high machining rates and low costs. Their machining parameters should be optimised to achieve high production rate under the specified machining requirements. However, the recent machine shops do not have time and cost enough to perform the cutting and grinding tests. Therefore, those removal processes should be studied in terms of the physical aspects as shown in Fig. 9.1. The removal volume is given geometrically by the machining parameters, first. Then, the removal forces may be estimated for the volume. Because of consuming the mechanical energy while removal processing such as shearing and friction, the energy, in turn, is transferred into the heat energy; and thus, the heat generation occurs in the machining point. Also the tool wear depends upon both the stress and the temperature on the tool face. Consequently, the tool wear is associated with the removal force and the heat generation.

The machining requirements are generally represented by the machining errors, surface finish and the affected layer with residual stress in subsurface. Those requirements are also associated with the force, temperature and the tool wear.

This Chapter describes the removal process in terms of the machining force, temperature and the cutting tool life or grinding wheel life in the basic cutting and grinding processes by using the analytical method. As widely known, the analytical method can facilitate to understand the essential feature of the mechanism, although we have certain limitations in its applicability depending upon the quality of the mathematical model. In fact, the mathematical model involves some assumptions to simplify the machining mechanism in practice (see Appendix A).

Importantly, there are a considerable number of prevailing and different machining methods as already shown in Chaps. 1 and 8, and at least we must understand drilling and milling in addition to basic cutting, i.e., turning, and grinding. These methods can be characterised as follows, and such characteristic features result in certain difficulties to apply the analytical method. It is thus worth suggesting to analyse them by using the numerical (computational) method, which

Y. Ito and T. Matsumura, *Theory and Practice in Machining Systems*,
DOI 10.1007/978-3-319-53901-0_9

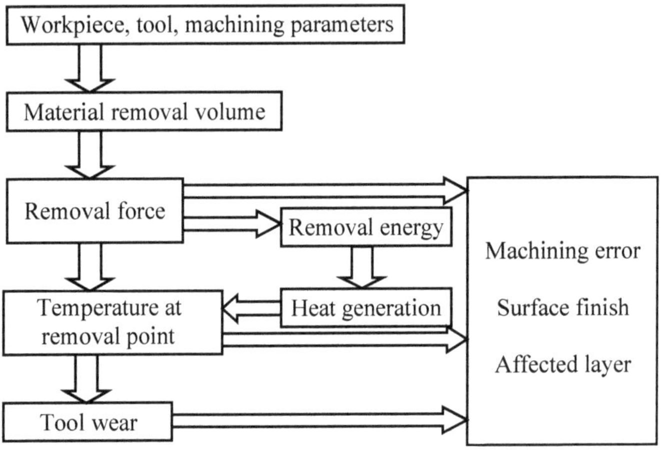

Fig. 9.1 Machining process

has been prevailed recently and becomes dominant in the research into the drilling and milling mechanisms.

(1) In drilling, the rake face changes continuously its angle from obtuse around the chisel to acute around the margin.
(2) In milling, the thickness of the swarf changes continuously with rotating the milling cutter, because the cutting edge moves as per the Trochoidal curve.

In this context, the grinding mechanism is at burning issue, although we may benefit considerably from the theoretical background as will be stated within this Chapter. In many respects, the actual states in grinding cannot be modelled in detail at present, but are too simplified, especially those for "Three-dimensional random distribution of abrasive grain" and "Self-sharpening" in time domain, which should be first quantified. In addition, the grain generates Trochoidal curve while grinding and is of obtuse rake angle. Admitting such difficulties, it is now urged to establish a new numerical model for grinding.

9.1 Cutting

9.1.1 Cutting Model

Several cutting operations are performed with three-dimensional swarf formation as shown in Fig. 9.2a. The material is removed by two cutting edges and the swarf formation depends upon the edge inclinations. In short, the cutting model changes along the edge. In order to understand quickly the cutting mechanism, thus, it is

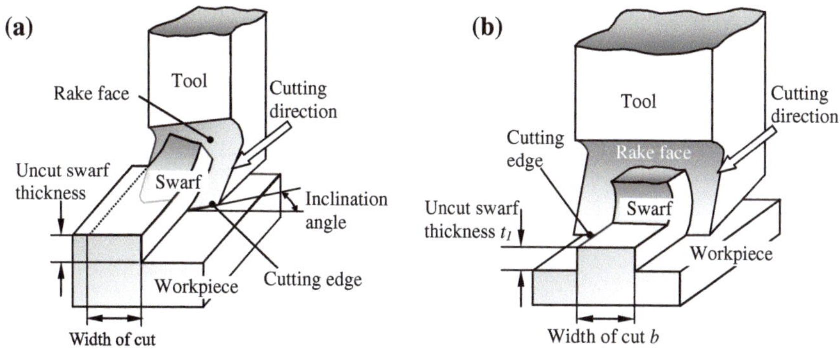

Fig. 9.2 Cutting manner: **a** three dimensional cutting and **b** orthogonal cutting

recommended that an orthogonal cutting manner shown in Fig. 9.2b will be discussed here.

As can be readily seen, the cutting edge is perpendicular to the cutting direction. As a result, the volume of rectangular section, which is specified by the uncut swarf thickness (depth of cut) t_1 and the width of cut b, is removed from the workpiece. Obviously, the swarf forms in the plane normal to the cutting edge, and therefore, the swarf formation does not change along the cutting edge.

Figure 9.3 shows an orthogonal cutting in any cross section containing both the cutting and the swarf flow directions. The material to be cut deforms in a plane strain state, when the cutting edge inclines at a rake angle α with respect to the normal direction to the workpiece surface. The clearance angle gives a space between the machined surface and the cutting edge. As will be clear from Fig. 9.3, the swarf is produced ahead of the tool by shearing the material continuously along the shear plane, which inclines at a shear angle ϕ with respect to the cutting direction.

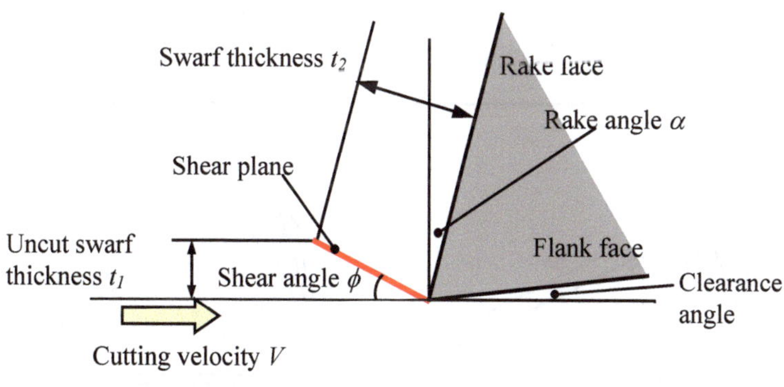

Fig. 9.3 Orthogonal cutting

In the orthogonal cutting, the cutting model is controlled by the shear angle shown in Fig. 9.4a. The shear angle is determined by t_1, α and the swarf thickness t_2:

$$\tan \phi = \frac{(t_1 / t_2) \cos \alpha}{1 - (t_1 / t_2) \sin \alpha} = \frac{r_c \cos \alpha}{1 - r_c \sin \alpha} \tag{9.1}$$

where, r_c is known as the cutting ratio:

$$r_c = \frac{t_1}{t_2} \tag{9.2}$$

In consequence, the cutting model is determined by measuring the swarf thickness, which is usually larger than the uncut swarf thickness; hence the value of r_c is less than unity.

The shear strain γ, which is defined as $\Delta s/\Delta y$ in Fig. 9.4b, can be expressed as:

$$\gamma = \frac{\Delta s}{\Delta y} = \frac{DD'}{AH} = \frac{DH}{AH} + \frac{HD'}{AH} = \cot \phi + \tan (\phi - \alpha) = \frac{\cos \alpha}{\sin \phi \cos (\phi - \alpha)} \tag{9.3}$$

Thus, large shear strains are associated with low shear angles and low or negative rake angles.

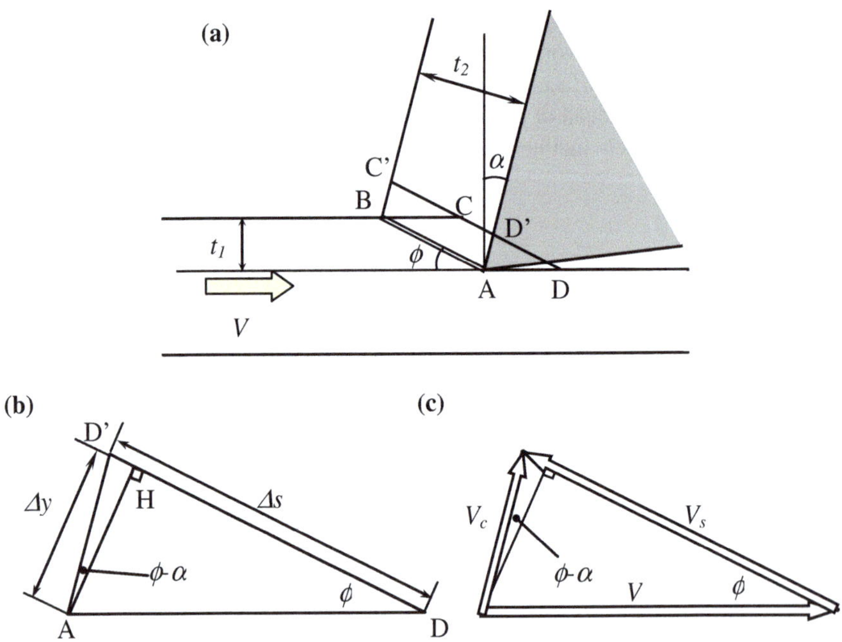

Fig. 9.4 Orthogonal cutting model: **a** shear plane cutting model, **b** shear strain and **c** velocities

Based on the principles of plasticity, vectors of the cutting velocity, the shear velocity and the swarf velocity must form a closed velocity diagram, as shown by the hodograph of Fig. 9.4c. The vector, sum of the cutting velocity and the swarf velocity, is equal to the shear velocity vector. The swarf velocity V_c and the shear velocity V_s are:

$$V_c = \frac{\sin\phi}{\cos(\phi - \alpha)} V = r_c V \tag{9.4}$$

$$V_s = V_c \sin(\phi - \alpha) + V\cos\phi = \frac{\cos\alpha}{\cos(\phi - \alpha)} V = \gamma \sin\phi\, V \tag{9.5}$$

Hence, the rate of the strain in cutting is:

$$\dot{\gamma} = \frac{\Delta s}{\Delta y \Delta t} = \frac{V_s}{\Delta y} \tag{9.6}$$

where, Δt is the time elapsed for the metal to travel a distance Δs along the shear plane, and Δy is the thickness of the shear zone.

9.1.2 *Cutting Force*

(a) *Orthogonal cutting*

The cutting force can be measured by the experimental set-up shown in Fig. 9.5, in which two-dimensional, i.e., orthogonal cutting is carried out. The cutting velocity is controlled with the diameter of the cutting point by the servomotor of the NC lathe. The tangential and the radial components, F_H and F_V, are measured by the piezoelectric dynamometers, where F_H and F_V are regarded as the principal and thrust components in the orthogonal cutting shown in Fig. 9.6. In consequence, the shear force F_s and the normal force F_{ns} applied to the shear plane are given by:

$$\left.\begin{aligned} F_s &= F_H \cos\phi - F_V \sin\phi \\ F_{ns} &= F_H \sin\phi + F_V \cos\phi \end{aligned}\right\} \tag{9.7}$$

The shear stress τ_s and the normal stress σ_s on the shear plane are:

$$\left.\begin{aligned} \tau_s &= \frac{F_s}{A_s} = \frac{(F_H \cos\phi - F_V \sin\phi)\sin\phi}{bt_1} \\ \sigma_s &= \frac{F_{ns}}{A_s} = \frac{(F_H \sin\phi + F_V \cos\phi)\sin\phi}{bt_1} \end{aligned}\right\} \tag{9.8}$$

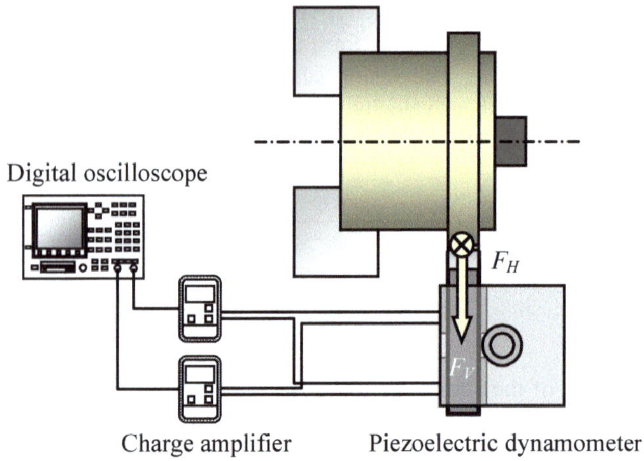

Fig. 9.5 Orthogonal cutting test

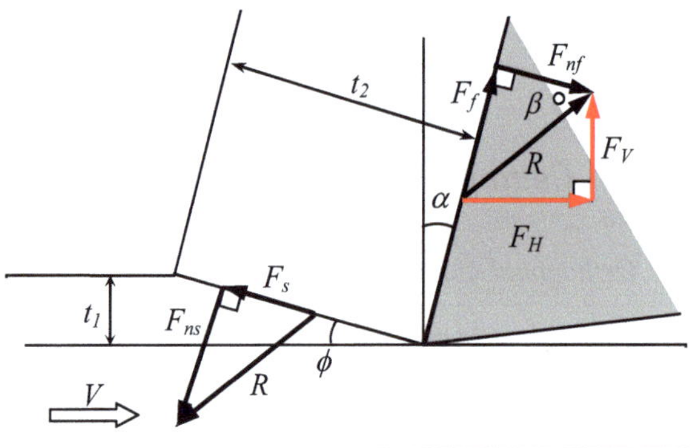

Fig. 9.6 Cutting force in orthogonal cutting

where, b is the width of cut, and A_s is area of the shear plane:

$$A_s = \frac{bt_1}{\sin\phi} \tag{9.9}$$

The friction and the normal forces at the rake face yields to:

$$\left.\begin{aligned} F_f &= F_H \sin\alpha + F_V \cos\alpha \\ F_{nf} &= F_H \cos\alpha - F_V \sin\alpha \end{aligned}\right\} \tag{9.10}$$

The coefficient of friction μ on the tool-swarf interface is thus written as:

$$\mu = \tan\beta = \frac{F_f}{F_{nf}} = \frac{F_V + F_H \tan\alpha}{F_H - F_V \tan\alpha} \tag{9.11}$$

where, β is the friction angle.

It is worth suggesting that the coefficient of friction in metal cutting generally is relatively large as compared with that at the sliding surface of the industrial machine.

The resultant cutting force R is also expressed as:

$$R = \frac{\tau_s b t_1}{\sin\phi\cos(\phi+\beta-\alpha)} \tag{9.12}$$

and the principal and thrust components are:

$$\left.\begin{aligned} F_H &= \frac{\tau_s b t_1 \cos(\beta-\alpha)}{\sin\phi\cos(\phi+\beta-\alpha)} \\ F_V &= \frac{\tau_s b t_1 \sin(\beta-\alpha)}{\sin\phi\cos(\phi+\beta-\alpha)} \end{aligned}\right\} \tag{9.13}$$

As can be readily seen, the shear angle is an important to estimate the cutting force according to Eq. (9.12). Within a cutting force context, thus, a facing problem is the accurate estimation of the shear angle, and one of the attempts to determine the shear angle so as to minimum the cutting force in an analysis made by Merchant (1945). The analysis is also based on the assumption that the maximum shear stress occurs in the shear plane. As a result, the shear stress on the shear plane is given by:

$$\tau = \frac{F_s}{A_s} = \frac{R\cos(\phi+\beta-\alpha)\sin\phi}{b t_1} \tag{9.14}$$

If β is independent of ϕ, the shear angle corresponding to the maximum shear stress is given by differentiating Eq. (9.14) with respect to ϕ and equating the resulting expression to zero:

$$\frac{d\tau}{d\phi} = \cos(\phi+\beta-\alpha)\cos\phi - \sin(\phi+\beta-\alpha)\sin\phi = 0 \tag{9.15}$$

Thus, the shear angle is written as:

$$\phi = \frac{\pi}{4} + \frac{\alpha}{2} - \frac{\beta}{2} \tag{9.16}$$

As will be clear from Eq. (9.16), when the rake angle decrease and/or the friction angle increases, the shear angle decreases. Importantly, we can obtain Eq. (9.16) using another derivation based on the cutting power, which is associated with Eq. (9.12). More specifically, Fig. 9.7 compares the theoretical values ϕ given by Eq. (9.16) with the measured data. Because the theoretical value is not in good agreement with the

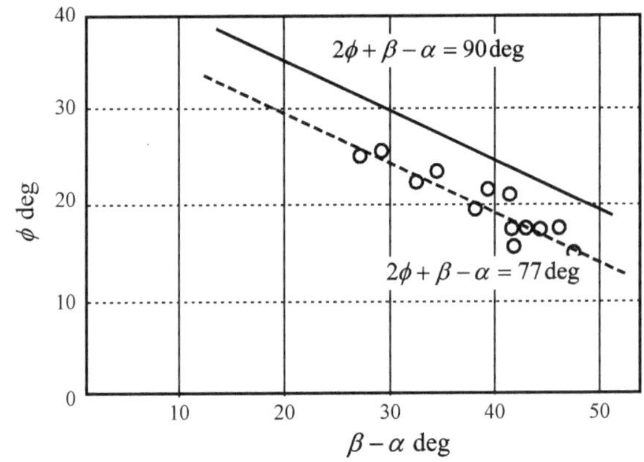

Fig. 9.7 Cutting equation with measurement data

measured result, Eq. (9.16) was modified by assuming that the shear stress τ_s would be influenced by the normal stress on the shear plane σ_s as follow:

$$\tau_s = \tau_0 + k_\tau \sigma_s \tag{9.17}$$

where, k_τ is a material constant.

The normal stress may be associated with:

$$\frac{\sigma_s}{\tau_s} = \frac{F_{ns}}{F_s} = \tan(\phi + \beta - \alpha) \tag{9.18}$$

From Eq. (9.17), the shear stress is given by:

$$\tau_s = \frac{\tau_0}{1 - k_\tau \tan(\phi + \beta - \alpha)} \tag{9.19}$$

The resultant cutting force is thus obtained by substituting Eq. (9.19) into Eq. (9.12):

$$R = \frac{\tau_0 b t_1}{\{1 - k_\tau \tan(\phi + \beta - \alpha)\} \sin\phi \cos(\phi + \beta - \alpha)} \tag{9.20}$$

By putting $dR/d\phi = 0$, the following equation is derived:

$$\phi = \frac{\cot^{-1} k_\tau}{2} + \frac{\alpha}{2} - \frac{\beta}{2} \tag{9.21}$$

where, $\cot^{-1} k_\tau$ is called machining constant.

When the machining constant is 77°, Eq. (9.21) agrees obviously with the measured data. However, Eq. (9.21) cannot always be applied to every material. In consequence, there have been a number of the additional attempts to predict the shear angle for a wide range of cutting conditions (Shaw 1984).

To this end, the effects of the cutting parameters on the cutting force are summarised as follows:

(1) The shear angle increases with the cutting speed, and thus the cutting force decreases with the cutting speed.
(2) The cutting force is almost proportional to the uncut swarf thickness. However, the specific cutting force, which is the cutting force loaded on the unit cutting area, is relatively large in small uncut swarf thickness. Therefore, the cutting force does not change proportionally to the uncut swarf thickness when the uncut swarf thickness is small.

(b) *Cutting force in turning*

The cutting force in turning, as shown in Fig. 9.8a, is discussed in this part. A typical single-point turning tool is shown in Fig. 9.8b, where the primary concern is the side cutting edge, while the secondary concern is the end cutting edge. Since the cutting force is usually associated with the cutting area, the specific cutting force is defined as the cutting force per unit area of material to be removed. Reportedly, the specific cutting force depends, in its first approximation, mainly upon the feed rate (feed speed) and the depth of cut in turning. In this context, Childs summarised values of the specific cutting force for representative engineering materials as shown in Table 9.1 (Childs et al. 2000). The cutting force is estimated by multiplying the cutting area by the specific cutting force, and in due course Kronenberg presented the following equation to estimate the cutting force F:

$$F = a \cdot p_s \tag{9.22}$$

where, a and p_s are the cutting area and specific cutting force, respectively.

The specific cutting force p_s is furthermore expressed as:

$$p_s = \frac{c_K}{{}^{\varepsilon K}\sqrt{a}} \tag{9.23}$$

where, c_K and ${}^{\varepsilon}K$ are constants, which are given by combination of the workpiece material and the tool material.

As can be readily seen, the specific cutting force can be used easily to estimate the cutting power in the machine shop. However, the specific cutting force depends upon the tool geometry, and thus Masuko presented another expression including the parameters of the tool geometry as shown in Table 9.2 (see also Appendix A).

(c) *Force model in micro cutting*

The principals of micro cutting are similar to those of milling of normal-sized parts. Swarf formation is generally associated with the uncut swarf thickness (Chae et al. 2006). When the uncut swarf thickness, h, is less than a critical minimum swarf thickness, h_m, as shown in Fig. 9.9a, elastic deformation occurs and the edge does not remove material. As the uncut swarf thickness approaches to the minimum

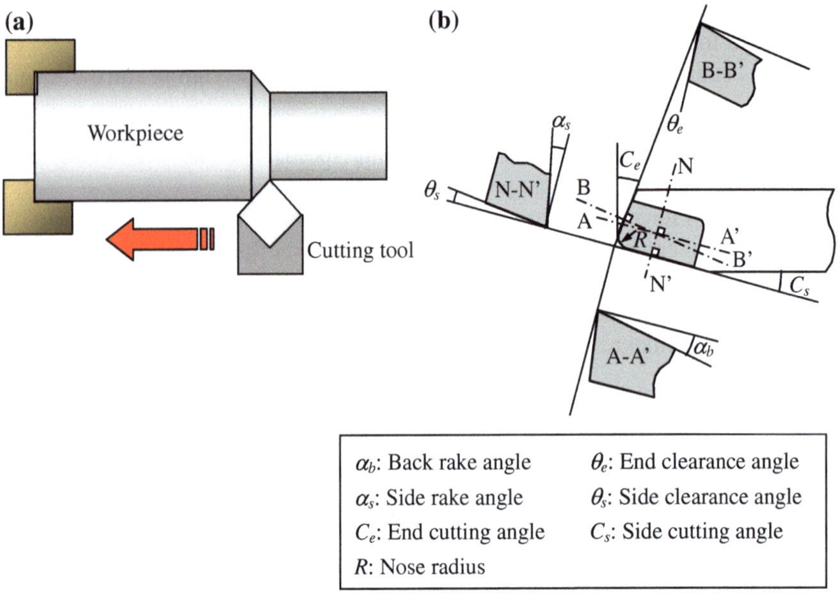

Fig. 9.8 Turning operation: **a** operation and **b** cutting tool

Table 9.1 Typical specific cutting forces for a range of engineering materials from Childs et al. (2000)

Material	Specific cutting force (GPa)
Aluminum alloys	0.5–1.0
Copper alloys	1.0–2.0
Cast irons	1.5–3.0
Carbon steels	2.0–3.0
Alloy steels	2.0–5.0

swarf thickness, swarfs are formed in shearing, whereas certain elastic deformation still occurs as illustrated in Fig. 9.9b. Therefore, the removal depth is less than the specified depth. When the uncut swarf thickness increases beyond the minimum swarf thickness, the elastic deformation phenomena decrease significantly and the entire depth of cut is removed as a swarf as shown in Fig. 9.9c.

In conventional cutting, the rake angle is defined as the inclination angle of the rake face. Swarf formation is controlled by the rake angle in a large uncut swarf thickness. However, when the uncut swarf thickness reduces to an order of micrometer, the edge radius can regulate the actual rake angle. As a result, cutting is performed at a large negative rake angle, which affects the magnitude of the ploughing and shearing forces. The minimum swarf thickness is associated with the edge radius and the material. Figure 9.10 shows the change in the swarf load/cutting force with the uncut swarf thickness (Liu et al. 2004). A sudden change in thrust forces occurs at a shifting from ploughing/sliding dominant mode to shearing dominant mode.

Table 9.2 Specific cutting forces with the parameters of the tool geometry

Nose radius	Carbon steel	Cast iron
$r = 0$	$p_s = 230\left(\frac{\sigma_z}{588}\right)^{0.5}\left(\frac{\delta}{76}\right)^{0.66}\left(\frac{60}{\kappa}\right)^{0.2}\left(\frac{n}{5}\right)^{0.12}\cdot\frac{1}{a^{0.1}}$	$p_s = 230\left(\frac{\sigma_z}{588}\right)^{0.5}\left(\frac{\delta}{76}\right)^{0.66}\left(\frac{r}{1.74}\right)^{0.1}\left(\frac{n}{5}\right)^{0.07}\cdot\frac{1}{a^{0.15}}$
$r > 0$	$p_s = 120\left(\frac{H_B}{160}\right)^{1.2}\left(\frac{\delta}{82}\right)^{0.9}\left(\frac{60}{\kappa}\right)^{0.17}\left(\frac{n}{5}\right)^{0.13}\cdot\frac{1}{a^{0.15}}$	$p_s = 120\left(\frac{H_B}{160}\right)^{1.2}\left(\frac{\delta}{82}\right)^{0.9}\left(\frac{r}{1.74}\right)^{0.1}\left(\frac{n}{5}\right)^{0.13}08\cdot\frac{1}{a^{0.2}}$

$\delta = 90$—rake angle [deg] $\kappa = 90$—side cutting angle [deg] $n = d/f$
r nose radius [mm], d depth of cut, f feed rate

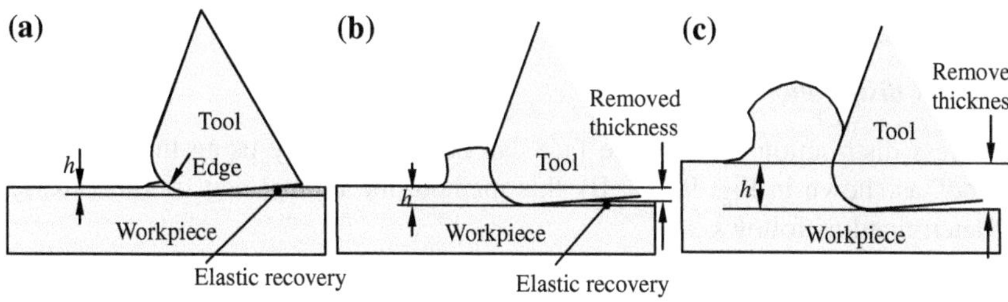

Fig. 9.9 Swarf formation in micro cutting: **a** uncut swarf thickness $h <$ minimum swarf thickness h_m, **b** $h \cong h_m$ and **c** $h > h_m$ (from Chae et al. 2006). h is the uncut chip thickness appeared in the original literature. h is regarded as t_1 in this text

In short, the cutting force is derived from the ploughing force around the edge radius and also the shearing force for swarf formation as described above. In the micro cutting, the ploughing effect is relatively large to the shearing. Therefore, the ploughing force cannot be ignored and the force in the micro cutting may be expressed as:

$$F = p_p \cdot b + p_s \cdot b \cdot t_1 \tag{9.24}$$

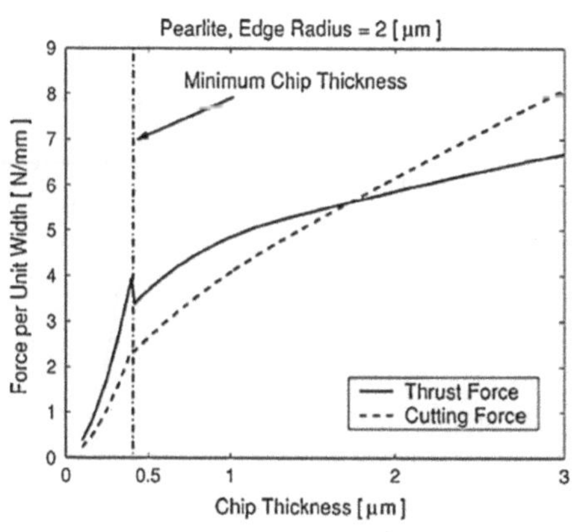

Fig. 9.10 Change in swarf load/cutting force with cutting thickness. (from Liu et al.): "Chip Thickness" in the original literature is regarded as "Swarf Thickness"

where, p_p and p_s are the ploughing force loaded on a unit edge length and the shearing force loaded on a unit cutting area, respectively. b and t_l are width of cut and the uncut swarf thickness.

When t_l is large enough to ignore the first term, the cutting force may be approximated as:

$$F \cong p_s \cdot b \cdot t_1 \tag{9.25}$$

Because the cutting area is given by $b \cdot t_1$, Eq. (9.22) regards as the second term in Eq. (9.24). In addition, the second term in Eq. (9.24) depends upon t_l, the first term becomes relatively large to the second term in a small uncut swarf thickness.

(d) *Stress distribution in cutting*

The stress distribution on the rake face can be measured by using the "Split tool method" as shown in Fig. 9.11a. By this method, the normal and friction stresses are determined as follows.

$$\left.\begin{aligned} \sigma_t &= \frac{N_{f1} - N_{f2}}{\Delta A_c} = \frac{N_{r2} - N_{r1}}{\Delta A_c} \\ \tau_t &= \frac{F_{f1} - F_{f2}}{\Delta A_c} = \frac{F_{r2} - F_{r1}}{\Delta A_c} \end{aligned}\right\} \tag{9.26}$$

where, ΔA_c is the differential cutting area.

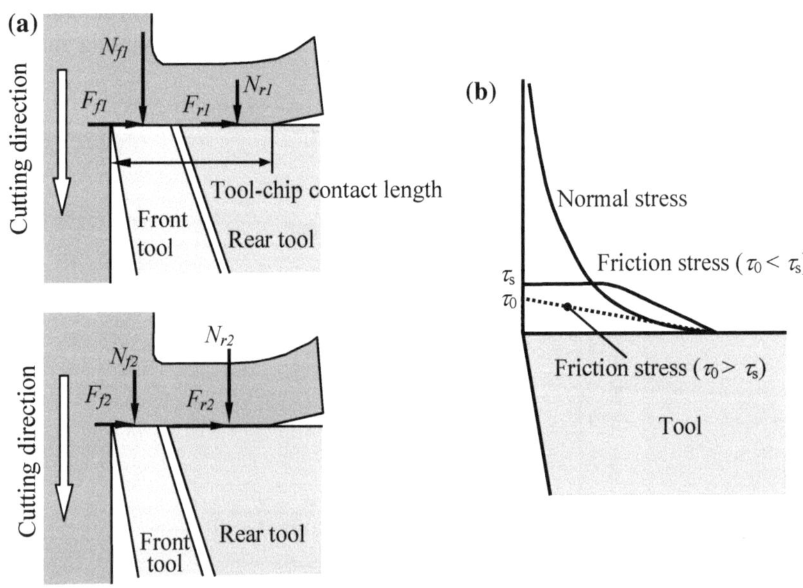

Fig. 9.11 Stress distributions on rake face: **a** measurement of stress distributions and **b** stress distributions

According to the measurement, the normal stress is approximated as an exponential distribution as shown in Fig. 9.11b, whereas the friction stress is distributed as a triangular or a trapezoidal function. When the friction stress is approximated as triangular distribution and its magnitude at the tool tip is larger than the shear stress on the shear plane, the trapezoidal distribution is applied.

9.1.3 *Cutting Temperature*

Within a cutting temperature context, we must discuss the measuring method and analysis.

(a) *Temperature measurement*

The cutting temperature can be measured by using various methods as shown in Fig. 9.12. Figure 9.12a shows the measurement of the average temperature at the cutting point by the thermoelectric couple. In this case, the junction between the workpiece and the tool is established at the cutting position, where the temperature rises. In contrast, the temperature distribution can be measured with a split tool illustrated in Fig. 9.12b, in which the platinum wire and the swarf establish the

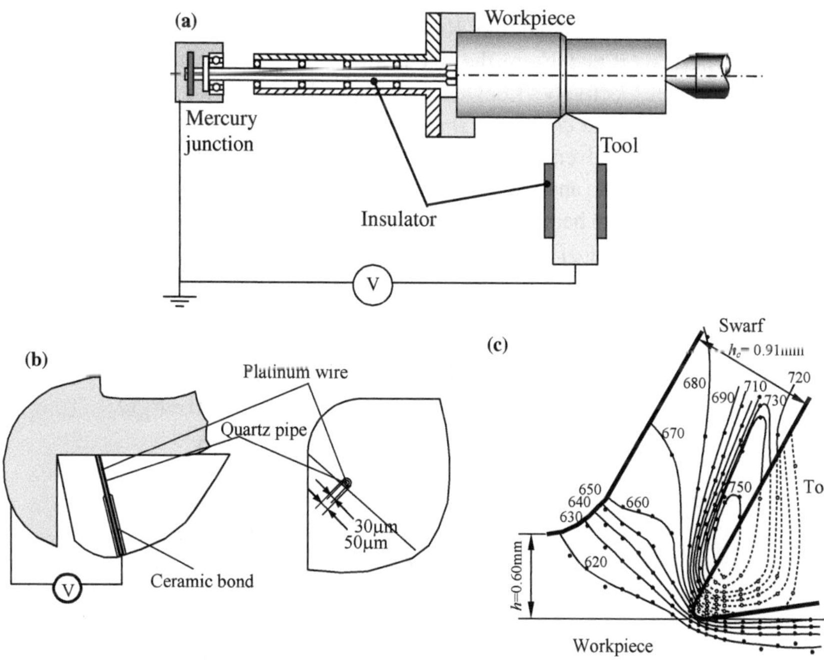

Fig. 9.12 Measurement of cutting temperature: **a** thermoelectric couple (average temperature), **b** thermoelectric couple (temperature distribution) and **c** infrared photograph

junction of the thermoelectric couple, and then the position of the junction can be moved by grinding the flank face. As a result, the temperature distribution is obtained with repeating measurement and grinding of flank face. Importantly, the temperature distribution is also detected by the non-contact measuring devices such as the radiation pyrometer and infrared photograph. These methods are very helpful to visualise the temperature distribution at a glance, and Boothroyd measured the temperature distribution as shown in Fig. 9.12c with an infrared photograph (Boothroyd 1961).

(b) *Analysis of the cutting temperature*

The cutting power is determined by the cutting velocity V and the cutting direction component of the cutting force F_H as $F_H \times V$. The power is consumed into the following energies:

(1) Shear energy on shear plane
(2) Friction energy on rake face
(3) Friction energy on flank face (on the worn tool)
(4) Momentum energy due to momentum change associated with metal as it crosses the shear plane
(5) Surface energy induced by the surface generation
(6) Residual energies in the subsurface and the swarf

Most of cutting energy is consumed in the shear and the friction energies, and as reported elsewhere, the energies (4), (5) and (6) are enough small to be ignored.

Lawen and Shaw analysed the cutting temperature in the model shown in Fig. 9.13, where t_1, b and V are the uncut swarf thickness, width of cut and the cutting velocity, respectively (Loewen and Shaw 1954). In this model, the shear plane inclines at an angle ϕ and the sharp tool without the flank wear land is used, and also the mechanical energies consumed in the cutting process are the shear energy on the shear plane and the friction energy on the rake face. All the mechanical energies convert into the heat generation, and the intensity of heat generation on the shear plane per unit area q_1 is associated with the shear energy as follows:

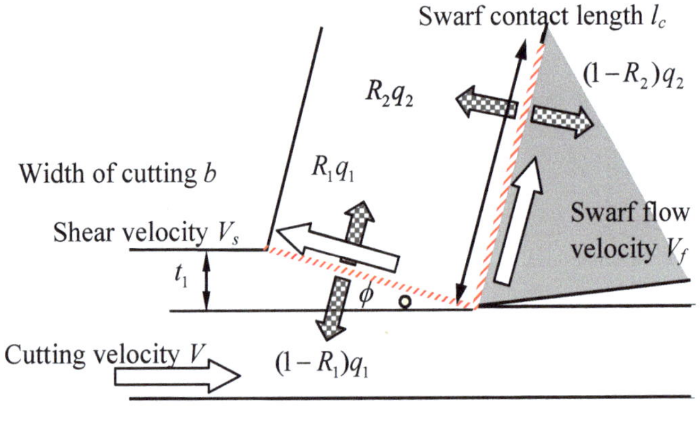

Fig. 9.13 Analysis model of cutting temperature

$$q_1 = \frac{F_s V_s}{J b t_1 \operatorname{cosec} \phi} \tag{9.27}$$

where, J is mechanical equivalent of heat. F_s and V_s are the shear force and the shear velocity, respectively.

In consideration that a part of the heat generation is consumed in the temperature rise in the swarf, and when the ratio of the heat conduction into the swarf is assumed to be R_1, $(1 - R_1)q_1$ is conducted into the workpiece. Based on the heat conduction into the swarf, the temperature $\bar{\theta}_s$ on the shear plane is given by:

$$\bar{\theta}_s = R_1 q_1 \frac{t_1 b \operatorname{cosec} \phi}{V t_1 b \rho_w c_w} + \theta_0 \tag{9.28}$$

where, θ_0 is the room temperature.

ρ_w and c_w are the density and the specific heat capacity of the workpiece.

The temperature rise on the shear plain is also obtained by the heat conduction into the workpiece. According to Jaeger's theoretical analysis, the mean and the maximum surface temperatures, $\bar{\theta}$ and $\theta_{\max}$, caused by the heat source moving at a velocity of V shown in Fig. 9.14a are:

$$\begin{aligned}
&L > 5 \quad \left.\begin{aligned} \bar{\theta} &= \frac{4}{3\sqrt{\pi}} \frac{ql}{k\sqrt{L}} = 0.752 \frac{ql}{k\sqrt{L}} \\ \theta_{\max} &= \frac{2}{\sqrt{\pi}} \frac{ql}{k\sqrt{L}} - 1.128 \frac{ql}{k\sqrt{L}} = 1.5\bar{\theta} \end{aligned}\right\} \\
&L < 5 \quad \left.\begin{aligned} \bar{\theta} &= \frac{2}{\pi} \frac{Kq}{kV} \bar{f} = 0.636 \frac{Kq}{kV} \bar{f} \\ \theta_{\max} &= \frac{2}{\pi} \frac{Kq}{kV} f_m = \frac{f_m}{\bar{f}} \bar{\theta} \end{aligned}\right\} \\
&K = \frac{k}{\rho c} \\
&L = \frac{Vl}{2K}
\end{aligned} \tag{9.29}$$

where, q is the intensity of heat generation and $2l$ is the width of the heat source. f_m and $\bar{f}$ are shown in Fig. 9.14b.

In order to carry out the temperature analysis on the shear plane due to the heat conduction into the workpiece, a continuous plane is assumed with extending the shear plane at an angle ϕ as shown in Fig. 9.14c. Because L in Eq. (9.29) is generally larger than 5 in the cutting process, the temperature on the shear plane is obtained by:

$$\bar{\theta}_s = \frac{0.752(1 - R_1) q_1 t_1 \operatorname{cosec} \phi}{2 k_w \sqrt{L_1}} + \theta_0 \quad L_1 = \frac{V_s t_1 \operatorname{cosec} \phi}{4 K_1} \quad K_1 = \frac{k_w}{\rho_w c_w} \tag{9.30}$$

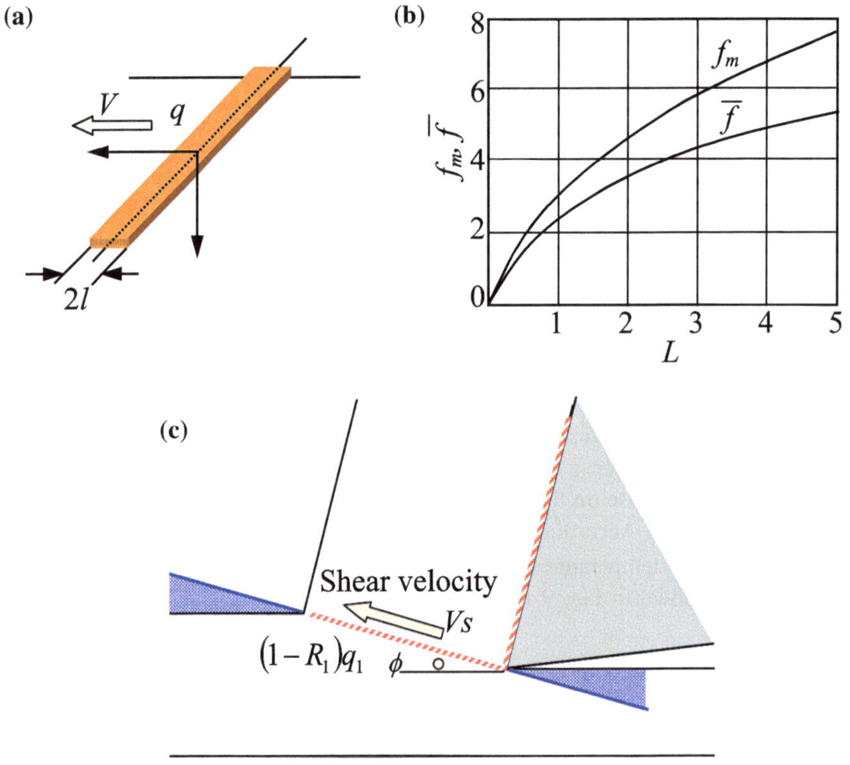

Fig. 9.14 Temperature analysis on shear plane: **a** moving heat source conduction, **b** parameters $\bar{f}$ and f_m and **c** analysis

where, k_w, ρ_w and c_w are thermal conductivity, density and specific heat capacity of the workpiece, respectively.

Because Eqs. (9.28) and (9.30) are the temperature on the shear plane, R_1 is acquired by:

$$R_1 = \frac{1}{1 + \frac{0.664\gamma}{\sqrt{L_1}}} = \frac{1}{1 + 1.328\sqrt{\frac{K_1 \gamma}{V t_1}}} \quad \gamma = \frac{V_s}{V \sin \phi} \tag{9.31}$$

Thus, $\bar{\theta}_s$ is acquired by substituting the determined R_1 into Eq. (9.28) or (9.30).

The intensity of the heat generation per unit area induced by the friction energy on the rake face q_2 is:

$$q_2 = \frac{F_f V_f}{J l_c b} \tag{9.32}$$

where, F_f, V_f and l_c are the friction force, the swarf velocity and the tool-swarf contact length, respectively.

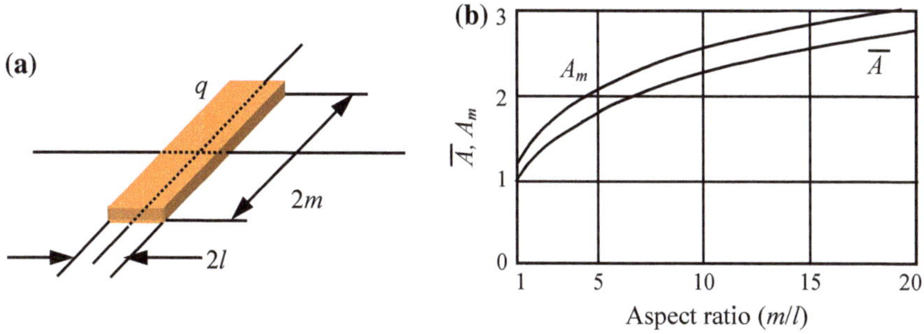

Fig. 9.15 Temperature analysis on rake face: **a** heat source and **b** parameters $\overline{A}$ and A_m

When the ratio of the heat conduction into the swarf is assumed to be R_2, $(1 - R_2)q_2$ is conducted into the tool. The heat generation on the rake face is regarded as the heat source moving at a swarf velocity on the back of the swarf. Based on the heat conduction into the swarf, the temperature rise is obtained by:

$$\bar{\theta}_t = \bar{\theta}_s + \frac{0.752 R_2 q_2 l_c}{2k_c\sqrt{L_2}} \quad L_2 = \frac{V_f l_c}{4K_2} \quad K_2 = \frac{k_c}{\rho_c c_c} \tag{9.33}$$

where, k_c, ρ_c and c_c are thermal conductivity, density and specific heat capacity of the swarf, respectively.

The heat source does not move on the rake face in terms of the tool and the heat conduction continuously occurs. In Jaeger's theoretical analysis, the mean and the maximum surface temperatures, $\bar{\theta}$ and $\theta_{\max}$, caused by the heat source shown in Fig. 9.15a are:

$$\left.\begin{aligned} \bar{\theta} &= \frac{ql}{k}\overline{A} \\ \theta_{\max} &= \frac{ql}{k}A_m \end{aligned}\right\} \tag{9.34}$$

where, q is heat generation per unit area and k is the thermal conductivity. $\overline{A}$ and A_m are parameters associated with aspect ratio of the heat source m/l in Fig. 9.15b.

The temperature on the rake face is given by the heat conduction into the tool:

$$\bar{\theta}_t = \frac{(1 - R_2)q_2 l_c}{k_t}\overline{A} + \theta_0 \tag{9.35}$$

where, k_t is thermal conductivity of the tool.

Because Eqs. (9.33) and (9.35) are the temperature on the rake face, R_2 is acquired by:

$$R_2 = \frac{q_2\dfrac{l_c\overline{A}}{k_t} - \bar{\theta}_s + \theta_0}{q_2\dfrac{l_c\overline{A}}{k_t} + q_2\dfrac{0.377 l_c}{k_c\sqrt{L_2}}} \tag{9.36}$$

Thus, $\bar{\theta}_t$ is obtained by substituting the determined R_2 into Eq. (9.33) or (9.35).

9.1.4 Tool Wear

Within a tool wear context, we must discuss the tool material and tool life equation. Of these, the tool life equation is extremely important to determine the cutting condition in practice from the economisation viewpoint.

(a) *Tool materials*

Several materials have been applied to the tool for cutting. In retrospect, the tool steel and high-speed steel were prevailed, although the tool steel and high-speed steel should be hardened by heat treatment. Thus, these tool materials cannot be used at the high cutting temperature. Nowadays, the cemented carbide, ceramics, cBN and diamond can commonly facilitate the cutting tool material, and each material can be characterised as follows:

(1) Cemented carbide tools consist of hard grains made of WC, TiC and/or TiN, and are manufactured by sintered process with the bonding materials such as Co.
(2) Ceramic tools are made of Al_2O_3, TiC, TiO, TiN, Cr_2O_3 and/or ZrO_2, and are also manufactured by sintered process with catalysts.
(3) cBN (cubic boron nitride) has high hardness and high thermal conductivity, and in general cBN tools are usually used for cutting of hard material such as high carbon steels.
(4) Diamond is, as well known, the hardest material made of carbon. The diamond tools are used for cutting of non-carbon materials such as aluminum alloys.

More importantly, the functional requirements of the cutting tools are high wear resistance and high fracture resistance. The tool wear occurs on the surface of the tool, whereas the tool fracture is caused by low toughness of the tool material. Thus, coating tools have recently been developed as a hybrid type, in which the hard materials are deposited physically or chemically on the tough material. TiC, TiN and/or Al_2O_3 are coated by the chemical vapor deposition, while TiAlN or TiSiN is by the physical vapor deposition.

(b) *Tool life equation*

The tool wear occurs in the cutting operations to a various extent as illustrated in Fig. 9.16, and the crater and flank wears are representative in the determination of the tool life. The crater wear generally occurs on the rake face caused by sliding of the swarf at high temperatures. The crater wear can be regulated by the depth K_T. In contrast, the flank wear occurs on the flank face by sliding between the finished surface and the clearance surface of tool, and can be regulated by the width of the land V_B. The grooving wear or the notched wear occurs at the boundary of cutting area, and it is reported that the grooving wear is caused by a large strain hardening and the chemical interaction with atmosphere.

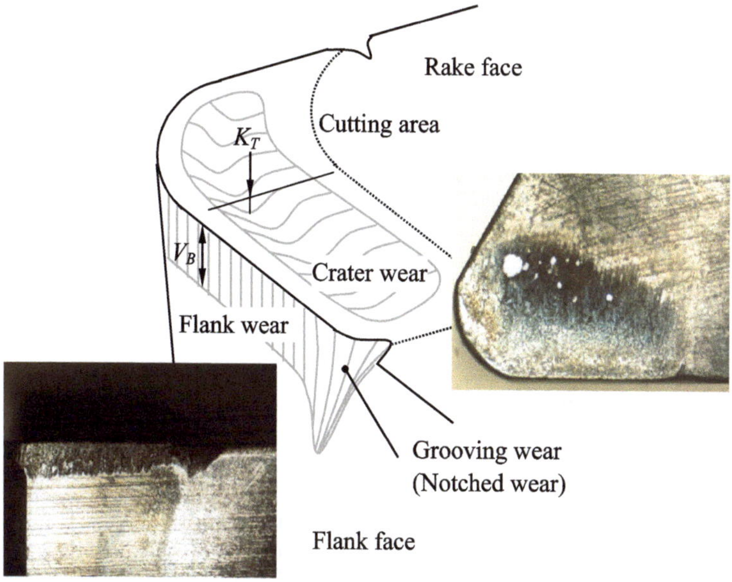

Fig. 9.16 Tool wears

Figure 9.17 illustrates the wear progresses with the cutting time. As can be readily seen, the crater wear increases linearly with the time. Meanwhile, the progress of the flank wear consists of the initial wear, the steady wear and the rapid wear. More specifically, the initial wear increases at a high rate, because of the unstable contact of the tool with the workpiece surface. Succeedingly, the steady wear increases at a relatively lower rate, and its progress depends upon the combination of materials, the stress and the temperature on the tool face. Then, the rapid wear occurs at a high temperature due to a great frictional energy on the flank wear land.

Normally the critical wear is a specified index to determine the tool life. Because the tool wear largely depends upon the cutting temperature, the wear rate increases with the cutting velocity. Figure 9.17b shows the relation between the cutting velocity V and the tool life T. Based on the linear relation in the logarithmic chart, the tool life equation, which is called Taylor's equation, is derived as follows:

$$VT^n = C \tag{9.37}$$

where, n and C are the parameters given by combination of the tool and the workpiece.[1]

[1]Expression of parameter n and C are the conventional use in Taylor's equation.

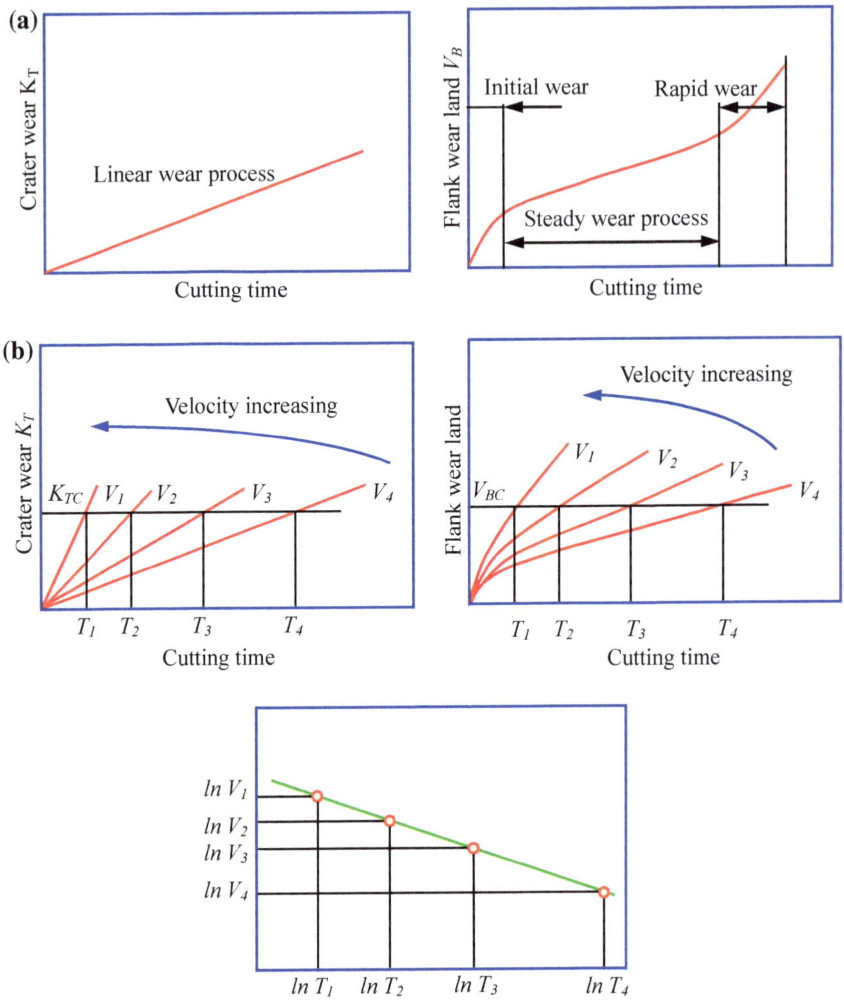

Fig. 9.17 Tool wear progress and tool life: **a** tool wear progress and **b** tool life

Similarly, the following equation associates the tool life with the cutting parameters.

$$V^{\xi_V} f^{\xi_f} d^{\xi_d} T^n = C \tag{9.38}$$

where, ξ_V, ξ_f, ξ_d, n and C are parameters, respectively.

9.1.5 *Surface Quality*

We have often used the term "Surface quality" so far; however, the surface quality is not defined quantitatively as yet. As a result, we have certain difficulties in consideration of to what extent the related attributes range. Admitting such a situation surrounding the surface quality, we will discuss it by taking the machining error, surface finish and affected layer as the representative attributes.

(a) *Machining accuracy*

The machining error is controlled by those related to the machine tool factors, the operation factors and the process factors. Because the cutter paths are controlled numerically, the machining error, i.e., geometrical accuracy in narrow scope, occurs due to the interpolation error in NC. In contrast, the motion accuracy, the dynamic response and the thermal deformation of the machine tool have influences on the machining accuracy in general. In addition, the machining error also largely depends upon the clamping accuracy of the workpiece and the tool controlled by the operator's skill or the performance of automated devices.

Of these various factors, the process factors associated closely with the machining error are the deformation of the workpiece and the tool, and the tool wear in cutting. The cutting force induces the mechanical deformation of the workpiece and the tool. The dynamic responses of them should also be considered in the machining accuracy. Importantly, the thermal deformation is caused by the heat generation of the cutting point, and the thermal expansion of the tool largely depends on the cutting temperature. The swarf control should not be ignored when the heated swarfs are left on the workpiece and the tool. More importantly, the flank wear leads to the geometrical error due to the change in the cutting edge shape, and the cutting force and the cutting temperature increase with the flank wear land. Thus, the mechanical and the thermal deformations increase with the tool wear progress.

(b) *Surface finish*

The control factors of the surface finish are classified into the geometrical profile, the material behaviour, the tool wear and the rigidity of the machine tool. The geometrical profile is controlled by the cutting parameters. For example, the surface is profiled by the tool geometry and the feed per revolution f in turning as shown in Fig. 9.18a. Reportedly elsewhere, the maximum surface roughness is approximately given by:

$$R_{\max} \cong \frac{f^2}{8r} \tag{9.39}$$

where, r is the nose radius of the tool.

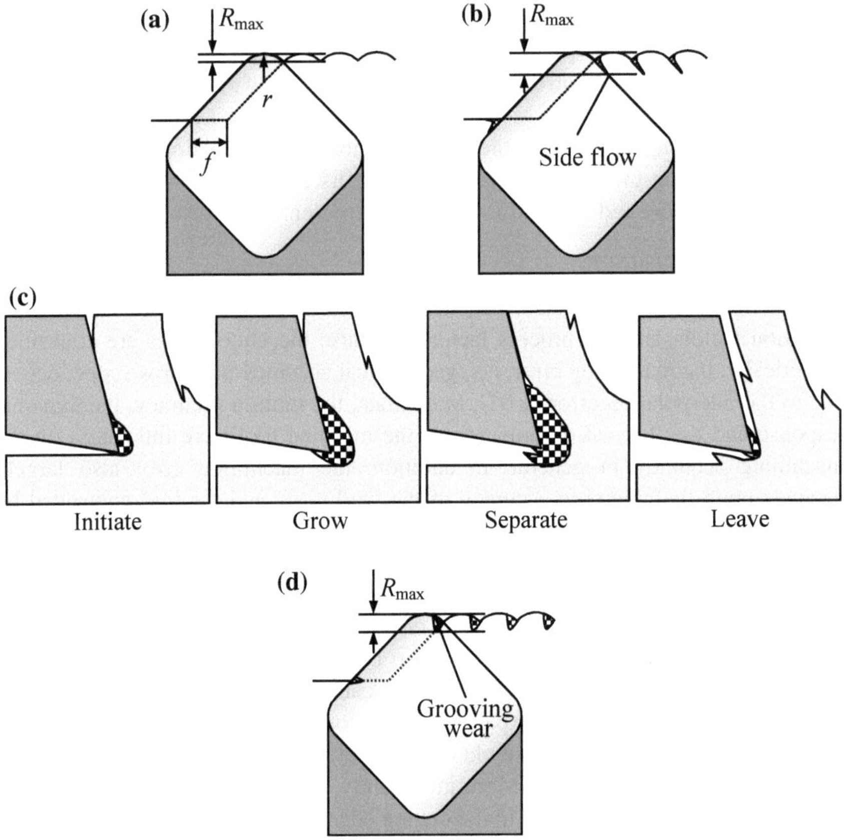

Fig. 9.18 Surface roughness in turning process: **a** geometrical roughness, **b** side flow, **c** built-up edge and **d** tool wear

As for the material behaviour, the side flow occurs around the boundary between the cutting and the non-cutting area and increases the surface roughness as shown in Fig. 9.18b. In addition to the side flow of the material, the built-up edge has a large influence on the surface roughness. The built-up edge is formed in accordance with the process schematically shown in Fig. 9.18c and induces the excessive cutting depth and the vibration, resulting in the deterioration of the surface roughness.

The grooving wear (the notched wear) on the end cutting edge increases the height of surface roughness as illustrated in Fig. 9.18d. Furthermore, when the workpiece is machined by the tool with large flank wear, the surface finish is deteriorated by a large friction between the tool and finished surface.

As can be readily seen, the vibrations in the machine tool-workpiece-cutting tool system in the cutting process have influences on the surface finish. The dynamic responses of them should be thus measured to improve the surface finish and to determine the appropriate cutting parameters.

(c) *Affected layer*

The affected layer changes the material properties not only on surface but also in subsurface. The residual stress is caused by the localised deformation, change in the localised volume due to the thermal expansion and the crystal transformation occurring in subsurface. The temperature rise in subsurface induces the crystal transformation of the material. Then, the material properties change with the cutting temperature. With the growing importance of the surface quality of the part, the affected layer becomes a crucial problem in the academia. In fact, the subsurface with less residual stress is demanded in the automobile part to improve the fuel efficiency.

9.2 Modelling of Grinding

9.2.1 Grinding Wheel and Stone

Grinding is generally performed as a post-process of cutting. Figure 9.19 shows the grinding process, and the grinding wheel consists of the three leading elements:

(1) Abrasive grains
(2) Bond
(3) Pores

The abrasive grains, the size of which are usually around or smaller than 0.1 mm, remove the material by playing the role of the cutting edges. Thus, the removal volume is much less than that of cutting. The material of the abrasive grains should be chosen according to the workpiece material, and Al_2O_3, Al_2O_3 +Cr_2O_3, SiC, ZrO_2, cBN and diamond are usually used for the abrasive grains. Importantly, the surface roughness of the finished workpiece depends upon the size of the grains, and the density of grains in a wheel also controls the grinding performance.

More importantly, the abrasive grains are bonded onto the grinding wheel body, and thus the bonding forces are regarded as the clamping force of grains.

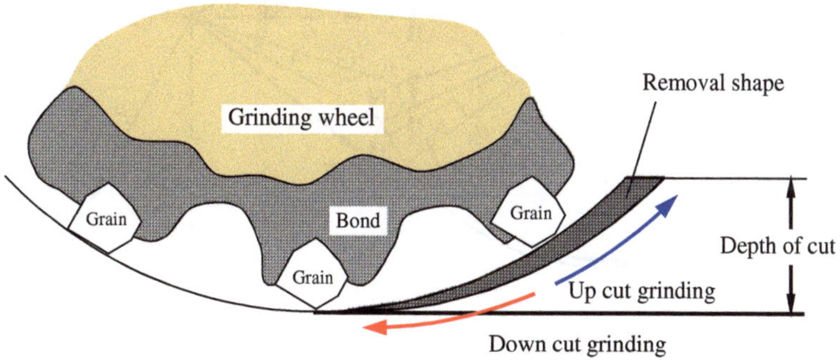

Fig. 9.19 Grinding process

In consequence, the bonding stiffness, which depends on the bonding material, has an influence on the grinding performance as well as the abrasive grain. Sodium silicate, MgO, $MgCl_2$, metal, Resin and rubber are used for bonding the grains.

The pores evacuate the swarf in the removal area by forming the swarf pockets. When the swarf pockets are clogged, dressing should be conducted to improve the removal performance.

Summarising, the different points of grinding from cutting are:

(1) The sizes and the orientations of the abrasive grains are not uniform.
(2) Dressing operation is required for forming of the wheel and the removing the pores clogged with the swarf and the worn grains. If the self-dressing occurs, dressing is not be required. More specifically, the self-sharpening of the abrasive grain, one of the self-dressing effects, facilitates the enhancement of the grinding performance, and furthermore be characterised the grinding processes.

9.2.2 Grinding Model

In discussing the grinding mechanism, it is very convenient to consider the similarity with the milling processes, although the cutting edge should be regarded as the very small blade.

Figure 9.20 shows the removal process in plunge grinding when the removal depth is Δ. In this model, the workpiece and the grinding wheel rotate at angular

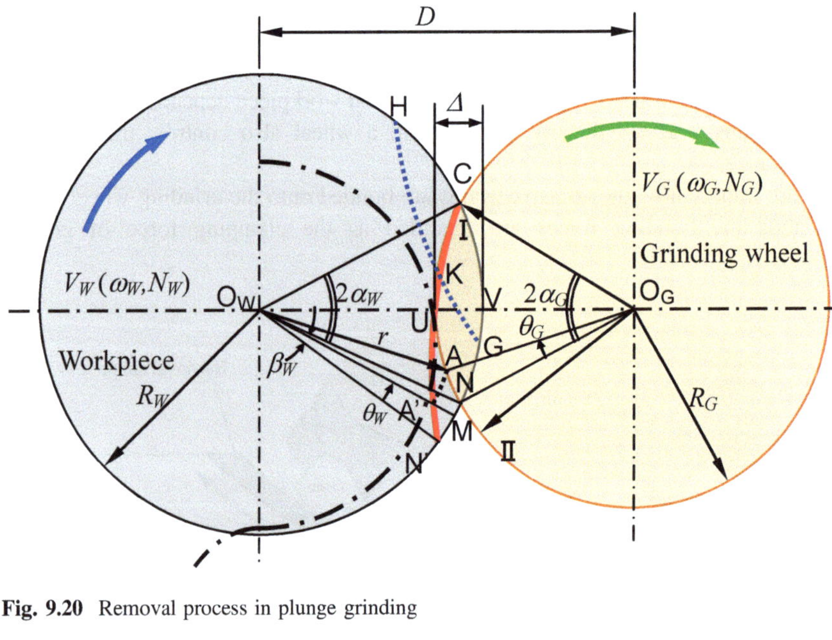

Fig. 9.20 Removal process in plunge grinding

velocities of ω_W and ω_G, their radii are R_W and R_G and then the circumferential speeds are V_W and V_G. As can be seen from Fig. 9.20, the angles of contact area are α_W and α_G, and consequently, the removal time t and the rotation angle β_W of the workpiece during t are:

$$\left.\begin{array}{l} t = 2\alpha_G / \omega_G \\ \beta_W = t\omega_W = 2\alpha_G K_W \end{array}\right\} \tag{9.40}$$

The removal shape is controlled by the ratio of both of the radii, the angular velocities, the circumferential speeds and the angles of the contact area:

$$\left.\begin{array}{l} K_R = R_W / R_G \\ K_W = \omega_W / \omega_G \\ K_V = V_W / V_G = K_W \cdot K_R \\ K_\alpha = \alpha_W / \alpha_G \end{array}\right\} \tag{9.41}$$

A grain engages in the workpiece at the point N and rotates at an angular of α_G and then exits at the point C. The workpiece also rotates at an angular velocity of ω_W during the removal process. Thus, Point N on the workpiece rotates to N′ when the grain finishes removal. The locus of the grain traces along N′A′KC on the workpiece. In order to obtain the locus, the radial depth y = MA′ is associated with the tangential length x = MN′.

The length of the locus L_W is:

$$L_W = R_W(2\alpha_W + \beta_W) = 2\alpha_W R_W(1 + K_W / K_\alpha) = 2\alpha_G R_G(1 + K_V) \tag{9.42}$$

From a geometrical consideration, the following expressions are obtained:

$$\left.\begin{array}{l} R_G \sin\alpha_G = R_W \sin\alpha_W \\ D = R_G + R_W - \Delta = R_G \cos\alpha_G + R_W \cos\alpha_w \end{array}\right\} \tag{9.43}$$

The angle of the contact area is approximately given by:

$$\alpha_G = \sqrt{\frac{K_R}{1 + K_R}\frac{2\Delta}{R_G}} \tag{9.44}$$

When the grain rotates at an angle of θ_G after the grain's engagement, Point A is removed. Then, Point A rotates to Point A′ at the exit of grain from the workpiece. The θ_G and the angular delay θ_W with respect to N′ are:

$$\left.\begin{array}{l} \theta_W = \dfrac{x}{R_W} = \dfrac{2\alpha_W + \beta_W}{2\alpha_G}\theta_G = (K_\alpha + K_W)\,\theta_G \\ \theta_G = \dfrac{x}{R_W(K_\alpha + K_W)} = \dfrac{x}{R_G(1 + K_V)} = c_\theta x \end{array}\right\} \tag{9.45}$$

The radius r designated by O_wA or O_wA' is given by the cosine formula:

$$\begin{aligned} r &= \sqrt{R_G^2 + D^2 - 2R_G D\cos(\alpha_G - \theta_G)} \\ &\cong (D - R_G)\left\{1 + \frac{R_G D}{2(D - R_G)^2}(\alpha_G - \theta_G)^2\right\} \\ &= (R_W - \Delta) + \frac{R_G + R_W}{2K_R}(\alpha_G - c_\theta x)^2 \end{aligned} \tag{9.46}$$

Consequently, the locus of the grain is obtained by:

$$y = R_W - r = \Delta - \frac{R_G + R_W}{2K_R}(\alpha_G - c_\theta x)^2 \tag{9.47}$$

On the basis of Fig. 9.21a, the successive cutting edge spacing is defined as the average distance between the grains on a line. When the abrasive grains I and II are aligned at a successive cutting edge spacing δ as shown in Fig. 9.21b, OO′DBA is the removed area and L_O = AB is:

$$L_0 = K_V \cdot \delta \tag{9.48}$$

Both of the grains traces along:

$$\left.\begin{aligned} y_1 &= \Delta - \frac{R_G + R_W}{2K_R}(\alpha_G - c_\theta x)^2 \\ y_2 &= \Delta - \frac{R_G + R_W}{2K_R}[\alpha_G - c_\theta(x - L_0)]^2 \end{aligned}\right\} \tag{9.49}$$

Figure 9.21c shows the removal area, and in due course, the cutting depth is expressed as follows:

$$\left.\begin{aligned} t_c &= y_1 = \Delta - \frac{R_G + R_W}{2K_R}(\alpha_G - c_\theta x)^2 & (0 \le x \le L_0) \\ t_c &= y_1 - y_2 = \frac{1}{2}\frac{1 + K_R}{K_R}\frac{K_V}{(1 + K_V)^2}\frac{\delta}{R_G}(L + L_0 - 2x) & (L_0 \le x \le L_c) \end{aligned}\right\} \tag{9.50}$$

The maximum and the mean thickness, $t_{\max}$ and t_m, are:

$$t_{\max} = \frac{1}{2}\frac{1 + K_R}{K_R}\frac{K_V}{(1 + K_V)^2}\frac{\delta}{R_G}(L - L_0) \tag{9.51}$$

$$t_m = \frac{1}{4}\frac{1 + K_R}{K_R}\frac{K_V}{(1 + K_V)^2}\frac{\delta}{R_G}(L - L_0) \tag{9.52}$$

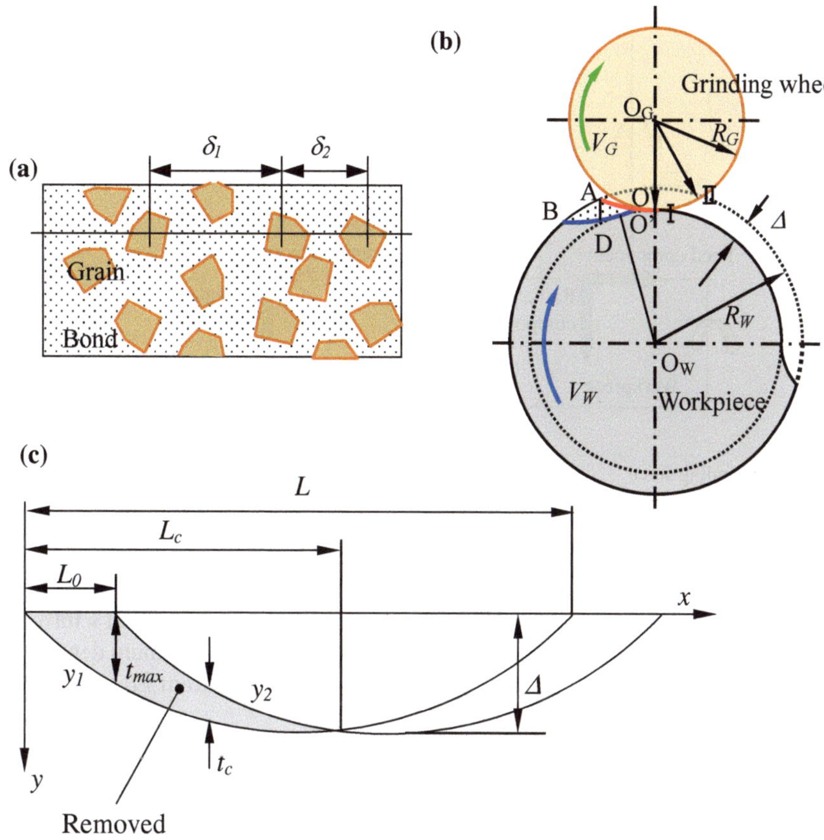

Fig. 9.21 Removal model in grinding process: **a** successive cutting edge spacing, **b** model of abrasive grains alignment and **c** removal area

The removal length L_c is:

$$L_c = \frac{1}{2}(L + L_0) \tag{9.53}$$

9.2.3 Grinding Force

The grinding force is decomposed into the tangential, the radial and the feed components as shown in Fig. 9.22. The force F is associated with the cutting parameters as follows:

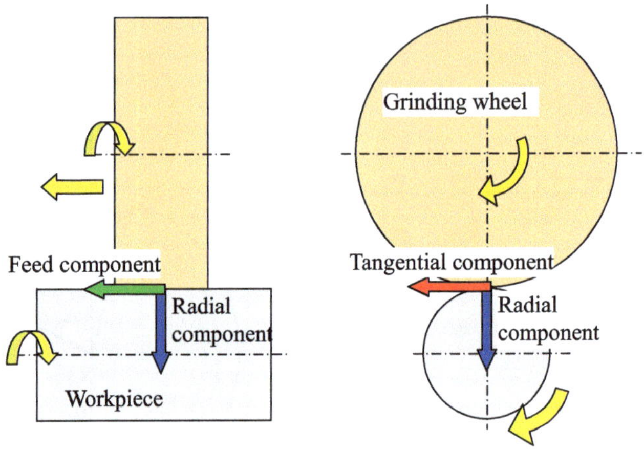

Fig. 9.22 Grinding force

$$F = \xi_k \Delta^{\xi_\Delta} V_G^{-\xi_G} V_W^{\xi_W} f_T^{\zeta} \tag{9.54}$$

where, f_T is the traverse feed when the grinding operation is conducted in a traverse manner. ξ_k, ξ_Δ, ξ_G, ξ_W and ζ are constants. By substituting the maximum depth $t_{\max}$ obtained from Eq. (9.51) and the removal length L_c obtained from Eq. (9.53), the removal size is approximately given by:

$$t_{\max} \cdot L_c = 2K_V \cdot \Delta \cdot \delta \tag{9.55}$$

where, δ is the successive cutting edge spacing. Thus, Eq. (9.54) can be written as:

$$P = \xi_k (K_V \cdot \Delta)^{\xi} f_T^{\zeta} = \xi_k \Delta^{\xi} V_G^{-\xi} V_W^{\xi} f_T^{\zeta} \tag{9.56}$$

where, ξ is constants, and in plunge grinding, the term f_T in Eq. (9.56) should be removed.

When the grain shape is assumed to be a cone shape shown in Fig. 9.23, the radial and the tangential components, p_g and q_g, for the single grain are:

$$\left.\begin{array}{l} p_g = \sigma_g t_m^2 \sin\varphi_g \tan\varphi_g \\ q_g = \frac{\pi}{4}\sigma_g t_m^2 \sin\varphi_g \end{array}\right\} \tag{9.57}$$

where, σ_g and φ_g are the constant and the vertical angle of the cone, respectively. The mean thickness is given by:

$$t_m \cong \frac{\delta}{\sqrt{2}} \frac{K_V}{1+K_V} \sqrt{\frac{R_G + R_W}{R_G R_W}\Delta} \tag{9.58}$$

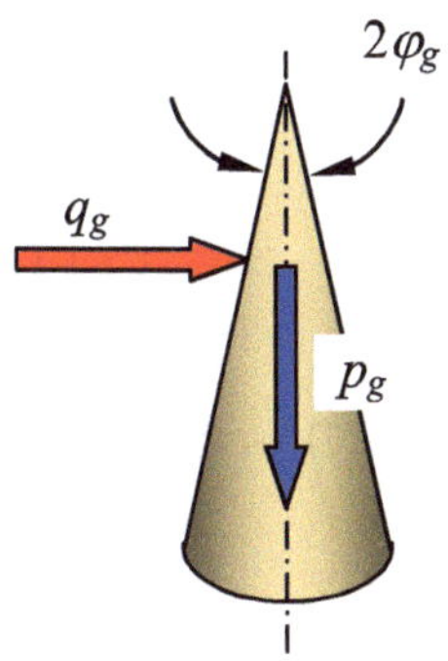

Fig. 9.23 Abrasive grain model

Because the removal length on the wheel is UC on the workpiece in Fig. 9.20, the length of the removal area is:

$$l_G = R_G \alpha_G \tag{9.59}$$

When the grinding width is B, the contact area is:

$$A = Bl_G = \sqrt{2}B\sqrt{\frac{R_G R_W}{R_G + R_W}\Delta} \tag{9.60}$$

The number of the simultaneously cutting edges n is associated with the mean spacing of the grains w:

$$n = \frac{A}{w^2} = \sqrt{2}\frac{B}{w^2}\sqrt{\frac{R_G R_W}{R_G + R_W}\Delta} \tag{9.61}$$

The radial and the tangential components are estimated as:

$$\left.\begin{array}{l} P = \dfrac{\sigma_g B}{\sqrt{2}}\left(\dfrac{\delta}{w}\right)^2\left(\dfrac{K_V}{1+K_V}\right)^2\left(\dfrac{R_G+R_W}{R_G R_W}\right)^{0.5}\Delta^{1.5}\tan\varphi_g \sin\varphi_g \\ Q = \dfrac{\pi\sigma_g B}{4\sqrt{2}}\left(\dfrac{\delta}{w}\right)^2\left(\dfrac{K_V}{1+K_V}\right)^2\left(\dfrac{R_G+R_W}{R_G R_W}\right)^{0.5}\Delta^{1.5}\sin\varphi_g \end{array}\right\} \tag{9.62}$$

9.2.4 *Grinding Temperature*

Within the grinding temperature context, we must discuss the measurement method and theoretical analysis for the individual and overall problems, e.g., those at abrasive grains, grinding wheel as a whole and workpiece.

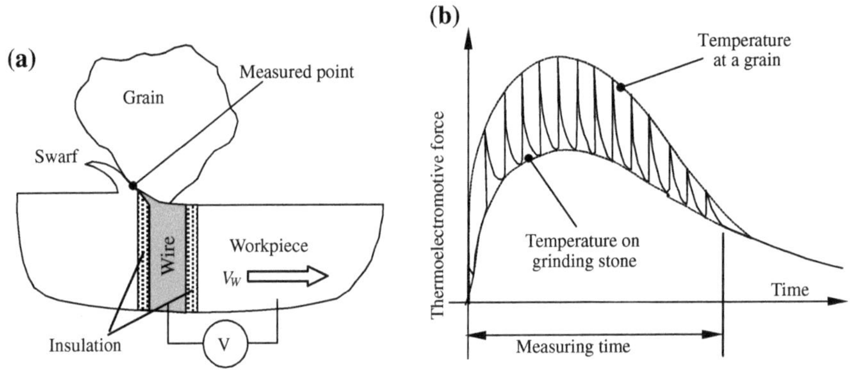

Fig. 9.24 Measurement of temperature at abrasive grains: **a** thermoelectric couple and **b** change in thermoelectromotive force

(a) *Temperature measurement*

Figure 9.24a shows a measurement method of temperatures at abrasive grains by a thermoelectric couple. The wire is embedded in the workpiece with an electrical insulating pipe. An electrical junction facilitates the measurement of the thermoelectromotive force at a contact point of a wire with the material in the removal process as shown in Fig. 9.24b. Thus, we can evaluate the temperatures at abrasive grains and on the grinding wheel by the peaks in the thermoelectromotive force and its envelope curve, respectively.

Figure 9.25a shows measurement of the temperature distribution in the workpiece by a thermoelectric couple. The relative distances between the measurement position and the wheel contact area change with giving the grinding depth and feeding the workpiece. The thermoelectromotive forces are measured sequentially in the depth from the surface as shown Fig. 9.25b. Then, the temperature distribution is finally evaluated as shown in Fig. 9.25c.

(b) *Temperature at an abrasive grain*

Figure 9.26 shows an analysis model for temperature at an abrasive grain. In this model, the removal energy is:

$$W = V \cdot t_1 \cdot b \cdot u_S \tag{9.63}$$

where, V, t_1 and b are the removal velocity, the removal depth and the removal width, respectively.

In consideration that u_s is the specific removal energy per volume, the heat generation on the shear plane per unit area is given by:

$$q = \frac{W}{Jbt_1 \operatorname{cosec} \phi} = \frac{u_S V}{J \operatorname{cosec} \phi} \tag{9.64}$$

where, J is mechanical equivalent of heat and ϕ is the shear angle.

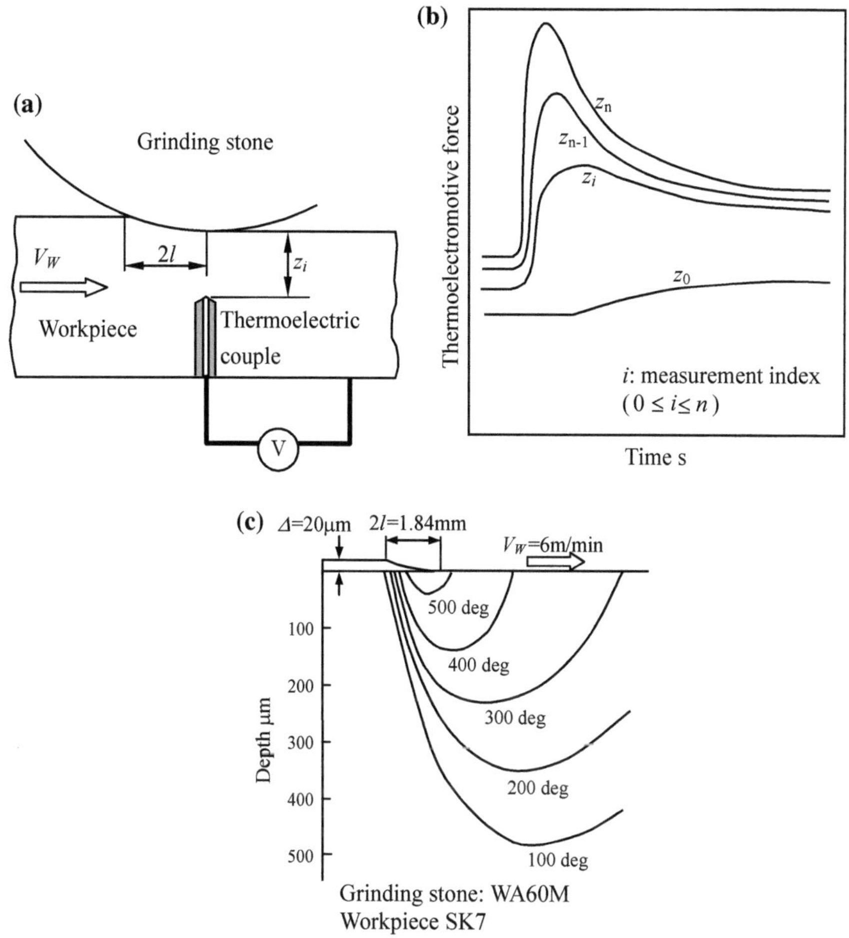

Fig. 9.25 Measurement of temperature distribution in workpiece: **a** thermoelectric couple, **b** temperature change and **c** temperature distribution in workpiece

When the ratio of the heat conduction into the swarf, R_1, is assumed, $(1 - R_1)q$ is consumed in the temperature rise in the workpiece. Based on the strength of the heat conduction into the swarf, the temperature rise on the shear plane $\bar{\theta}_g$ is given by:

$$\bar{\theta}_g = \frac{R_1 q b t_1 \operatorname{cosec} \phi}{c \rho V t_1 b} = \frac{R_1 q \operatorname{cosec} \phi}{c \rho V} \tag{9.65}$$

where, ρ and c are the density and the specific heat capacity of the workpiece, respectively.

Based on the heat conduction into the workpiece, Eq. (9.29) is applied to analysis of the temperature rise on the shear plane $\bar{\theta}_g$, and thus:

$$\left.\begin{aligned} \bar{\theta}_g &= 0.752\frac{(1-R_1)qt_1\operatorname{cosec}\phi}{2k\sqrt{L}} \quad (L \geq 5) \\ \bar{\theta}_g &= 0.636\tfrac{K(1-R_1)q}{kV}\bar{f} \quad (L<5) \end{aligned}\right\} \tag{9.66}$$

$$K = \frac{k}{\rho c}, L = \frac{Vt_1\operatorname{cosec}\phi}{4K}$$

where, k is thermal conductivity and $\bar{f}$ is obtained from the diagram already shown in Fig. 9.14b.

Because Eqs. (9.65) and (9.66) are the temperature rise on the shear plane, R_1 is acquired by:

$$\left.\begin{aligned} R_1 &= \frac{1}{1+0.665\operatorname{cosec}\phi/\sqrt{L}} \quad (L \geq 5) \\ R_1 &= \frac{1}{1+1.572\operatorname{cosec}\phi/\bar{f}} \quad (L<5) \end{aligned}\right\} \tag{9.67}$$

As a result, $\bar{\theta}_g$ is obtained by substituting R_1 into Eq. (9.65) or (9.66).

(c) *Temperature on grinding wheel*

On the strength of the heat conduction theory, the temperature on the grinding wheel is analysed using the mathematical model illustrated in Fig. 9.27. In the model, the arc AB can be replaced by the line AB, because the grinding depth Δ is small. When the tangential component of the grinding force Q is loaded on the wheel, the heat generation q is:

$$q = \frac{1}{J}\frac{QV_G}{2lB} \tag{9.68}$$

where, V_G and B are the circumferential speed and the grinding width, respectively.

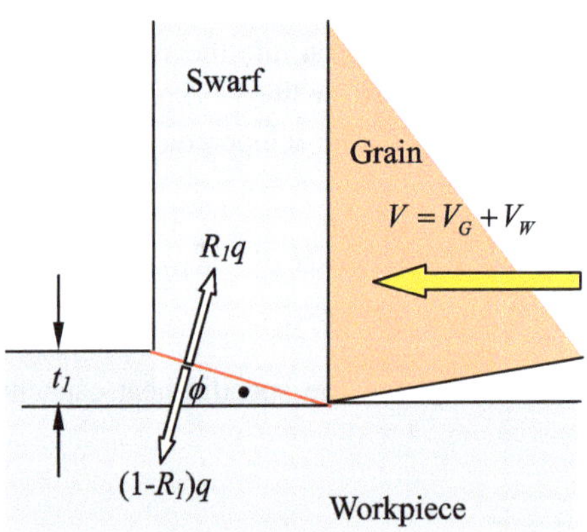

Fig. 9.26 Temperature analysis at an abrasive grain

The contact length $2l$ is:

$$2l = l_G = R_G \alpha_G = \sqrt{\frac{K_R}{1+K_R} 2\Delta R_G} \tag{9.69}$$

The ratio of the heat conduction into the workpiece is assumed to be R_w. Based on the moving heat source analysis, the temperature rise on the contact area is given by:

$$\bar{\theta}_w = 0.752 \frac{R_w q l_G}{2k\sqrt{L}}, \quad L = \frac{V_W l_G}{4K} > 5, \quad K = \frac{k}{\rho c} \tag{9.70}$$

where, k, ρ and c are the thermal conductivity, the density and the specific heat capacity of the workpiece, respectively, and also V_W is the feed rate of the workpiece.

The temperature rise on the contact area is also given by the following equation with assuming the ratio of the heat conduction into the grinding wheel, R_g:

$$\bar{\theta}_g = 0.752 \frac{R_g q l_G}{2k_g\sqrt{L_g}}, \quad L_g = \frac{V_G l_G}{4K_g} > 5, \quad K_g = \frac{k_g}{\rho_g c_g} \tag{9.71}$$

where, k_g, ρ_g and c_g are the thermal conductivity, the density and the specific heat capacity of the grinding wheel, respectively.

When the ratio of the heat conduction into the swarf is R_c, the temperature rise on the contact area is:

$$\bar{\theta}_c = R_c q B l_G \,/\, \rho c (V_W \Delta B) = \frac{R_c q l_G}{\rho c V_W \Delta} \tag{9.72}$$

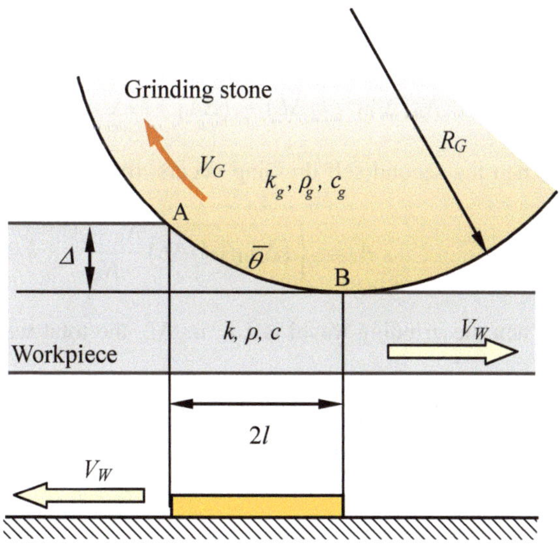

Fig. 9.27 Temperature on grinding wheel

R_w, R_g and R_c are determined to satisfy the following equations:

$$\left.\begin{array}{r}\bar{\theta}=\bar{\theta}_g=\bar{\theta}_c\\ R_w+R_g+R_c=1\end{array}\right\} \tag{9.73}$$

Consequently, those ratios are:

$$\begin{array}{l} R_w=\dfrac{1}{\Sigma}\quad R_g=\dfrac{1}{\Sigma}\dfrac{k_g}{k}\sqrt{\dfrac{L_g}{L}}\quad R_c=\dfrac{1}{\Sigma}\dfrac{0.752V_W\Delta}{2K\sqrt{L}}\\ \Sigma=1+\dfrac{k_g}{k}\sqrt{\dfrac{L_g}{L}}+\dfrac{0.752V_W\Delta}{2K\sqrt{L}}\end{array} \tag{9.74}$$

Then, temperature rise is, as can be readily seen, obtained by substituting these ratios into Eqs. (9.70), (9.71) or (9.72).

9.2.5 *Wheel Life*

The grinding wheel life is defined as the period from a dressing to the succeeding dressing. In short, the wheel life is analysed using a narration-like mathematical model described below:

(1) N_0 grains on the grinding wheel remove the material simultaneously and the average wear area of a grain is ΔA_0 at the beginning of grinding.
(2) The flank wear of a grain increases in proportion to the grinding travel, where the proportionality constant is k_f.
(3) Then, the grinding force increases with the flank wear.
(4) The self-dressing occurs in an incremental travel of ΔL and the wheel wear area reduces by the average wear area of a grain.

When the first self-dressing occurs, the total wear area changes to:

$$A_1=N_0\left(\Delta A_0+k_f\Delta L\right)-\left(\Delta A_0+k_f\Delta L\right)=(N_0-1)\left(\Delta A_0+k_f\Delta L\right) \tag{9.75}$$

When the second self-dressing occurs, the total wear area changes to:

$$A_2=\left\{\left(\Delta A_0+k_f\Delta L\right)\frac{N_0-1}{N_0}+k_f\Delta L\right\}(N_0-1) \tag{9.76}$$

When the grinding travel is $L=n\cdot\Delta L$, the total wear area A_n is:

$$A_n=(N_0-1)\left[\Delta A_0\left(\frac{N_0-1}{N_0}\right)^{n-1}+k_f\Delta L N_0\left\{1-\left(\frac{N_0-1}{N_0}\right)^n\right\}\right] \tag{9.77}$$

The average grain life in the removal area L_f is obtained by:

$$L_f = \frac{1}{N_0}(\Delta L + 2\Delta L + \ldots. + N_0 \Delta L) = \frac{(N_0+1)\Delta L}{2} \cong \frac{N_0 \Delta L}{2} \tag{9.78}$$

Thus, the grinding travel until the nth self-dressing, L, is rewritten using L_f.

$$L = n \cdot \Delta L = \frac{2nL_f}{N_0} \quad or \quad \frac{n}{n_0} = \frac{L}{2L_f} \tag{9.79}$$

When n approach infinity, $\left(\frac{N_0-1}{N_0}\right)^n$ in Eq. (9.77) is approximated by $\exp\left(-\frac{n}{N_0}\right)$. Because N_0 is a large number, the total wear area A_T is expressed as:

$$\begin{aligned} A_T &= \Delta A_0 N_0 \exp\left(-\frac{L}{2L_f}\right) + 2k_f L_f N_0 \left\{1 - \exp\left(-\frac{L}{2L_f}\right)\right\} \\ &= 2N_0 k_f L_f \left\{1 - (1-\alpha)\exp\left(-\frac{L}{2L_f}\right)\right\} \end{aligned} \tag{9.80}$$

where, α_w is the ratio of the initial wear area to wear area at infinity L:

$$\alpha_w = \Delta A_0 \,/\, 2k_f L_f. \tag{9.81}$$

The flank wear ratio a_f in the removal area S is defined as $a_f = A_T \,/\, S$. The density of grains λ is given by $N_0 \,/\, S$. Thus, the flank wear ratio a_f is expressed as:

$$a_f = 2\lambda k_f L_f \left\{1 - (1-\alpha)\exp\left(-\frac{L}{2L_f}\right)\right\} \tag{9.82}$$

When the critical wear ratio a_c is given to conduct the dressing operation, the wheel life L_T is determined by:

$$L_T = -2L_f \cdot \ln\left\{\left(1 - \frac{a_c}{2\lambda k_f L_f}\right) \cdot \frac{1}{1-\alpha}\right\} \tag{9.83}$$

When $2\lambda k_f L_f \,/\, a_c$ is less than unit, the wheel life is not determined because self-dressing occurs on all of grains before the wear ratio a_f reaches a_c.

9.2.6 Surface Quality

The grinding operation is performed to finish the workpiece after cutting and heat treatment. Thus, the micro/sub-micro order of surface roughness is required for the

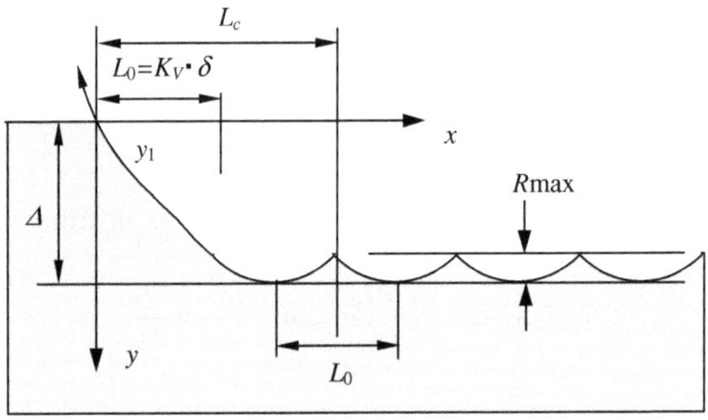

Fig. 9.28 Surface profile machined by abrasive grains

grinding process. Based on Fig. 9.28 and by referring to Eq. (9.47), a profile curve removed by grains is given by:

$$y_1 = \Delta - \frac{R_G + R_W}{2K_R}(\alpha_G - c_\theta x)^2 \tag{9.84}$$

The geometrical surface roughness is obtained by:

$$R_{\max} = \Delta - (y_1)_{x=L_c} = \frac{1}{8}\frac{1 + K_R}{K_R}\frac{K_V^2}{(1 + K_V)^2}\frac{\delta^2}{R_G} \tag{9.85}$$

However, the surface roughness measured in the grinding operations is usually larger than the geometrical roughness obtained theoretically. This is due to localised plastic deformation, adhesion of swarfs, the displacement of the wheel and the workpiece and randomness of the grain alignment.

In addition, (1) when the temperature is relatively high on the finished surface, the grinding burn occurs resulting in the change in the surface colour, (2) the hardness in subsurface also changes with the crystal transformation, and (3) the grinding cracks induced by the thermal stress deteriorate the material strength. Equally, the residual stress is caused by the localised mechanical deformation, the localised thermal stress and the localised volume expansion due to the crystal transformation.

9.3 Machinability

Machinability of a material is defined as how much more easy to remove an unnecessary allowance than other materials. Within the machinability context, we focus on machinability in cutting to be clear the machinability measures. Major interests for machinability are thus

(1) cutting force,
(2) cutting temperature,
(3) tool life,
(4) quality such as the surface finish and the geometrical accuracy, and
(5) swarf control.

The cutting performance depends not only upon the workpiece material but also on the cutting tool, cutting conditions, workpiece shape and the machine tool characteristics. Thus, machinability should be evaluated in terms of the required cutting performance while carrying out the operation. For example, (1) when the cutting operations are conducted on a machine tool with low stiffness, the cutting force would be important, (2) when the thin cylinders are machined, the cutting temperature would be also critical as well as the cutting force, (3) when the high machining rates and the continuous operation are required for the NC operations, the tool life and the swarf control would be considered, and (4) when conducting the finishing, the surface quality would be evaluated.

Admitting that the machinability has not been defined with wide acceptability, and that should be defined in consideration of various factors, primary concerns at present are the tool life and the cutting force.

(a) *Tool life*

When machinability is evaluated in terms of the tool performance, the tool lives are determined with changing the cutting velocity, provided that other conditions such as the workpiece shape, the tool material, the feed rate and the depth of cut keep constant. The paramcters n and C in the tool life equation, Eq. (9.37), are determined to estimate the cutting velocity for the standard tool life such as 30 or 60 min in the machine shop. Then, the cutting velocities for workpiece materials are compared to evaluate machinability.

Machinability rating is useful to evaluate how much easy the material is machined. The American Iron and Steel Institute (AISI) has determined machinability ratings for a wide variety of materials in running the turning tests, where the 160 Brinell B1112 steel, sulphite-free cutting steel, is used as the standard material to be compared. More specifically, the tool material HSS (high speed steel) is used in the cutting tests, and the tool life is determined in cutting of B1112 steel at a cutting velocity of 54 m/min (180 surface feet per minute). Then, it is arbitrarily assigned that a machinability rating of B1112 steel is 100%, and the cutting speed V_i is acquired so that the tool life of the sample material is the same as that of B1112. In short, the machinability rating is:

$$MR = \frac{V_i}{54} \times 100 \tag{9.86}$$

When the cemented carbide is used in the tests, the cutting velocity for a tool life of 20 min in cutting of sample material is compared with that of B1112:

$$MR = \frac{V_{20,\,sample}}{V_{20,\,B1112}} \times 100 \tag{9.87}$$

A material with a machinability rating less than 100% would be more difficult to machine than B1112, whereas a material with a value more than 100% would be easier. For example, the machining rating of 0.4% carbon steel is 57; that of austenite stainless steel is 35; and those of nickel super alloy and Inconel are 13.

(b) *Cutting forces and power consumption*

The cutting forces are directly related to the powers consumed. As publicised elsewhere, the cutting forces is often estimated by the specific cutting force or the specific cutting energy u. Because the specific cutting force is the cutting force per unit area, the higher value indicates the lower machinability. The specific cutting force is, in general, available at the technical data issued by the research divisions of the companies (see Appendix A).

The specific cutting energy u is defined as the cutting energy for removing the unit volume of the material, and is written as:

$$u = \frac{F_p V}{V b t_1} = \frac{F_p}{b t_1} \tag{9.88}$$

where, F_p is the principal cutting force and also V, b and t_1 are the cutting velocity, the width and the depth of cut, respectively.

For further convenience, the specific cutting energies are summarised in Table 9.3 (Shaw 1984) and although the table is obtained in the specified cutting parameters with continuous swarf and no built-up edge, F_p is:

$$F_p = u b t_1 \tag{9.89}$$

Table 9.3 Approximate values of specific energies: rake angle, 0 deg; uncut swarf thickness, 0.25mm from Shaw (1984)

Material	Specific energy Jm^{-3}
Aluminum alloy	7.02×10^8
Gray cast iron	10.53×10^8
Free-machining brass	10.53×10^8
Free-machining steel (AISI 1213)	17.55×10^8
Mild steel (AISI 1018)	21.06×10^8
Titanium alloy	35.10×10^8
Stainless steel (18–8)	40.14×10^8
High temperature alloys (Ni or Co base materials)	40.14×10^8

To a first approximation, the feed component F_f is assumed to be one half of F_p:

$$F_f = \frac{F_P}{2} \tag{9.90}$$

The horsepower consumed is estimated as:

$$W = \frac{F_p V}{1000} \quad [\text{kW}] \tag{9.91}$$

References

Boothroyd G (1961) Photographic technique for the determination of metal cutting temperatures. Br J Appl Phys 12:238–242

Chae J, Park S-S, Freiheit T (2006) Investigation of micro-cutting operations. Int J Mach Tools Manuf 46:313–332

Childs T-H-C, Maekawa K, Obikawa T, Yamane Y (2000) Metal machining: theory and applications. Arnold, London 2000

Loewen E-G, Shaw M-C (1954) On the analysis of cutting-tool temperatures. Trans. ASME 76:217–231

Liu X, DeVor R-E, Kapoor S-G, Ehmann K-F (2004) The mechanics of machining at the microscale: assessment of the current state of the science. Trans ASME J Manuf Sci Eng 126:666–678

Merchant M-E (1945) Mechanics of the metal cutting process. II. plasticity conditions in orthogonal cutting. J Appl Phys 16:318–324

Shaw M-C (1984) Metal cutting principles. Oxford Science Publication, New York

Chapter 10
Numerical Simulation for Metal Cutting and Grinding

The numerical simulation has recently been increased with the advance of computer technologies. The simulation software is implemented to understand machining phenomena and control the machining processes. Consequently, the simulation is beneficial to reduce the time and cost for development and improvement in machining. The requirements of the simulation are:

(1) Simulation accuracy for quantitative analysis,
(2) Computational time,
(3) Versatility to simulate many cases and phenomena, and
(4) Database management in house.

Although FEM (Finite Element Method) has been most commonly used for the engineering analysis, many issues have left in terms of the above four requirements. Therefore, not only FEM but also the other methods should be applied and examined for phenomena to be discussed.

This chapter reviews numerical modelling in cutting and grinding. Because the simulation technologies have progressed to demonstrate as the actual processes especially in the cutting field, the modelling and the applications are mainly focused on the cutting simulation here. Then, in grinding works, some differing remarks from those in cutting are described with reviewing some contributions.

10.1 Cutting Simulation

10.1.1 Outline of Simulation Models

The cutting simulation has been developed with progress of the computer technology. Because all the phenomena in the cutting process are associated with the cutting force in terms of energy consumption and tribological aspect, as described

Y. Ito and T. Matsumura, *Theory and Practice in Machining Systems*,
DOI 10.1007/978-3-319-53901-0_10

in introduction of Chap. 9, many researches have been done on the cutting force analysis or prediction. The presented cutting force models are classified into:

(a) Cutting force coefficients method,
(b) FEM, and
(c) Cutting energy method.

Characteristic features of these methods are shown in Table 10.1. Because each modelling has own advantages and disadvantages, the simulation model should be taken into account for the simulation performance to be required.

10.1.2 Cutting Force Coefficients Method

The cutting force coefficient method associates the cutting force F with the specific coefficients, the edge length S and the cutting area A as:

$$F = p_e S + p_s A \tag{10.1}$$

where, p_e and p_s are the indentation force loaded on a unit edge length and the shearing force generating the swarf loaded on a unit cutting area, respectively. Here, p_e and p_s are called "specific indentation force" and "specific shearing force."

Because the cutting force is generally divided into three components according to the coordinate system discussed in the cutting process, Eq. (10.1) is applied to estimate each component. For example, in turning, the force components are given by:

$$\left.\begin{aligned} F_t &= p_{te} S + p_{ts} A \\ F_r &= p_{re} S + p_{rs} A \\ F_a &= p_{ae} S + p_{as} A \end{aligned}\right\} \tag{10.2}$$

where, F_t, F_r and F_a are the tangential, radial and the axial force components loaded on the workpiece; and (p_{te}, p_{re}, p_{ae}) and (p_{ts}, p_{rs}, p_{as}) are the specific indentation and the specific shearing forces in each component, respectively.

In the steady turning process, where the cutting parameters don't change, Eq. (10.2) is applied directly corresponding to the feed, the depth and the tool geometry.

In milling, the non-steady cutting process is performed with rotation of the cutter, where the uncut swarf thickness changes with the rotation angle of the edge. Therefore, the time series analysis should be conducted to estimate the periodical change in the cutting force. Because the cutting edges on the end mill are generally formed helically, the uncut swarf thickness also changes with the cutter height. In order to consider the change in the uncut swarf thickness with the rotation angle and the cutter height, the cutting edges on the end mills are divided into discrete small segments as shown in Fig. 10.1. Equation (10.2) may be applied to estimate the differential force loaded on the edge segments as:

Table 10.1 Cutting force model

	Model	Output	Advantages and disadvantages
Cutting force coefficient method	Cutting fore is associated with the specific force parameters, the cutting edge length and the cutting area	Cutting force	Advantages: Computational time is short enough to apply to the chatter vibration analysis
			Disadvantages: Many cutting tests are required for database management depending on the cutting parameters and the tool geometry
Finite element method	Material flow is simulated numerically with strain, stress and other physical variable in plastic mechanics	Cutting force, stress, strain, temperature in the material and the tool	Advantages: Swarf formation is controlled by the simulation. Workpiece and tool damage may be evaluated with stress, strain and temperature
			Disadvantages: Three-dimensional complex analyses require long computational times. Database management of constitutive equation and friction model requires for accurate simulation
Cutting energy method	Three-dimensional swarf flow is modelled by piling up the orthogonal cuttings. The swarf flow direction is determined to minimise the cutting energy	Cutting force, swarf flow direction and cutting energy	Advantages: Swarf flow can be determined to control swarf. The simulation may be applied to tool design because cutting force can be estimated for any tool geometry using only the orthogonal cutting data. Database management is relatively easy
			Disadvantages: Computational times are relatively long to determine the swarf flow direction compared to cutting force coefficient model

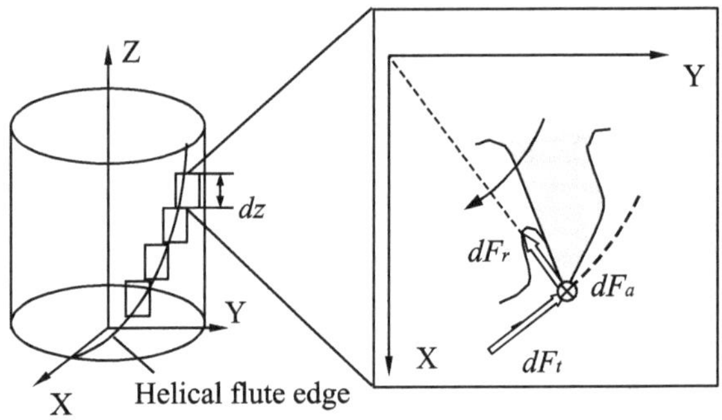

Fig. 10.1 End milling model

$$\left.\begin{aligned} dF_t &= p_{te}dS + p_{ts}dA \\ dF_r &= p_{re}dS + p_{rs}dA \\ dF_a &= p_{ae}dS + p_{as}dA \end{aligned}\right\} \tag{10.3}$$

where, dF_t, dF_r and dF_a are the tangential, radial and the axial force components loaded on a segmented edge; and dS and dA are the edge length and the cutting area of the segment, respectively. The change in the cutting force is given by sum of the differential forces along the cutting edges removing the material. This method is called "instantaneous rigid force model."

Many mechanistic force models in milling have been proposed since Martellotti's work (Martellotti 1941). The cutting forces in slab milling and face milling were associated with the cutting force coefficients in a mechanistic model (Koenigsberger and Sabberwal 1961). The swarf loads in millings were predicted in the mechanistic model of milling (Kline et al. 1982). Then, oblique cutting model was applied to predict the cutting force (Armarego and Deshpande 1989, 1990). The cutting force in the ball end milling was presented using the oblique cutting model (Yang and Park 1991). Furthermore, the mechanistic force model was applied to the profile milling cutter (Bayoumi et al. 1994). A semi-mechanistic analytical model was presented to predict the cutting force of the ball end mills (Yucesan and Altintas 1996). Then, the oblique cutting model was applied to the force prediction (Lee and Altintas 1996). As more practical application, the cutter runout was introduced to the mechanistic model (Feng and Menq 1994a, b). A thermomechanical oblique cutting model was applied to ball end milling (Fontaine et al. 2007). Those works were reviewed in some literatures (Smith and Tlusty 1991; Ehmann et al. 1997).

In drilling, the specific force parameters (p_{te}, p_{re}, p_{ae}) and (p_{ts}, p_{rs}, p_{as}) in the cutting force model expressed as Eq. (10.3) are not constant because the rake angle changes along an edge. Furthermore, the local tool geometries of the chisel and the lips are largely different. The cutting speeds are quite low around the chisel, in which the rake angle is negative. Therefore, the indentation may be dominant contribution to the cutting force in drilling around the chisel. The diameter of the indentation zone was determined and a force model was presented for the lips

(Williams 1974). The oblique cutting model was applied to estimate the cutting force in drilling with flat face drills (Armarego and Cheng 1972).

The cutting force coefficients method estimates the cutting force loaded on the tool or the workpiece in a short time due to the only geometrical calculation is performed to get the edge length and the cutting area. Therefore, this model may be used for checking the excessive cutting force and the chatter vibration with CAM software. Because the specific parameters depend upon not only the cutting parameters but also the tool geometry, database management requires many cutting tests with changing those parameters. As for the swarf flow, the force models based on the oblique cutting determine the swarf flow angles based on the local inclination angles at the segmented edges. They introduce Stabler's law or the implicit equation.

10.1.3 Finite Element Method

FEM has been widely applied to analyse engineering problems such as structure, fluid and thermodynamics. The FE analysis has also been conducted to simulate swarf formation, stress and strain in the tool and the material, the cutting force and the temperature distribution since 1970s. Many contributions in the FE analysis for the cutting simulation are classified into modelling and applications. As initial works on the FE analysis from 1970s to 1990s, many models were presented to simulate the swarf formation as actual one under limitations of the computer capacities. The material behaviour has been regarded as elastic-plastic or rigid-plastic to formulise the stress-strain relation. In modelling of the material flow during cutting, Eulerian and Lagrangian views were applied to the finite elements in the analysis. Eulerian fixes the elements in space and allows the material flow through them; while Lagrangian fixes the elements to the flowing material, so that they convect with the material. In the FE analysis of the cutting process, criterion is required to separate the swarf from the workpiece around the cutting edge. Some criteria have been applied to represent the swarf formation or morphology. With the discussion of the swarf/workpiece separation criterion, the mesh control is also critical issue to save the computational time with high accuracy in the simulation. The structured mesh has been shifted to remeshing and adaptive mesh in 1990s. Those works are reviewed well in literatures (Heisel et al. 2009; Childs et al. 2000).

The earliest works appeared to analyse the shear zone in a simplified model (Zienkiewicz 1971; Kakino 1971). They demonstrated the plastic yield along the primary shear zone in a small strain elastic-plastic analysis with assuming the shape of the swarf. The model neglected friction between the swarf and the tool, the strain rate and the temperature dependency in the material flow stress.

Then, iterative convergence method was presented to simulate the swarf formation (Shirakashi and Usui 1976). The method assumes the pre-formed swarf with the shear plane angle and the flow lines, which is to be modified; calculates the distribution of stress and strain by FEM; calculates strain rate, temperature and the total strain with a small tool advance; and modifies the flow stress in the elements.

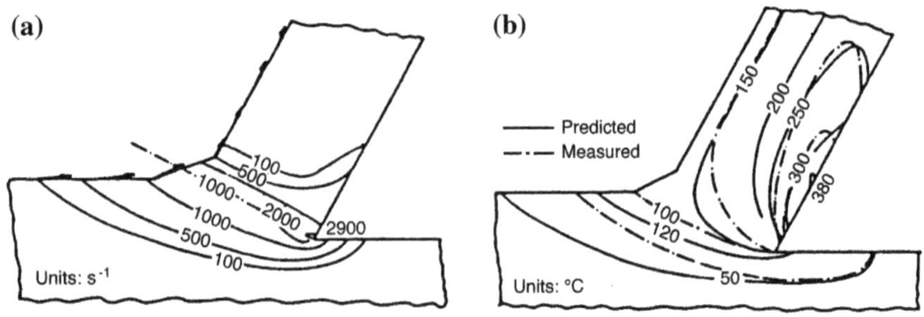

Fig. 10.2 Swarf formation simulation in iterative convergence method: **a** srain rate and **b** temperature, from Shirakashi and Usui (1976)

The procedure repeats until the generated plastic flow is consistent with the assumed shape. The presented method considered the swarf-tool friction and the material flow stress variation with strain, strain rate and temperature acquired in high strain rate Hopkinson bar tests. Figure 10.2 shows an example of the swarf shape, equivalent strain rate and temperature distribution in machining of α-brass.

Rigid-plastic modelling was applied to simulate the steady cutting using software developed from metal forming analyses (Iwata et al. 1984). The modelling also adjusts an initially assumed flow field to bring it into agreement with the computed field.

Updated Lagrangian elastic-plastic analysis was first introduced to simulate the non-steady swarf formation analyses, in which the swarf formations are demonstrated from the contacts of tools with the workpiece. Then, the criterion of the swarf/workpiece separation at the cutting edge, shown in Fig. 10.3, is discussed in terms of strain, energy and displacement. Figure 10.4 shows an example of the simulations following a strain-based separation criterion (Strenkowski and Carrol 1985). Updated Lagrangian elastic-plastic simulation demonstrated discontinuous swarf formation in β-brass at a low cutting speed in which a swarf separation was controlled by a displacement controlled criterion with crack nucleation and growth as shown in Fig. 10.5 (Obikawa and Usui 1996).

The earlier works on FE analysis were limited by computer capacities until 1990s. Most of FE simulations demonstrated the swarf formations in orthogonal cuttings. With progress of the computer technology, three-dimensional elastic-plastic

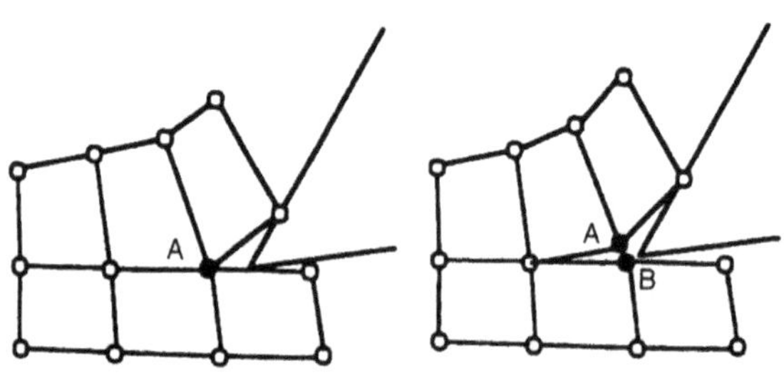

Fig. 10.3 Swarf/Workpiece separation in FE model

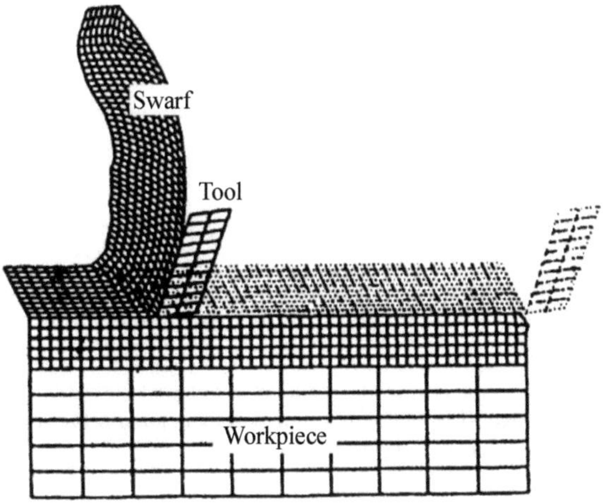

Fig. 10.4 Simulation of non-steady cutting process from Strenkowski and Carrol (1985)

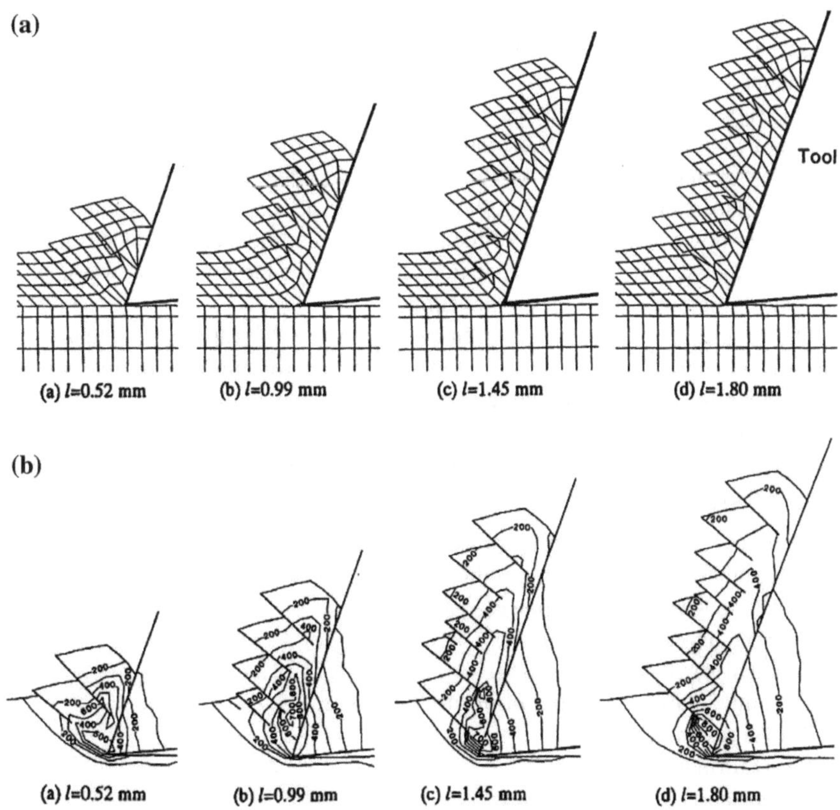

Fig. 10.5 Simulation of discontinuous swarf formation and temperature: **a** Swarf formations and **b** Temperature distribution from Obikawa and Usui (1996)

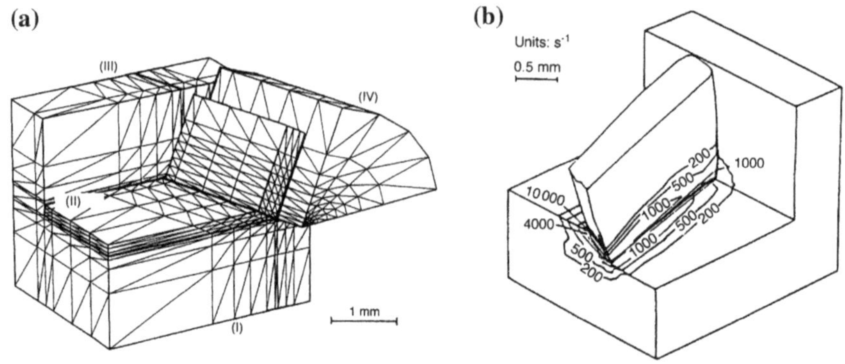

Fig. 10.6 Three-dimensional FE analysis: **a** Simulation model and **b** Equivalent strain rate with swarf formation from Maekawa and Maeda (1993)

analysis was first done to simulation the swarf formed at the primary and secondary cutting edges, in which a displacement controlled criterion was employed for the swarf/workpiece separation in the iterative convergence method (Maekawa and Maeda 1993). Figure 10.6 shows an example of the three-dimensional swarf formations.

The rigid-plastic FE analyses were also applied to three-dimensional swarf formations with a shift from Eulerian to Lagrangian modelling. Then, the material behaviour at the cutting edge was expressed to separate the swarf from the workpiece by remeshing (Ueda et al. 1996). By means of adaptive mesh refinement in non-steady flows, concentration of the mesh at the cutting edge introduces a crack following the swarf/workpiece separation. The adaptive meshing, then, was applied to rigid-plastic and elastic-plastic modelling.

The commercial software have been available since AdvantEdge™ (http://www.thirdwavesys.com/) and Deform™ (http://www.deform.com/) were released by Third Wave Systems and Scientific Forming Technologies Corporation. They have been commonly used for simulation in the 2-D and 3-D cutting processes. Since then, the recent works have widely applied to analyse the cutting phenomena. The modelling and applications of the FE analyses were also classified into (Mackerle 2003):

- Material removal and cutting process in general,
- Computational models for specific machining processes,
- Effects of geometric and process parameters,
- Thermal aspects in machining,
- Residual stresses in machining,
- Dynamic analysis and control of machine tools,
- Tool wear and failure,
- Swarf formation mechanism,
- Optimisation, and other topics.

The computational power has recently been increasing to simulate three-dimensional cutting processes such as milling and drilling. Nevertheless, those computational times are still long to finish the calculation. The FE analysis also requires the constitutive equation representing the material properties, which give the flow stresses obtained in split Hopkinson pressure bar tests. However, the constitutive equations for the materials that software users cut are not fully equipped in the database of the commercial software. Also the realistic friction models depending upon the tool surface and the workpiece are not included and the software user determines the representative value. Although the FE analysis is powerful tool to understand the cutting process including the surface characteristics after finishing, the following issues should be discussed to implement the software in the actual researches and developments:

(1) Computational time to finish the analysis of three-dimensional cutting process,
(2) Database of the constitutive equation and customisation to adapt the material to be cut by the software user,
(3) Realistic friction model and its parameters corresponding to the combination of the tool and the workpiece material, and
(4) The correct selection of the model parameters and of the assumptions made.

10.1.4 Cutting Energy Method

The model predicts the cutting force by piling up the orthogonal cuttings in three-dimensional swarf flow (Usui et al. 1978). The swarf flow direction is predicted to minimise the cutting energy with the cutting forces. Recently, dry cutting or semi-dry cutting has been carried out to reduce the environmental impact and the cost for the waste fluid processing. In such operations, the swarf flow should be controlled by the tool geometry and cutting parameters. The advantages of this model are that not only the cutting force but also the swarf flow direction can be predicted in a short time by only the orthogonal cutting data associated with a few cutting parameters. Therefore, the model may be used for the tool design and the optimisation of the cutting parameters.

In the model, the swarf flow in turning is interpreted as a piling up of the orthogonal cuttings in the planes containing the cutting velocity and the swarf flow velocity as shown in Fig. 10.7a. The orthogonal cutting model is determined by:

$$\left.\begin{array}{l}\phi = F(\alpha, V, t_1)\\ \tau_s = G(\alpha, V, t_1)\\ \beta = H(\alpha, V, t_1)\end{array}\right\} \quad (10.4)$$

where, ϕ, τ_s and β are the shear angle, shear stress on the shear plane and the friction angle in the orthogonal cutting process. In addition, α, V and t_1 are the rake angle, the cutting velocity and the uncut swarf thickness.

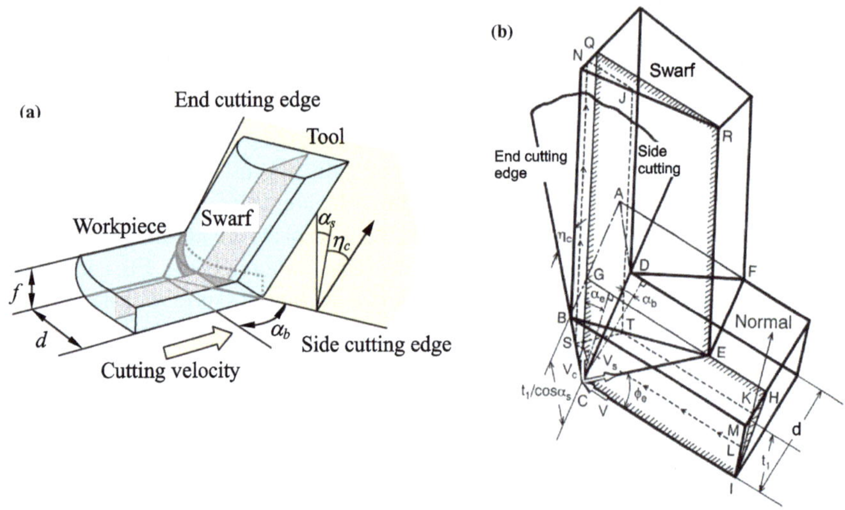

Fig. 10.7 Cutting model in turning: **a** Swarf flow in turning and **b** Swarf flow model

Figure 10.7b shows an orthogonal cutting model in the swarf flow. When the swarf flow angle η_c is assumed, the rake angle in the plane containing the cutting velocity and the swarf flow velocity, α_e, is given by:

$$\alpha_e = \sin^{-1}(\sin\alpha_s \cos\alpha_b \cos\eta_c + \sin\eta_c \sin\alpha_b) \tag{10.5}$$

where, α_s is the side rake angle and α_b is the back rake angle. Thus, the cutting model changes with the swarf flow angle as can be imagined from Eq. (10.4). The shear energy consumed in the shear plane U_S is:

$$U_S = \tau_s V_s A_s = \tau_s \frac{\cos\alpha_e}{\cos(\phi_e - \alpha_e)} V A_s \tag{10.6}$$

where, A_s is the area of the shear plane CEFD; V and V_s are the cutting velocity and the shear velocity; and ϕ_e is the shear angle in the orthogonal cutting plane.

The friction energy on the rake face U_f is:

$$U_f = F_t V_c = F_t \frac{\sin\phi_e}{\cos(\phi_e - \alpha_e)} V \tag{10.7}$$

where, V_c is the swarf flow velocity and F_t is the friction force:

$$F_t = \frac{\tau_s \sin\beta}{\sin\phi_e \cos(\phi_e + \beta - \alpha_e)} \frac{\cos\alpha_e}{\cos\alpha_b \cos\alpha_s} fd \tag{10.8}$$

where, d is the depth of cut and f is the feed rate per revolution.

Because the most of the cutting power is consumed in the shear energy and the friction energy, the cutting energy is expressed as:

$$U = U_s + U_f = J(\eta_c,\ \alpha_b,\ \alpha_s,\ d,\ f,\ V) \tag{10.9}$$

Obviously, α_s, α_b, d, t_1 and V are the cutting parameters in the operation, and thus the cutting energy is controlled by the swarf flow angle η_c. In short, η_c is an important determinant to minimise the cutting energy.

More specifically, the cutting force can be predicted by the swarf flow model, which is based on the minimum cutting energy. The principal component of the cutting force $(F_p)_{Cs}$, which is the force in the rotating coordinate system at an angle of the side cutting angle C_s, is duly associated with:

$$U = U_s + U_f = \left(F_p\right)_{Cs} V \tag{10.10}$$

Thus, $(F_p)_{Cs}$ is obtained by:

$$\left(F_p\right)_{Cs} = \frac{\tau_s \cos\alpha_e}{\cos(\phi_e - \alpha_e)}\left[A_s + \frac{t_1 d \sin\beta}{\cos(\phi_e + \beta - \alpha_e)\cos\alpha_b\cos\alpha_s}\right] \tag{10.11}$$

On the basis of Fig. 10.8, the normal force loaded on the rake face N_t is given by:

$$N_t = \frac{(F_p)_{Cs} - F_t \sin\alpha_e}{\cos\alpha_s \cos\alpha_b} \tag{10.12}$$

The feed component $(F_f)_{Cs}$ and radial component $(F_d)_{Cs}$ are:

$$\left.\begin{aligned} \left(F_f\right)_{Cs} &= -N_t \sin\alpha_s + F_t \cos\eta_c \cos\alpha_s \\ \left(F_d\right)_{Cs} &= -N_t \cos\alpha_s \sin\alpha_b + F_t \sin\eta_c \cos\alpha_b \\ &\quad - F_t \cos\eta_c \sin\alpha_s \sin\alpha_b \end{aligned}\right\} \tag{10.13}$$

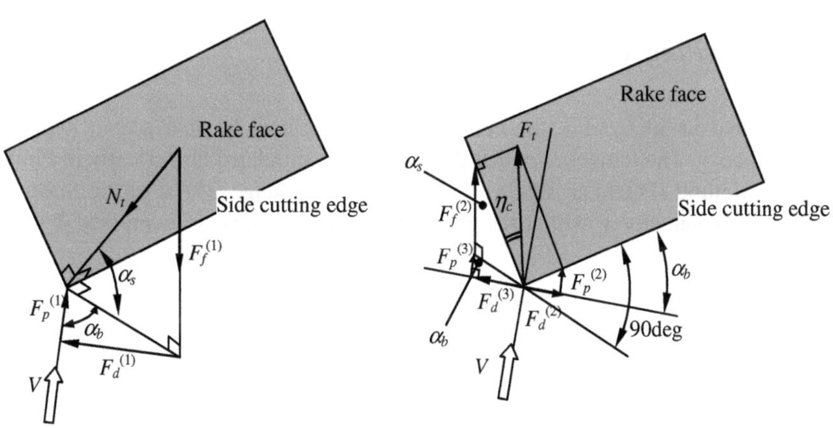

Fig. 10.8 Cutting force loaded on tool

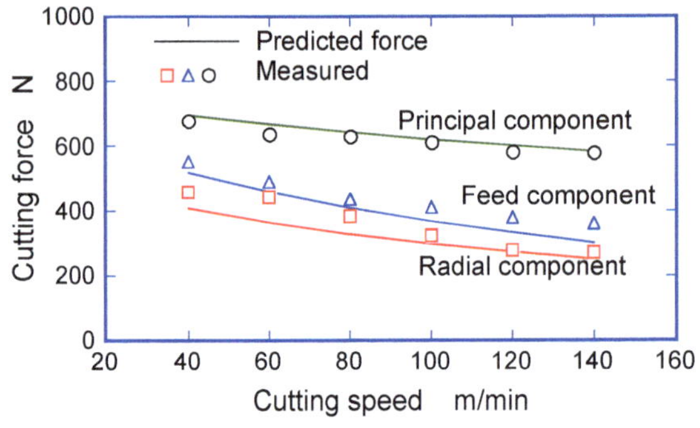

Fig. 10.9 Cutting force simulation: material cut, Inconel X750; tool, carbide K10 (−5, −5, 5, 5, 15, 15, 0.8); depth of cut, 1.0 mm; feed rate, 0.2 mm/rev; and lubrication, dry

When the turning tool inclines at a side cutting angle C_s, each component is converted by the following expressions:

$$\left.\begin{aligned} F_p &= (F_p)_{Cs} \\ F_f &= (F_f)_{Cs}\cos C_S - (F_d)_{Cs}\sin C_S \\ F_d &= (F_d)_{Cs}\cos C_S + (F_f)_{Cs}\sin C_S \end{aligned}\right\} \quad (10.14)$$

Figure 10.9 shows an example of the cutting force prediction in turning of Inconel X750. The change in the principal, the feed and the radial components are in good agreement with the experimental data.

The force model was extended to plane milling (Hirota et al. 1978) and ball end milling (Tsai and Liao 2008). Then, the model was applied to simulation with more complex tools such as roughing end mills (Matsumura and Usui 2010). In drilling, the cutting edges consist of the chisel and the lips. The different swarf formation occurs on an edge. The model was first applied to the cutting process of the lips (Hirota and Kasahara 1981). Then, the cutting force from the chisel engagement to the lips disengagement were simulated (Matsumura and Leopold 2010).

10.1.5 Cutting Temperature Simulation

As described in Sect. 9.1.3, the cutting temperature may be analysed in the heat transfer theory. The temperature distribution can be simulated numerically in Finite Volume Method (Patankar 1980). The temperature change $\partial\theta/\partial t$ of the material moving at a velocity $\boldsymbol{v}$ with heat generation q per volume is expressed by the following equation:

$$\rho c\frac{\partial\theta}{\partial t} = div(k \cdot grad\theta) - div(\rho c\mathbf{v}\theta) + q \quad (10.16)$$

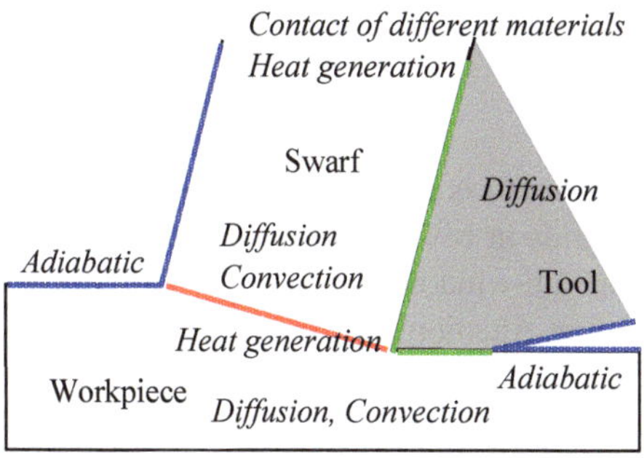

Fig. 10.10 Numerical analysis model of cutting temperature

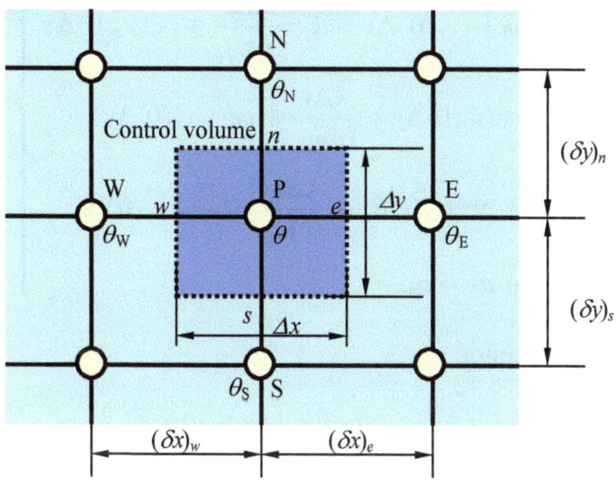

Fig. 10.11 Control volume in two-dimensional grid

where, k, ρ and c are thermal conductivity, density and specific heat capacity, respectively. Obviously, the temperature change is caused by thermal diffusion, thermal convection and heat generation, and thus the cutting temperature is analysed in the orthogonal cutting illustrated in Fig. 10.10 by assuming the following preconditions.

(1) In the analysis model, only thermal diffusion occurs in the tool,
(2) Thermal diffusion and thermal convection simultaneously occurs in the workpiece and the swarf,
(3) Heat generation occurs on the shear plane and the rake face corresponding to the shear energy and the friction energy, and also
(4) Because heat conduction into the surrounding air is much lower than conduction in the material, the surfaces of the workpiece and the swarf is regarded as adiabatic boundaries.

When the cutting temperature is analysed for the steady state of orthogonal cutting, Eq. (10.16) is converted into the following differential equation by substituting $\partial\theta/\partial t$:

$$\frac{\partial(\rho c v_x \theta)}{\partial x} + \frac{\partial(\rho c v_y \theta)}{\partial y} = \frac{\partial}{\partial x}\left(k\frac{\partial \theta}{\partial x}\right) + \frac{\partial}{\partial y}\left(k\frac{\partial \theta}{\partial y}\right) + S \tag{10.17}$$

where, vector $\boldsymbol{v}$ in Eq. (10.16) is decomposed into X and Y components (v_x, v_y).

In consideration of grids in two-dimensional analysis shown in Fig. 10.11, and also when applying an up-window scheme to thermal convection, Eq. (10.17) yields to the discretisation equation for Point P as follows:

$$a_P\theta_P = a_E\theta_E + a_W\theta_W + a_N\theta_N + a_S\theta_S + q\Delta x\Delta y$$

$$\left.\begin{aligned}
a_E &= \frac{k\Delta y}{(\delta x)_e} + \rho c\langle -v_x, 0\rangle\Delta y \quad a_W = \frac{k\Delta y}{(\delta x)_w} + \rho c\langle v_x, 0\rangle\Delta y \\
a_N &= \frac{k\Delta x}{(\delta y)_n} + \rho c\langle -v_y, 0\rangle\Delta x \quad a_S = \frac{k\Delta x}{(\delta y)_s} + \rho c\langle v_y, 0\rangle\Delta x \\
a_P &= \frac{k\Delta y}{(\delta x)_e} + \rho c\langle v_x, 0\rangle\Delta y + \frac{k\Delta y}{(\delta x)_w} + \rho c\langle -v_x, 0\rangle\Delta y \\
&\quad + \frac{k\Delta x}{(\delta y)_n} + \rho c\langle v_y, 0\rangle\Delta x + \frac{k\Delta x}{(\delta y)_s} + \rho c\langle -v_y, 0\rangle\Delta x \\
&= a_E + a_W + a_N + a_S
\end{aligned}\right\} \tag{10.18}$$

where, $\langle a, b\rangle$ is an operator:

$$\langle a, b\rangle = \begin{cases} a & (a > b) \\ b & (a < b) \end{cases} \tag{10.19}$$

Needles to say, Eq. (10.18) is established for each grids point, the simultaneous equations are solved by a successive relaxation method.

Figure 10.12a, b show an example of the analysis models, and also a temperature distribution. As will be clear from Fig. 10.12, (a) the position of maximum temperature on the rake face is away from the tool tip, and (b) most of heat generation on the shear plane conducts into the swarf due to the thermal convection in the material flow. Importantly, Eq. (10.17) for the rhombic grids is formulated in the same manner as those of Eq. (10.18).

10.1.6 Tool Wear Progress Simulation

(a) *Wear model*

The tool life equations have been associated with the cutting parameters such as the cutting velocity. However, it is well known that the tool wear progress also changes with the tool geometry. In other words, the wear generally depends upon stress and

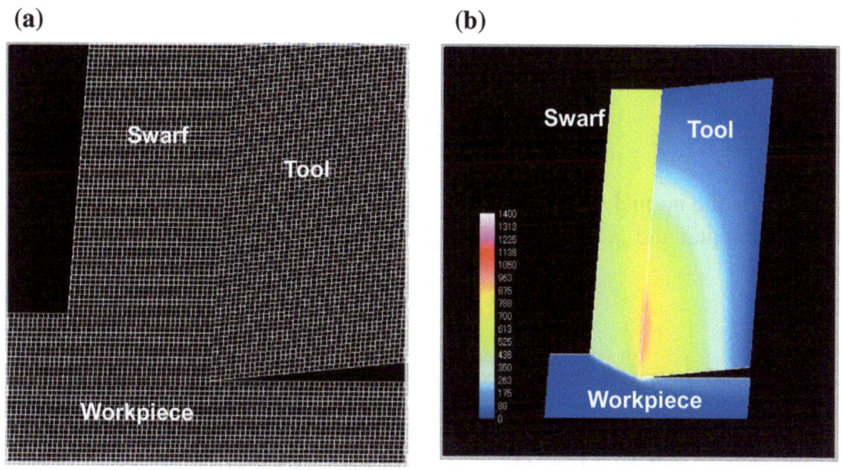

Fig. 10.12 Example of FDM analysis: **a** analysis model and **b** temperature distribution in a cross section of cutting area

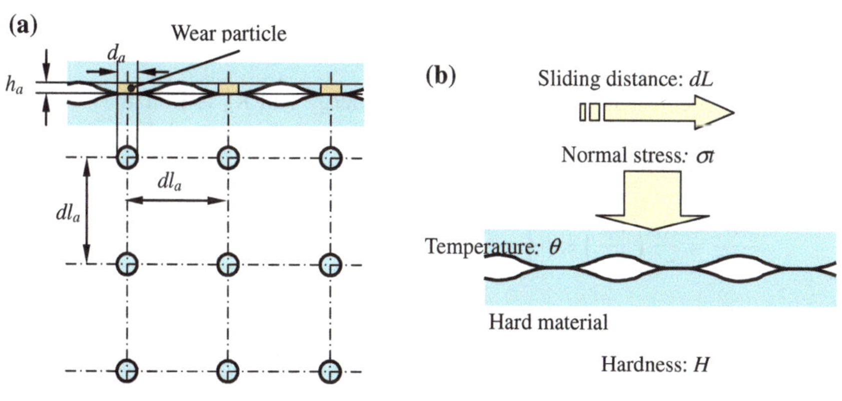

Fig. 10.13 Wear model: **a** Adhesion model and **b** Ploughing model

temperature on interface of materials, which are acquired in the force prediction and the temperature analysis, and thus the tool wear should be determined in association with stress and temperature on the tool face. For the wear model, we have so far various proposals, and of those, the adhesion model and the ploughing model will be introduced herein (Usui et al. 1984).

Figure 10.13a shows the interface of materials in the adhesion model. In this model, asperities with a diameter d_a exist at an incremental distance dl_a, and the wear particles, which are approximated as small cylinders with a height h_a, occur in a sliding distance ranged dL. When the material moves in a distance dL, thus, the asperities in a unit area encounter N_a asperities on the other material, and N_a yields to:

$$N_a = \left(\frac{dL}{dl_a}\right)\left(\frac{4\sigma_t}{H\pi d_a^2}\right) \tag{10.20}$$

where, σ_t is the normal stress. In addition, H is hardness of the asperities associated with the temperature θ, and given by:

$$H = \alpha_1 \exp\left(\frac{\alpha_2}{\theta}\right) \tag{10.21}$$

where, α_1 and α_2 are the material constants. The wear volume per unit area dW is:

$$dW = \left(\frac{dL}{dl_a}\right)\left(\frac{4\sigma_t}{H\pi d_a^2}\right)\frac{\pi}{4}d_a^2 h_a z = \frac{\sigma_t}{H}\left(\frac{h_a}{dl_a}\right) z dL \tag{10.22}$$

where, z is Holm's probability, i.e., probability of generation of the wear particles at a contact. Simultaneously, z is associated with the Boltzman distribution as follows:

$$z = K_c \exp\left(\frac{-\Delta E}{K_B\theta}\right) \tag{10.23}$$

where, K_c is a constant. ΔE is the activation energy and K_B is Boltzman constant controlled by the surface structure. The wear rate is obtained by substituting Eqs. (10.21) and (10.23) into Eq. (10.22):

$$\frac{dW}{\sigma_t dL} = \frac{K_c}{\alpha_1}\left(\frac{h_a}{dl_a}\right)\exp\left\{\frac{-(\Delta E + K_B\alpha_2)}{K_B\theta}\right\} = C_1 \exp\left(\frac{-C_2}{\theta}\right) \tag{10.24}$$

where, C_1 and C_2 are called wear characteristic constants.

In the ploughing model shown in Fig. 10.13b, the wear volume dW per sliding distance dL is characterised by the following equation (Rabinowicz 1961):

$$dW = K\frac{\sigma_t}{H}dL \tag{10.25}$$

where σ_t is the normal stress. In addition, H is hardness of the harder material in tool-work-surface contact, associated with the temperature θ and given by:

$$H = \alpha_1 \exp\left(\frac{\alpha_2}{\theta}\right) \tag{10.26}$$

where α_1 and α_2 are constants. K is expressed as:

$$K = \beta_1 \exp\left(\frac{-\beta_2}{\theta}\right) \tag{10.27}$$

where, β_1 and β_2 are constants. In short, the wear volume is obtained by substituting Eqs. (10.26) and (10.27) into Eq. (10.25):

$$\frac{dW}{\sigma_t dL} = C_1 \exp\left(\frac{-C_2}{\theta}\right) \tag{10.28}$$

where, C_1 and C_2 are wear characteristic constants.

Although ploughing wear model is characterised by the same formula as the adhesion model, the constants C_1 and C_2 are different. Adhesion is strongly controlled by temperature on the tool face. Thus, the parameter for the temperature dependency C_2 in the adhesion model is larger than that of the ploughing model.

(b) *Tool wear prediction*

The wear progress is predicted using the wear characteristic equation shown in Eqs. (10.24) and (10.28). More specifically, the wear progress can be clarified by the following procedures:

(1) The stress is distributed as already shown in Fig. 9.11b based on the cutting force, which is predicted in Sect. 10.1.4.
(2) Then, the temperature distribution on the tool face can be given in the numerical simulation described in 10.1.5.
(3) Consequently, the wear rate is determined by the wear characteristic equation.

Importantly, in the flank wear prediction, the wear rate should be re-calculated for the width of the flank wear land with modifying the stress distribution so that the wear rate is the same on all area of the flank wear land (Usui et al. 1984). Because the flank wear generally depends upon the mechanical effect, Eq. (10.28) based on the ploughing model may be applied to the wear prediction. The wear characteristic equation in Eq. (10.28) associates the wear rate in a unit area for a unit sliding distance with stresses and temperatures on the tool face. Therefore, Eq. (10.28) is transferred to Eq. (10.29) for the time series wear prediction:

$$\frac{dV_B}{dt} = C\sigma_f \exp\left(-\frac{\lambda}{\theta_f}\right)\left(\frac{1}{\tan\gamma} - \tan\alpha\right)V \tag{10.29}$$

where, V_B is the width of flank wear land; σ_f and θ_f are the stress and the temperature on the flank wear land; and α and γ are the rake angle and the clearance angle. C and λ are the wear characteristic constants depending upon the tool and the workpiece materials. As described in Sect. 9.1.4, the flank wear progress starts in the initial wear at high wear rates due to unstable contact of the tool and the workpiece. The initial wear rate depends upon not only the wear characteristics but

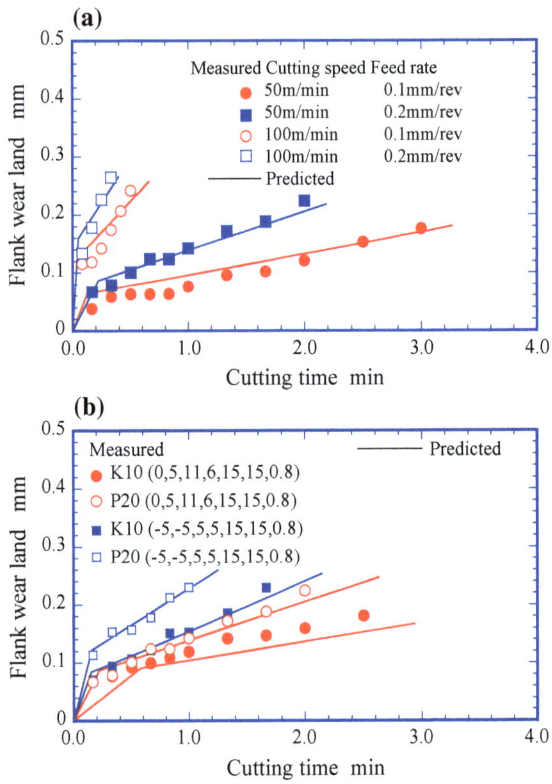

Fig. 10.14 Flank wear simulation: **a** Effect of cutting parameters in turning with P20 (0, 5, 11, 6, 15, 15, 0.8) and **b** Effect of tool geometry and material at a cutting speed of 50 m/min and feed rate of 0.2 mm/rev

also the other dynamic factors. Therefore, in the flank wear prediction, the initial wear offset V_{B0} is introduced at cutting time $T = 0$ as:

$$V_B(T) = V_{B0} + \int_0^T \left(\frac{dV_B}{dT}\right) dT \tag{10.30}$$

Figure 10.14 shows examples of the flank wear predictions compared to the measured widths of the flank wear lands.

10.2 Grinding Simulation

The numerical simulation in grinding has been increasing with scientific models in the removal process, grinding forces and energy, grinding temperature, affected layer with residual stress in subsurface since 1990s. Those works were reviewed well in literatures (Brinksmeier et al. 2006). The FE analyses have also been done on the modelling in two-dimensional (2D) and three-dimensional (3D) processes, and Doman has classified the due work into macro and micro models in terms of the scope size to be discussed (Doman et al. 2009). The macro-scale modelling has discussed the wheel and the workpiece interaction; while the material behaviour around the individual grains are focused in the micro-scale.

10.2.1 Micro-scale Modelling

The phenomena at interface of an abrasive grain on a material surface are classified into rubbing, ploughing and cutting modes, in which their transitions depend upon the removal volume. The major process in cutting is material removal with the swarf formation, while the grinding process consists of the above modes to be identified.

The rubbing occurs with local elastic deformation at small depths as shown in Fig. 10.15a. The mode shifts then to ploughing with higher depths, in which plastic upheaval of the material occurs ahead of and to the side of the grain as illustrated in Fig. 10.15b. Although the material is not physically removed, a groove or scratch is left in plastic footprint. When the grain penetrates into the material in depths over the critical value, the material is removed with the swarf formation as shown in Fig. 10.15c. These modes have been observed experimentally by several researchers. Therefore, many analytical works have been done on deformation and removal at the grain/workpiece interface (Durgumahanti et al. 2010). Furthermore, those modes occur simultaneously on a wheel with multiple contacts of the abrasive grains. Because several shapes and sizes of abrasive grains are also distributed on a wheel with random arrangement, a probabilistic model of the grinding process was presented (Stepien 2009). The grinding process has also been discussed in terms of the specific energy (Ghosh et al. 2008). The energy consumed into:

- Chip formation due to shearing,
- Primary rubbing,
- Secondary rubbing,
- Ploughing,
- Wear flat rubbing,

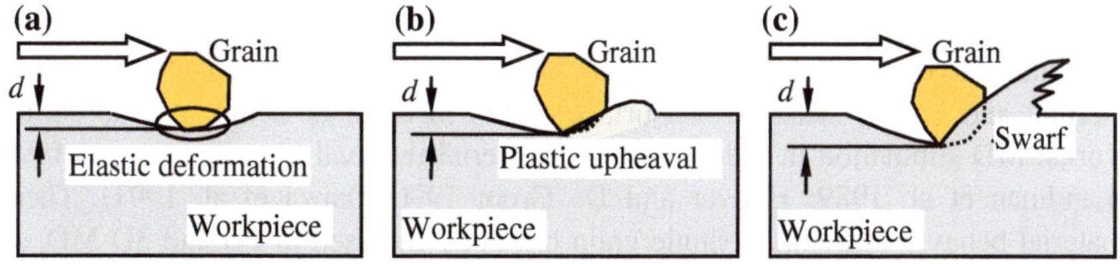

Fig. 10.15 Interaction of a grain and a material surface: **a** Rubbing, **b** Ploughing and **c** Cutting, from Doman et al. (2009)

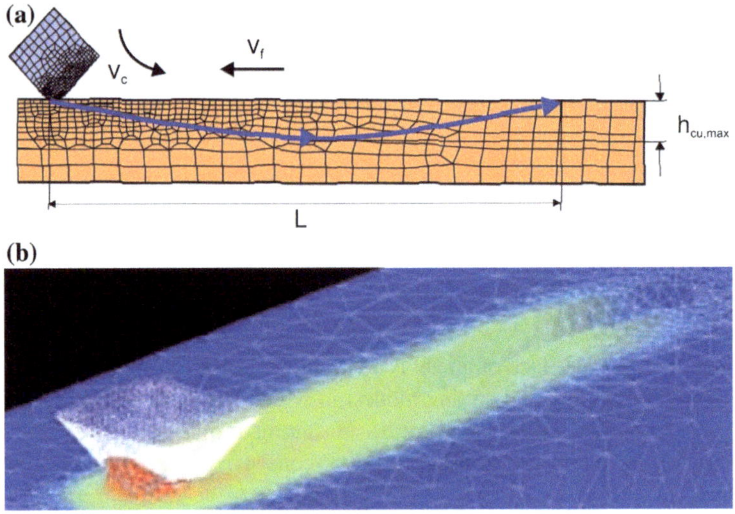

Fig. 10.16 A single scratching process in FE analysis: **a** FE model and **b** Removal process, from Klocke (2003)

- Friction between the loaded chip and workpiece,
- Friction between bond and workpiece, and so on.

Compared to cutting, many interactions of not only grains but also bond should be considered in discussion at the grains/workpiece interface.

The FE analyses have been carried out to demonstrate the above modes with applying the material models: elastic, elastoplastic and thermo-elastoplastic formulations. Most of the applications are focussed on the interaction of a single abrasive grain with the material. The sliding of an abrasive grain was simulated with the material deformation in elasticity theory (Ram et al. 2003). A multi-scale model was introduced to demonstrate the elastic contact of two rough surface (Yao et al. 2004). The rubbing and ploughing modes of an abrasive grain was simulated in the elastoplastic material model (Doman 2006) The thermal model was incorporated to analyse temperature and stress fields in scratching of the thermo-elastoplastic model (Klocke et al. 2002). The swarf formation process of a grain was simulated with friction and heat generation on the grain in the thermos-structural coupling analysis (Ohbuchi and Obikawa 2003). The realistic model shown in Fig. 10.16 has recently been demonstrated for a single grain in 3D FE analyses (Klocke 2003).

Molecular dynamics (MD) is a powerful tool to understand the mechanical, thermal and crystal states in subsurface of the material in detail. As the earlier works, MD simulation in indentation and material removal appeared around 1990 (Landman et al. 1989; Hoover and De Groot 1990; Ikawa et al. 1991). Then, material behaviour around a single grain has been analysed in 2D and 3D MD, as shown in Fig. 10.17 (Shimada et al. 1994; Rentsch and Brinksmeier 2005).

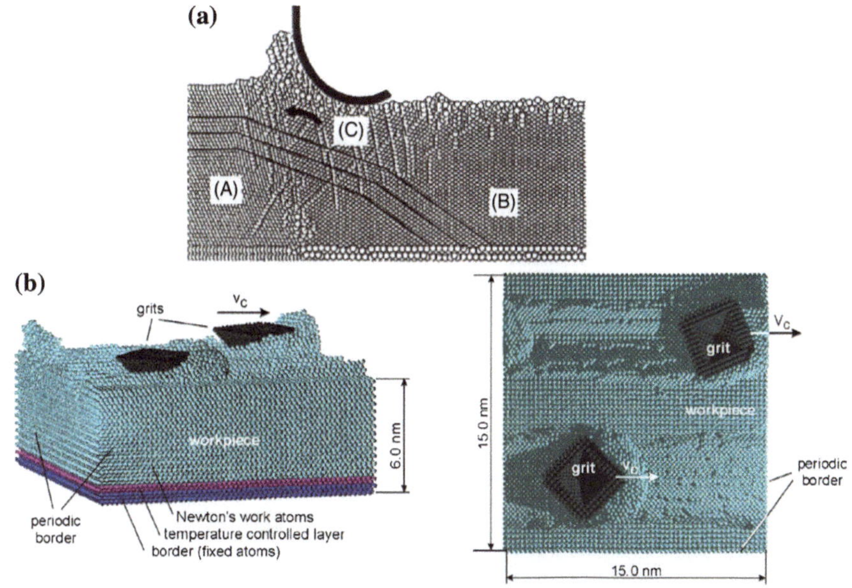

Fig. 10.17 Deformation and scratching in MD analysis: **a** Crystal orientation and deformation at grain boundary in 2D MD: (*A*) and (*B*) are original crystal, and new crystals appears in (*C*) from Shimada et al. (1994) and **b** Groove scratching with 2 grains in 3D MD from Rentsch and Brinksmeier (2005)

10.2.2 Macro-scale Modelling

The grinding operations are generally performed to finish the product shapes accurately with controlling the affected layers. The affected layer depends upon temperature and localised plastic deformation in subsurface. The macro-scale numerical models analyse the temperature distributions to associate the thermal deformation and the change in the material properties due to the phase transformation in subsurface. The temperature distribution can be analysed by finite difference method (FDM) or FEM. However, when the thermal deformation is associated with the temperature distribution, FEM may be applied to thermal-mechanical coupling analysis. The thermal FE analyses have been conducted in 2D and 3D models. In the FE modelling, the heat source should be given corresponding to the grinding parameters. The moving heat source is approximated as a uniform (rectangular) or triangular distribution. In 2D model, a two-dimensional semi-finite plate is assumed with insulted bottom surface; and convective cooling is applied to top surface when coolant is supplied (Biermann and Schneider 1997). Because the heat source is critical in the thermal analysis, the temperature distributions were analysed with some of the heat source profiles (Tian et al. 2009). Many 3D FE analyses have also been conducted to demonstrate more

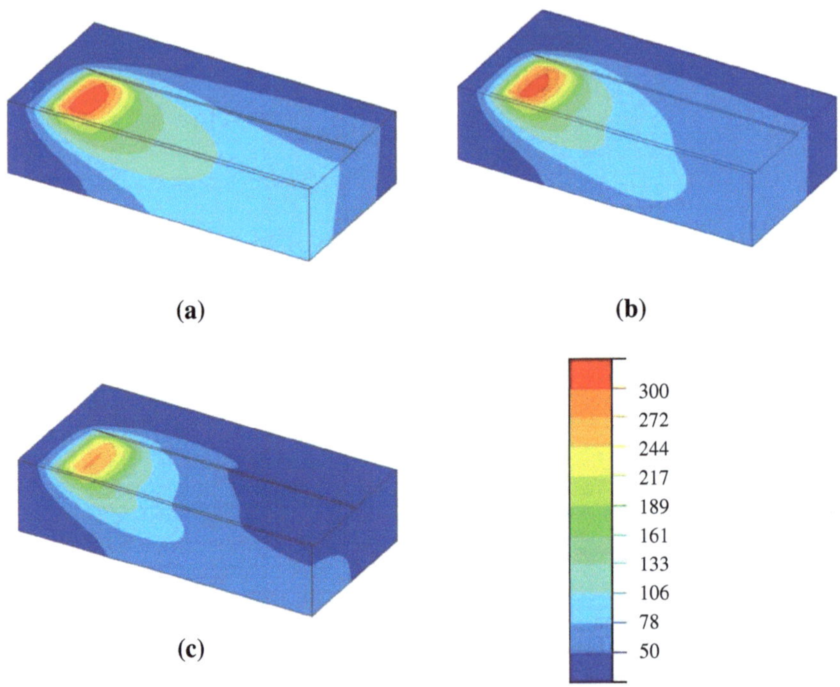

Fig. 10.18 Thermal analysis in 3D FEM: **a** Dry, **b** MQL and **c** N_2 at 0.7 bar, from Hesselbach et al. (2004)

actual grinding process than ever before. Figure 10.18 compares the temperature distributions of grinding in dry, MQL and N_2 at 0.7 bar in 3D thermal analysis (Hesselbach 2004). The FE simulation depicts the effect of the coolant supply on the surface temperature.

The thermal FE analysis may be incorporated with structure model. Thermal-structural analyses have been done to evaluate the internal state in subsurface. An explicit model was presented to demonstrate the grinding-induced residual stresses associated with coupling of thermal deformation, mechanical deformation and phase transformation by use of FE commercial code, ADINA (Moulik et al. 2001), in which a triangular heat source and traction pressure were applied onto the workpiece surface as shown in Fig. 10.19. The FE analyses are effective in understanding phase transformation, thermally induced stresses and work hardening of the material.

The macro-scale analysis is helpful to evaluate physical properties in subsurface. However, the simulation depends upon the boundary conditions such as heat source, heat convection and pressure to be given in the analysis. The heat source and pressure are associated with the abrasive grain/workpiece interface, in which the material undergoes deformation and removal simultaneously with multiple

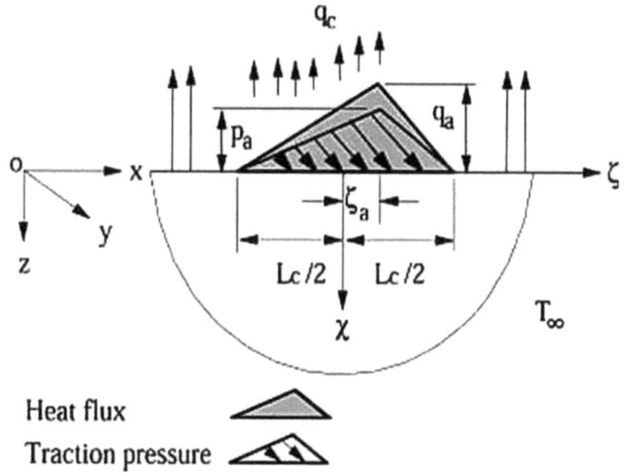

Fig. 10.19 Distribution of heat source and traction pressure in thermal mechanical analysis from Moulik et al. (2001) *Lc* is length of grinding zone; q_a and p_a are the peak values of the heat flux and pressure

contacts of the grains. Moreover, the contacts change with the wears of grains. Therefore, more realistic analysis requires of the multi-scale model with time-series change in the heat source and the pressure including self-dressing.

References

Armarego E-J-A, Cheng C-Y (1972) Drilling with flat rake face and conventional twist drill -I. theoretical investigation. Int J Mach Tool Des Res 12:17–35

Armarego E-J-A, Deshpande N-P (1989) Computerized predictive cutting models for forces in end milling including eccentricity effects. Ann CIRP 38:45–49

Armarego E-J-A, Deshpande N-P (1990) Computerized end milling force predictions with cutting models allowing eccentricity and cutter deflections. Ann CIRP 40:25–29

Bayoumi A-E, Yucesan G, Kendall L-A (1994) An analytical mechanistic cutting force model for milling operations: a theory and methodology. Trans ASME, J Eng Ind 116:324–330

Biermann D, Schneider M (1997) Modeling and simulation of workpiece temperature in grinding by finite element analysis. Mach Sci Technol 1:173–183

Brinksmeier E, Aurich J-C, Govekar E, Heinzel C, Hoffmeister H-W, Klocke F, Peters J, Rentsch R, Stephensong D-J, Uhlmann E, Weinert K, Wittmann M (2006) Advances in modeling and simulation of grinding processes. CIRP Ann-Manuf Technol 55:667–696

Childs T-H-C, Maekawa K, Obikawa T, Yamane Y (2000) Metal Machining: Theory and Applications. Arnold, London 2000

Doman D-A (2006) Rubbing and plowing phases in single grain grinding. Ph.D. thesis, Dalhousie University, Halifax, Nova Scotia, Canada

Doman D-A, Warkentin A, Bauer R (2009) Finite Element Modeling Approaches in Grinding. Int J Mach Tools Manuf 49:109–116

Durgumahanti U-S-P, Singh V, Rao P-V (2010) A new model for grinding force prediction and analysis. Int J Mach Tools Manuf 50:231–240

Ehmann K-F, Kapoor S-G, DeVor R-E, Lazoglu I (1997) machining process modeling: a review. Trans ASME, J Manuf Sci Eng 119:655–663
Feng H-S, Menq C-H (1994a) The prediction of cutting forces in the ball end milling process–I model formulation and model building procedure. Int J Mach Tool Des Res 34:697–710
Feng H-S, Menq C-H (1994b) The prediction of cutting forces in the ball end milling process–II cut geometry analysis and model verification. Int J Mach Tool Des Res 34:711–719
Fontaine M, Moufki A, Devillez A, Dudzinski D (2007) Modelling of cutting forces in ball-end milling with tool–surface inclination: part I: predictive force model and experimental validation. J Mater Process Technol 189:73–84
Ghosh S, Chattopadhyay A-B, Paul S (2008) Modelling of specific energy requirement during high-efficiency deep grinding. Int J Mach Tool Manu 48:1242–1253
Heisel U, Krivoruchko D-V, Zaloha W-A, Storchak M, Stehle T (2009) Die FEM-Modellierung als moderner Ansatz zur Untersuchung von Zerspanprozessen. ZWF Prozessmodellierung, Carl Hanser Verlag 104:604-016
Hesselbach J, Hoffmeister H-W, Maiz K, Machanova I (2004) Surface grinding with cryogenics. Prod Eng XI:1–4
Hirota A, Kasahara K (1981) Analytical prediction of chip formation and cutting forces in drilling operation (1st report)-Cutting model, energy method and preliminary results. J Jpn Soc Precis Eng 47:987–993
Hirota A, Usui E (1978) Analytical prediction of cutting forces in plain milling operation. J Jpn Soc Precis Eng 44:508–514 (in Japanese)
Hoover W-G, De Groot A-J (1990) Large-scale elastic-plastic indentation simulations via nonequilibrium molecular dynamics. Phys Rev A 10(42):5844–5853
Ikawa N, Shimada S, Tanaka H, Ohmori G (1991) An atomistic analysis of nanometric chip removal as affected by tool-work interaction in diamond turning. Ann CIRP 40:551–554
Iwata K, Osakada K, Terasaka Y (1984) Process modeling of orthogonal cutting by the rigind-plastic finite element method. Trans ASME, J Eng Mater Technol 106:132–138
Kakino Y (1971) Analysis of the mechanism of orthogonal machining by the finite element method. J Jpn Soc Precis Eng 37:503–508
Kline W-A, DeVor R-E, Lindberg J-R (1982) The prediction of cutting forces in end milling with application to cornering cuts. Int J Mach Tool Des Res 22:7–22
Klocke F (2003) Modelling and simulation in grinding. 1st European conference on grinding. In: Fortschritt-Berichte VDI Reihe 2 Fertigungstechnik pp 8.1–8.27 (Aachen, 6.–7. November)
Klocke F, Beck T, Hoppe S, Krieg T, Müller N (2002) Examples of FEM application in manufacturing technology. J Mater Process Technol 120:450–457
Koenigsberger F, Sabberwal A-J-P (1961) An investigation into the cutting force pulsations during milling operations. Int J Mach Tool Des Res 1:15–33
Landman U, Luedtke W-D, Nitzan A (1989) Dynamics of tip-substrate interactions in atomic force microscopy. Surf Sci 210:177–184
Lee P, Altintas Y (1996) Prediction of ball-end milling forces from orthogonal cutting data. Int J Mach Tool Des Res 36:1059–1072
Mackerle J (2003) Finite element analysis and simulation of machining: an addendum a bibliography (1996–2002). Int J Mach Tools Manuf 43:103–114
Maekawa K, Maeda M (1993) Simulation analysis of three-dimensional continuous chip formation processes (1st report)—FEM formation and a few results. J Jpn Soc Precis Eng 59:1827–1833
Matsumura T, Leopold J (2010) Simulation of drilling process for control of burr formation. J Adv Mech Des, Syst, Manuf 4:966–975
Matsumura T, Usui E (2010) Predictive cutting force model in complex-shaped end milling based on minimum cutting energy. Int J Mach Tools Manuf 50:458–466
Martellotti M-E (1941) An analysis of the milling process. Trans ASME 63:677–700
Moulik P-N, Yang H-Y-T, Chandrasekar S (2001) Simulation of thermal stresses due to grinding. Int J Mech Sci 43:831–851
Obikawa T, Usui E (1996) Computational machining of titanium alloy—finite element modeling and a few results. Trans ASME, J Manuf Sci Eng 118:208–215

Ohbuchi Y, Obikawa T (2003) Finite element modeling of chip formation in the domain of negative rake angle cutting. J Eng Mater Technol 125:324–332

Patankar S (1980) Numerical heat transfer and fluid flow. Hemisphere Publishing Co

Rabinowicz E, Dunn L-A, Russel P-G (1961) A study of abrasive wear under three-body condition. Wear 4:345–355

Ram A, Danckert J, Faurholdt T (2003) Finite element analysis of stresses due to normal and sliding contact conditions on an elastic surface. In: Fourth European LS-DYNA users conference 21–34

Rentsch R, Brinksmeier E (2005) Tribology aspects in state of the art MD cutting simulations. In: Proceedings of 8th CIRP International workshop on modeling of machining operations pp 401–408 (Chemnitz, Germany, May 2005)

Shimada S, Ikawa N, Tanaka H (1994) Structure of micromachined surface simulated by molecular dynamics analysis. Ann CIRP 43:51–54

Shirakashi T, Usui E (1976) Simulation analysis of orthogonal metal cutting process. J Jpn Soc Precis Eng 42:340–345

Smith S, Tlusty J (1991) An overview of modeling and simulation of the milling process. Trans ASME, J Eng Ind 113:169–175

Stepien P (2009) A probabilistic model of the grinding process. Appl Math Model 33:3863–3884

Strenkowski J-S, Carrol III-J-T (1985) A finite element model of orthogonal metal cutting. Trans ASME, J Eng Ind 107:349–354

Tian Y, Shirinzadeh B, Zhang D, Liu X, Chetwynd D (2009) Effects of the heat source profiles on the thermal distribution for ultraprecision grinding. Precision Eng 33:447–458

Tsai C-L, Liao Y-S (2008) Prediction of cutting forces in ball-end milling by means of geometric analysis. J Mater Process Technol 205:24–33

Ueda K, Manabe K, Nozaki S (1996) Rigid-plastic FEM of three-dimensional cutting mechanism (2nd Report)—simulation of plain milling process. J Jpn Soc Precis Eng 62:526–531

Usui E, Hirota A, Masuko M (1978) Analytical prediction of three dimensional cutting process—Part 1 basic cutting model and energy approach. Trans ASME, J Eng Ind 100:222–228

Usui E, Shirakashi S, Kitagawa T (1984) Analytical prediction of tool wear. Wear 100:129–151

Williams R-J (1974) A study of the drilling process. Trans ASME, J Eng Ind 96:1207–1215

Yang M-Y, Park H-D (1991) The prediction of cutting force in ball end milling. Int J Mach Tool Des Res 31:45–54

Yao Y, Schlesinger M, Drake GWF (2004) A multiscale finite-element method for solving rough-surface elastic-contact problems. Can J Phys 82:679–699

Yucesan G, Altintas Y (1996) Prediction of ball end milling force. Trans ASME, J Eng Ind 118:95–103

Zienkiewicz O-C (1971) The finite element method in engineering science, 2nd edn. MaGraw-Hill, London

Appendix A
Quick Calculation Method for Cutting Force

As well and widely known, the cutting force can be theoretically estimated on the basis of the orthogonal (two-dimensional) cutting model; however, the estimated value differs considerably from the measured one. This discrepancy is caused by some preconditions when establishing the orthogonal cutting model as will be discussed in the following.

Figure A.1 shows some visualised results for and the generalised illustration of the actual cutting state. In contrast, Fig. A.2 is the orthogonal cutting model widely accepted. By comparing these two, we can see the very simplicity of the orthogonal cutting model, and importantly, we must be aware of three leading assumptions in the model as follows.

(1) The work deforms suddenly at the shear plane from elastic to plastic states.
(2) The theoretical analysis is based on the "Principle of Volume Constant", which is the common rule in the plastic deformation, and thus applicable only to the swarf of flow type.
(3) In the removal of unnecessary allowance as the swarf, we can observe the *side flow*; however, the theoretical analysis is based on the plane stress field, which is applicable to the centre area in the width of swarf.

In addition, we must consider the size effect in actual cutting, by which the specific cutting force becomes larger with the decrease of depth of cut.

More importantly, a crucial issue lies in the determination of the shear angle φ in the model. In general, the shear angle is given by the "*Cutting Equation*", for example, those of Merchant and Lee-Shaffer; however, we have a handful of cutting equations depending upon the yield criteria of various materials, for example, maximum main stress theory, maximum shear stress theory, octahedral shear stress hypothesis, least energy principle and so on. Importantly, even now we have

Y. Ito and T. Matsumura, *Theory and Practice in Machining Systems*,
DOI 10.1007/978-3-319-53901-0

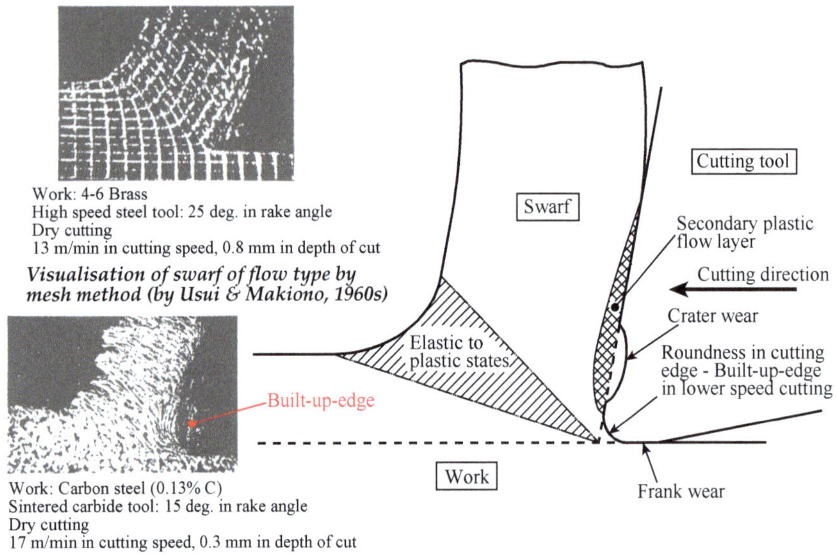

Fig. A.1 Visualisation and illustration of actual cutting state

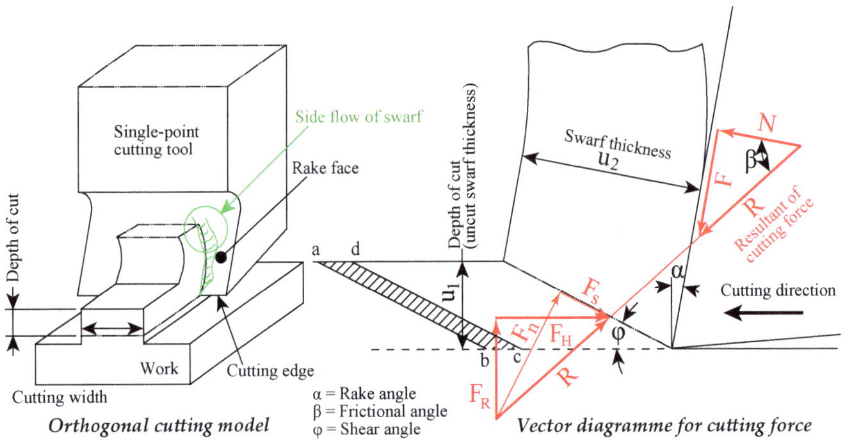

Fig. A.2 Concept of orthogonal cutting model and vector diagramme for cutting force

uncertainties in the yield criterion for each material, and thus cannot establish the acceptable one-to-one relationship between the material and its yield criterion.

As can be readily seen, the cutting force cannot be estimated correctly from the cutting model even when using the finite element method (FEM), and thus in nearly all cases, we must measure the cutting force (see Fig. 6.7). In contrast, we need to have a simplified and experience-based expression to estimate the cutting force on the factory floor.

In many respects, the resultant cutting force is decomposed into the three components as shown, in case of turning, in Fig. A.3. From the past, there are several expressions for the quick calculation of the principal component of the cutting force in turning, e.g., those of Kronenberg, ASME and Masuko, and of these, the expression proposed by Masuko is simple and with the ease of use, and is given by

$$F_H = p_s \times S_R \times t$$

where,

F_H = Principal component of cutting force (N),
p_s = Specific cutting force (N/mm^2),
S_R = Feed rate (mm/rev),
t = Depth of cut (mm)

In the expression, we need to know the specific cutting force to calculate the principal component of the cutting force. At present, some cutting tool manufacturers provide us with the specific cutting force as shown in Table A.1, although we can observe certain differences depending upon the manufacturer's data when the material to be cut is specified.

In general, the ratio of the feed component or back component to the principal component is nearly constant, when the work material and tool material are specified, and thus we may estimate duly the feed and back components on the basis of principal component.

Of special interest, in two-flute drilling to the rough (preparatory) hole generated by, for example, casting and forging, and also to pilot holes, one cutting edge

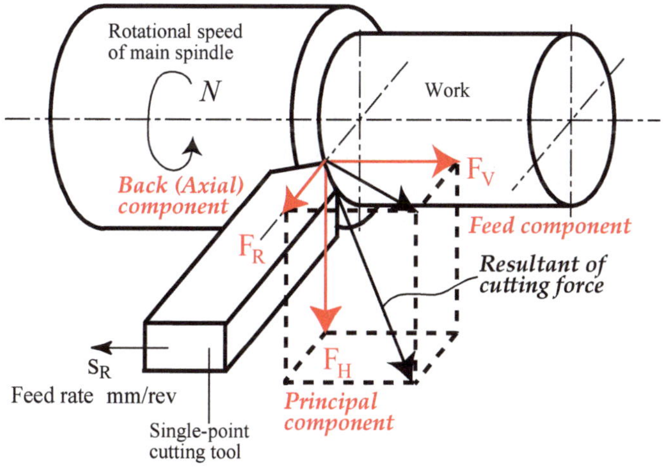

Fig. A.3 Three components of cutting force in turning

Table A.1 Specific cutting force in turning—2010s

By courtesy of Mitsubishi Materials Work materials	Tensile strength (MPa), or hardness	Specific cutting force (MPa) Feed rate 0.1 mm/rev	0.4 mm/rev
Carbon steel (SS400, S10C)	520	3,610	2,500
(S45C)	620	3,080	2,450
(S55C)	720	4,050	2,950
Tool steel (SKS)	770	3,150	2,450
Cr-Mo steel(SCM)	600	3,610	2,700
Ni-Cr-Mo steel (SNCM439)	HB 352	3,310	2,400
Cast iron (FC250)	HB 200	2,110	1,400
Meehanite cast iron (FC350)	360	2,300	1,600
By courtesy of Tungaloy	1 MPa ≒ 1 N/mm^2	Specific cutting force (N/mm^2)	
Al-alloy	HB 89	1,130	810
Mg-alloy		390	390
Brass		1,080	1,080
Carbon steel (S40C)	590	3,490	2,500

appears as to be a single-point cutting tool. The cutting edge can remove the undue proportion with a half feed rate as shown in Fig. A.4. In this case, the characteristic feature is in the continuously changing rake angle, i.e., from obtuse angle around the chisel to acute angle around the margin. In due course, we can obtain the cutting torque from the principal component, and the thrust from the feed component.

In the case of milling, the cutting edge moves as per the Trochoidal curve as shown in Fig. A.5. The swarf thickness changes thus continuously with the rotating position of the cutting edge (blade), and can be determined by the encircled region of both the loci, which are generated by the preceding (edge A) and being engaged (edge B) cutting edges. In fact, the thickness of the swarf can be approximately written as

$$h = s_z \sin \varphi$$

where,

$s_z = s/Nz$: Feed rate per tooth mm/rev
N: Rotational speed of milling cutter rev/min
z: Number of teeth
s: Feed speed of table mm/min

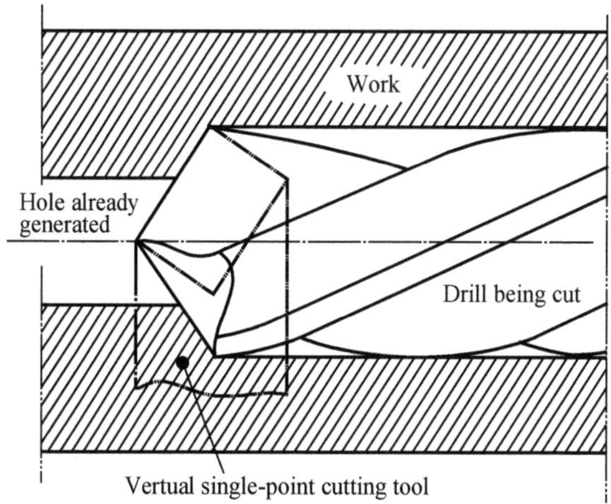

Fig. A.4 Model of drilling for already generated hole—Simulation by turning

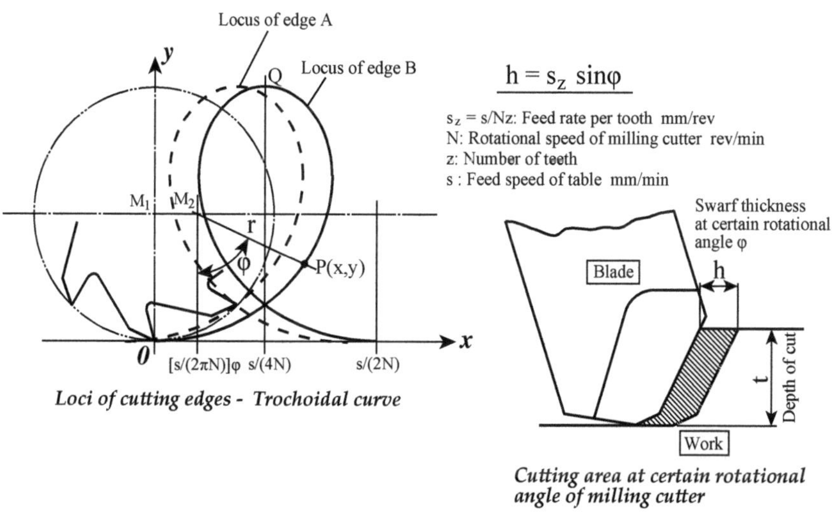

Fig. A.5 Swarf thickness in milling

In short, the characteristic feature is in the continuously changing thickness of the swarf with rotating the milling cutter, and in simplified calculation, the cutting force of milling is given roughly by using an average value of the thickness of the swarf.

Appendix B
Core Design Principles for Machine Body and Structural Body Component

In discussing the utilisation technology of the machine tool, it is not mandatory to know the design principle for the machine tool. In short, nearly all users expect that they can benefit from the machine tool purchased notwithstanding its design and manufacturing technologies.

Against to this context, it is somewhat useful to understand the basic knowledge of the design principle to plan and contrive the effective use of the machine tool, especially from the viewpoint of machine-attachment-tool-work system.

In short, it is desirable for the user to understand the two leading principles for the structural design: one is the "*Guiding Accuracy of Travelling and Rotating Structural Body Components*", and the other is the "*Allowable Deflection-based Design Principle*".

Guiding Accuracy of Structural Body Component

The utmost important design determinant in the machine tool is the realisation of the perfect form-generating movement under any dynamic and thermal loading. To fulfill such the determinant, the three guiding accuracies, i.e., pitching, yawing and rolling, should be in the best, i.e., minimum error magnitude as far as possible.

Figure B.1 illustrates three guiding accuracies, and in this context, the "*Principle of Narrow Guide*" should be always maintained. For example, the yawing becomes very small, when the length of the guideway is longer while maintaining the gap Δ at the width B in the constant.

In short, the magnitude of yawing yields to

Magnitude of yawing = Δ/L

In many respects, thus, the guideway structure and configuration are at fatal issue in the design. In general, the guiding accuracy is discussed on the basis of the geometrical accuracy error in each structural body component and also caused by

Y. Ito and T. Matsumura, *Theory and Practice in Machining Systems*,
DOI 10.1007/978-3-319-53901-0

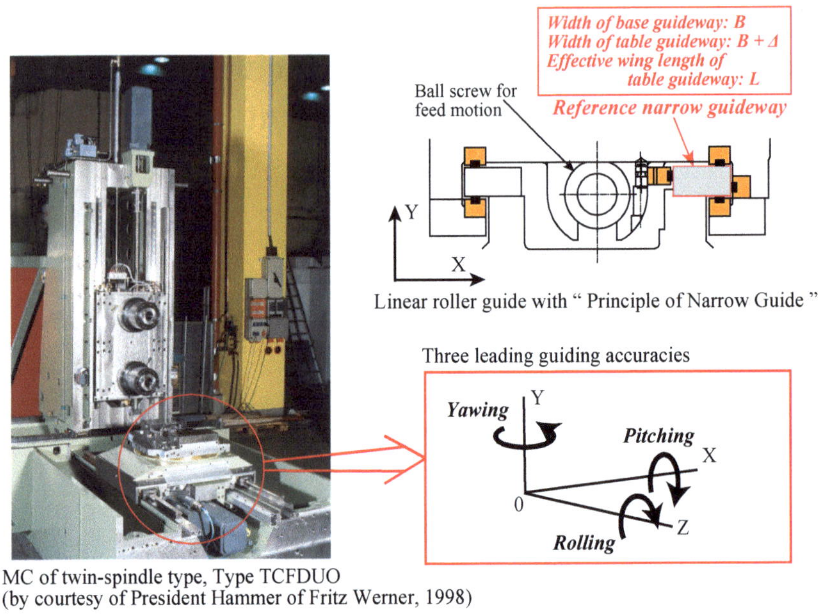

Fig. B.1 Definitions for three guiding accuracies and "Principle of Narrow Guide"

the assembly, but without considering the deflection caused by the dynamic and thermal loads. Paraphrasing, the guiding accuracy is, in principle, discussed, provided that the machine body is completely rigid showing any elastic deflection. It is thus necessary and inevitable to evaluate the guiding accuracy in running the machine tool under non-loading and loading, although such an evaluation is far from fruition. In addition, it is preferable that the individual machining error in the structural body component should be considered in the estimation of the assembly error by weighing its influence on the guiding accuracy.

Allowable Deflection-based Structural Design and Its Evaluation Index

In retrospect, the machine tool was defined as follows.

"Only the advanced industrial nation can produced both the machine tool and the ordnance"

This definition implies the utmost characteristic feature in the design principle of the machine tool. In fact, the machine body, i.e., assembly of structural body components, should be designed on the basis of the "*Allowable Deflection-based Principle*", whereas all other civil and defence supplies are designed on the basis of the "*Allowable Stress-based Principle*".

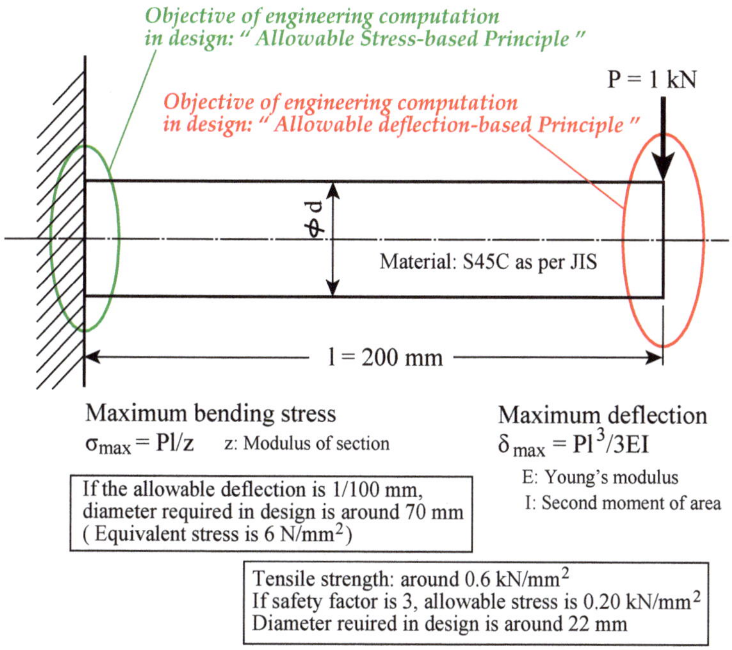

Fig. B.2 Comparison of "Allowable Deflection-based Principle" and "Allowable Stress-based Principle"

For the sake of understanding differing phases between these design principles, Fig. B.2 shows a typical mathematical model, i.e., round cantilever beam of fixed-end, under bending load and made of S45C as per JIS. In the case of the "*Allowable Stress-based Principle*", the maximum stress acts on the fixed end and should be lower than the allowable stress. As shown together in Fig. B.2, thus, the diameter of the beam should be more than 22 mm, if the safety factor is 3.

In the case of "*Allowable Deflection-based Principle*", the maximum deflection occurs at the tip of the beam, and it should be less than the allowable deflection. In general, the allowable deflection is 0.01 mm in maximum for the machine tool in consideration of the machining accuracy of the component, and in due course, the diameter yields to around 70 mm. In this case, the bending stress at the fixed-end is around 6 N/mm^2, which is around 1/30 of the allowable stress.

In short, the body of the machine tool should be, in design principle, larger than that of others, whereas we must be aware that people operate and handle even machine tool as same as others. This means that the machine tool is forced to be smaller than the size, which should be fulfilled in accordance with the design principle.

As will be clear from the above, the structural designer of the machine tool should overcome these fatal obstacles by nature. In general, the structural body component is provided with the stiffening rib, partition, double-wall, cell and so on

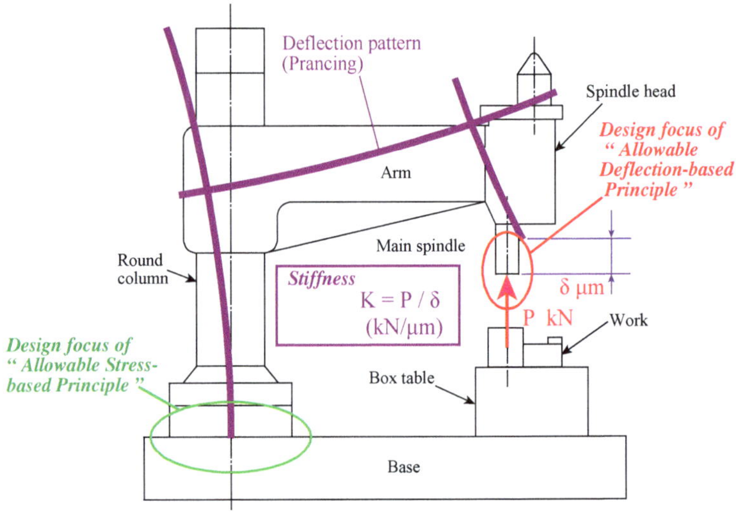

Fig. B.3 Definition of "Stiffness" representing characteristic feature in structural design of machine tool

to recover and maintain the satisfactory rigidity and damping under the severe dimensional constraint (see for example Fig. 3.16).

In due course, we must represent such the characteristic feature of the machine tool by the index, i.e. stiffness, as shown in Fig. B.3.

MIX
Papier aus verantwortungsvollen Quellen
Paper from responsible sources
FSC® C105338

If you have any concerns about our products,
you can contact us on
ProductSafety@springernature.com

In case Publisher is established outside the EU,
the EU authorized representative is:
Springer Nature Customer Service Center GmbH
Europaplatz 3, 69115 Heidelberg, Germany

Printed by Libri Plureos GmbH
in Hamburg, Germany